California Civil Surveying Reference Manual

Second Edition

George M. Cole, PhD, PE, PLS

PPI®

PPI2PASS.COM

A KAPLAN COMPANY

Report Errors for This Book

PPI is grateful to every reader who notifies us of a possible error. Your feedback allows us to improve the quality and accuracy of our products. Report errata at **ppi2pass.com**.

CALIFORNIA CIVIL SURVEYING REFERENCE MANUAL
Second Edition

Current release of this edition: 4

Release History

date	edition number	revision number	update
Sep 2020	2	2	Minor corrections.
Jan 2022	2	3	Minor corrections.
Dec 2022	2	4	Minor corrections.

PPI
ppi2pass.com

ISBN: 978-1-59126-656-3

Topics

Mathematics Basics

Field Data Acquisition

Plane Survey Calculations

Land Planning/ Development

Support Material

Table of Contents

Appendix
Table of Contents

Preface and Acknowledgments

The surveying aspects of civil engineering practice encompass an extremely broad range of knowledge and skills. Moreover, it's a field that is in a continuing state of flux, with advances in surveying technology and changes in associated areas such as law and land use practices. While the challenges of dealing with the breadth and the dynamic nature of surveying are what makes it such an exciting part of civil engineering practice, it requires significant time and effort by practicing engineers and exam candidates to keep up with the rapidly changing scope of the profession. This manual is designed to assist in that effort. The *California Civil Surveying Reference Manual* is based on the venerable *Surveyor Reference Manual*; the contents of that manual have been carefully reviewed, updated, and tailored to more completely cover the topics important to modern civil surveying practice in California, as well as those covered by the California Civil Surveying Exam.

I would also like to thank the PPI editorial and production staff, including Meghan Finley, content manager; Richard Iriye, typesetter; Sam Webster, product data operations manager; Tom Bergstrom, cover designer and technical illustrator; Cathy Schrott, editorial operations manager; and Grace Wong, director of editorial operations. Thank you for your hard work and patience in helping to put this book together.

Finally, when you have finished going through this book, I encourage you to assist in its improvement by notifying PPI of any comments or errors through the online errata submission form at **ppi2pass.com**. Revisions to the next edition of this book will be greatly influenced by readers' comments on this one. Please take the time to help those who will follow you in the surveying profession by letting us know what you liked and disliked about this book, and what we can do to make the next edition even better. Thank you.

George M. Cole, PhD, PE, PLS

About the Author

George M. Cole, PhD, PE, PLS, is a practicing surveyor and engineer whose assignments have ranged from surveying remote areas of the Arctic to participating in the first worldwide satellite geodesy program. He holds a bachelor of science degree from Tulane University and a master of science degree and doctorate from Florida State University. In addition to authoring the *Fundamentals of Surveying Practice Exam* and the *Principles and Practice of Surveying Practice Exam*, Dr. Cole has authored books on water boundaries, land tenure and cadastral systems, and numerous technical papers and periodical articles. He currently serves as an adjunct professor at both the University of Puerto Rico and Florida State University.

Introduction

This book is intended to serve multiple purposes. First, as the title suggests, it is intended to serve as a comprehensive reference for practicing civil engineers on the surveying aspects of their practice. If you are such a practicing professional, you will find it to be an invaluable addition to your library. Second, because of its comprehensive coverage, it is an ideal resource for those preparing for the California Civil Surveying Exam. (It is also an ideal book for instructors teaching review courses for that exam.)

1. ABOUT THE CALIFORNIA CIVIL SURVEYING EXAM

Administered as a computer-based test (CBT), the California Civil Surveying Exam may be taken at any Prometric test center, even outside of California. The exam tests the entry-level competency of a candidate to practice civil engineering within the profession's acceptable standards for public safety. The exam is open book and is administered over a 2.5 hour period. The exam contains 55 multiple-choice problems derived from five primary content areas as outlined in the board-adopted Engineering Surveying Test Plan.

- Topographic Surveys (35%)
- Construction Surveys (35%)
- Accuracy and Error Analysis (10%)
- Preparation of Reports and Maps (20%)

For the California Civil Surveying Exam, the board-adopted test plan lists the content areas of the exam and their assigned scoring percentages. The percentages assigned to each content area are the approximate proportion of total test points; however, the test plan does not reveal the total test points in advance (it varies from exam to exam). This makes it difficult to anticipate the exact number of problems for each content area of the exam.

For each problem, you will be asked to select the best answer from four choices. While you don't have to move sequentially through the exam, you can if you wish to. You can navigate to a specific problem. You can skip a problem or flag it for later review. You can return to any problem and change your answer. The onscreen index (listing, directory, etc.) can be used to see the problems that have been answered, flagged for review, or skipped. In some cases, you may have to toggle back and forth between a problem and an on-screen illustration if the illustration takes up too much space. A timer on the screen indicates how much time remains.

The official exam is graded against a "cut score"—a predetermined minimum passing score that varies from exam to exam. Historically, if you score above 60% of the total examination point value, you have a chance of passing.

On the official exam, after initial scoring, any problem that the board finds to be flawed may be deleted. In the event of deletion, the point value of the deleted problem becomes zero and the total number of points possible on that exam is adjusted accordingly.

2. PREPARING FOR THE EXAM

Plan Your Approach

You should consider preparation for the California Civil Surveying Exam to be a long-term project and plan carefully. The exam is comprehensive and fast-paced. Rapid recall, discipline, stamina, and mastery of the subject areas covered are all essential to success. Development of these skills may require considerable preparation in addition to the years of academic study and work practice you need to qualify. Therefore, it is important to plan your preparation for the exam as you would plan for a large engineering project.

These steps can help prepare you.

step 1: Answer the practice problems at the end of each chapter of this book. For future reference, prepare a concise outline as you work through each area. Your review should be on a rigorous schedule to help you develop the discipline and stamina necessary to do well on the exam. In the review, pay special attention to the following topics, which are fundamental to surveying and form a necessary core of knowledge: coordinate geometry (COGO) functions (such as inverse, side shot, bearing-bearing intersection, traverse, areas), differential leveling, horizontal and vertical curves, earthwork volumes, and datums.

step 2: For any areas in which you are not comfortable, read additional reference material as you work through each chapter. Tab pages where frequently used or hard-to-find information is located.

step 3: Take a sample exam, such as *Practice Exams for the California Civil Surveying Exam*, available from PPI, to evaluate your readiness for the exams.

step 4: Work on any weak areas revealed by the sample exam. To bolster your preparedness in those areas, use *Solved Problems for the California Civil Surveying Exam*, available from PPI, to practice until you feel confident. More problems covering the content areas in the exam are available in *Fundamentals of Surveying Practice Exam* and *Principles and Practice of Surveying Practice Exam*, also available from PPI.

step 5: Conduct a final review of your notes.

Learning to use your time wisely is one of the most important things that you can do during your review. You will undoubtedly encounter review problems that take much longer than you expect. You may cause some delays yourself by spending too much time looking through reference books for the information that you need. Learning to recognize such situations more quickly will help you make intelligent decisions during the exams.

Additional Reference Material

You will find that this book is an excellent starting point for preparing for your exam. However, additional references may be helpful, especially in areas in which you are uncomfortable. There are countless texts available that cover the various topics in depth. Listed here are several personal favorites of the author which offer coverage of the areas to be tested on the exams.

Bureau of Land Management. *Manual of Instruction for Surveys of Public Lands.* Washington, DC: Government Printing Office.

Cole, George M. *Water Boundaries.* Hoboken, NJ: John Wiley & Sons.

Colley, Barbara C. *Practical Manual of Land Development.* New York, NY: McGraw-Hill.

Davis, Foote, Anderson, and Mikhail. *Surveying Theory and Practice.* New York, NY: McGraw-Hill.

Hickerson, Thomas F. *Route Location and Design.* New York, NY: McGraw-Hill.

Van Sickle, Jan. *GPS for Land Surveyors.* Boca Raton, FL: CRC Press.

3. WHAT TO TAKE TO THE EXAM

You should come to the exam prepared with the following materials. Be sure to set them out ahead of time and devise a convenient way to carry them into the examination room with you.

- government-issued photo ID
- reference materials (only one box)
- two of the following four measuring devices: ruler, protractor, architect scale, engineer scale
- two non-QWERTY calculators

4. WHAT TO DO AT THE EXAM

The California Civil Surveying Exam is a difficult exam and requires thorough preparation in all areas, including multiple-choice test-taking techniques. Easy and difficult questions, with variable point values, are distributed throughout the exam. Besides contending with the nature and difficulty of the exam itself, many examinees spend too much time on difficult problems and leave insufficient time to answer the easy ones. You should avoid this. This test-taking strategy can be beneficial to you during the exam.

step 1: Work on the easy questions immediately and record your answers.

step 2: Work on questions that require minimal calculations and record your answers.

step 3: When you get to a question that looks "impossible" to answer, go ahead and guess.

step 4: When you face a question that seems difficult but solvable, you may need considerable time to search for relevant information in your books, references, or notes. Continue to the next question.

step 5: When you come to a question that is solvable but you know requires lengthy calculations or is time-consuming, leave it for later. For this question, you know exactly where to look for relevant information in your books, references, or notes.

The intent of this test-taking strategy is to save you precious time. Based on the number of exam questions and allotted time, you should not spend more than an average of 2.5 minutes per question. Thus, a lengthy or "time-consuming" question is one that will take you more than 2.5 minutes to answer.

After you have gone over the entire exam, it will be clear which questions you have already answered and which still require your attention. You should now go over the exam a second time, with the following approach in mind.

step 1: Go back and tackle the problems that require lengthy calculations; the problems that require time to search for relevant information in books, references, and notes; and the problems that seemed impossible.

step 2: Recheck your work for careless mistakes.

step 3: Set aside the last few minutes of your exam period to fill in a guess for any unanswered problems. There is no penalty for guessing. Only questions answered correctly will be counted toward your score.

Now that you know all there is to know about the exams and how to prepare for them, the rest is up to you. Plan your approach, and get to work. The very best of luck to you!

Topic I: Mathematics Basics

Basic Geometry

1. MEASURE OF ANGLES

In geometry, an *angle* is defined as the space between two lines diverging from a common point (the *vertex*).[1] Angles are most commonly measured in *degrees*, defined as $1/360$ of a complete circle and represented by a degree symbol (°). If a circle is divided into 360 equal arcs, with radii connecting the ends of each arc, the angle formed by any two adjacent radii measures 1°. (See Fig. 1.1.)

For closer measurement, a degree is divided into 60 equal arcs, called *minutes*, and each minute is divided into 60 *seconds*. Minutes are represented by a single prime ('), and seconds by a double prime ("). For example, an angle measuring thirty-six degrees, twenty-four minutes, and fifty-two seconds is written as 36°24'52".

Angles can also be measured in *radians*. A *radian* is the angle subtended by an arc on the perimeter of a circle with length equal to the radius of the circle. A radian is equal to $180°/\pi$ (approximately $57.296°$).

A protractor is used to measure angles on paper; in the field, a transit and a theodolite are used.

2. TYPES OF ANGLES

Angles are referred to using different names, depending on the degree of the angle. A *right angle* is an angle of 90°; an angle of 180° is a *straight angle*. An angle of less than 90° is an *acute angle*, and an angle of more than 90° and less than 180° is an *obtuse angle*. (See Fig. 1.2.)

Figure 1.1 *Angle Measurements*

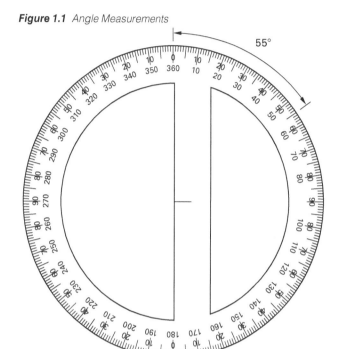

Figure 1.2 *Angles*

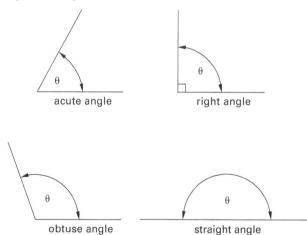

Two angles are said to be *complementary* if their sum is 90°, and *supplementary* if their sum is 180°. Figure 1.3 and Fig. 1.4 illustrate these pairs of angles.

---

[1] The definition of an angle may vary in other disciplines. In trigonometry, an angle can also be defined as the amount of rotation required to bring one line into coincidence with another. (See Chap. 7.) In surveying, an angle can also be defined as the difference in direction of two intersecting lines.

Figure 1.3 *Complementary Angles*

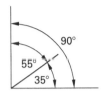

Figure 1.4 *Supplementary Angles*

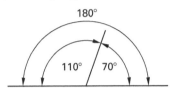

3. TRANSVERSALS AND ALTERNATE ANGLES

A *transversal* is a line that cuts two or more lines. When two parallel lines are cut by a transversal, the *alternate interior angles* and *alternate exterior angles* are equal. (See Fig. 1.5 and Fig. 1.6.)[2]

Figure 1.5 *Alternate Interior Angles*

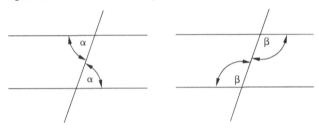

Figure 1.6 *Alternate Exterior Angles*

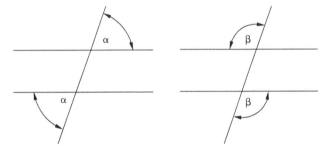

4. ADDING AND SUBTRACTING ANGLES

In surveying, it is often necessary to find the sum of or difference between the measurements of two or more angles. A good example is when determining the sum of the interior angles of a closed transverse to determine the angular error of closure. The measurements of angles can be added or subtracted like any other numerical value; these calculations are typically laid in three columns (degrees, minutes, and seconds), and each column is added separately and recorded. When necessary, degrees can be converted to minutes, and minutes can be converted to seconds, or vice versa.

Example 1.1

What is the sum of angles measuring $32°46'32''$ and $14°22'44''$?

Solution

$$32°46'32'' + 14°22'44'' = 46°68'76''$$
$$= 46°69'16''$$
$$= 47°09'16''$$

Example 1.2

What is the difference between two angles measuring $96°08'14''$ and $52°33'50''$?

Solution

For ease of calculation, convert one degree of the wider angle to minutes and one minute of the wider angle to seconds, for an equivalent measurement of $95°67'74''$.

$$95°67'74'' - 52°33'50'' = 43°34'24''$$

Example 1.3

What is the sum of the angle measurements shown?

$93°18'22''$

$65°13'8''$

$218°19'30''$

$67°05'20''$

$96°04'50''$

Solution

$$93°18'22'' + 65°13'08'' + 218°19'30'' = 539°59'130''$$
$$+67°05'20'' + 96°04'50''$$
$$= 539°61'10''$$
$$= 540°01'10''$$

5. AVERAGE OF MEASUREMENTS OF AN ANGLE

Because the measurement of angles is subject to both human and mechanical error, angles are often measured repeatedly. The results of these readings are then summed into a single value (the *accumulated angle*) and averaged, providing a value closer to the true

---

[2] Interior angles are those inside a shape or containing lines. Exterior angles are those outside a shape or containing lines. The sum of interior and exterior angles is 180°. When two lines are crossed by another line, the pairs of angles on opposite sides of the crossing line, but inside the two lines, are called alternate interior angles.

measurement of the angle than any individual reading. For example, an angle of $36°30'30''$ might be measured on the first reading as $36°30'$, but after six readings, the accumulated angle might be $219°03'00''$. Dividing the accumulated angle by six gives an average measurement of $36°30'30''$, the true measurement. How many times an angle is measured, and how the different measurements are taken, varies depending on the measuring device used.

Angles are averaged using the same concepts as used in addition and subtraction of angles. The angles are divided like any other numerical value, treating degrees, minutes, and seconds as separate numbers, and converting between the three where necessary.

Example 1.4

An angle is measured twice for an accumulated angle of $314°13'$. What is the final measurement of the angle?

Solution

$$\frac{314°13'}{2} = 157°06'30''$$

Example 1.5

An angle is measured six times for an accumulated angle of $318°03'$. What is the average measurement of the angle?

Solution

$$\frac{318°03'}{6} = \frac{318°00'180''}{6} = 53°00'30''$$

6. DEGREES AND DECIMALS OF A DEGREE

In some situations (e.g., tables of trigonometric functions), angles are expressed in degrees and decimals of a degree rather than degrees, minutes, and seconds. To convert degrees and minutes to degrees and decimals of a degree, divide the minutes by 60 and add the resulting decimal value to the degrees.

To convert degrees, minutes, and seconds to degrees and decimals of a degree, first divide the seconds by 60 to convert them to decimals of a minute. Add the decimals of a minute to the minute value, and divide this sum by 60 to convert the value to decimals of a degree. Add the decimals of a degree to the degrees.

To convert degrees and decimals of a degree to degrees, minutes, and seconds, multiply the decimal fraction by 60 and add the product (in minutes and decimals of a minute) to the degrees. Then multiply the decimal fraction by 60 and add the product to the degrees and minutes. If there is still a decimal fraction remaining, it is left as part of the seconds value.

Example 1.6

Convert a measurement of $46°24'15''$ to degrees and decimals of a degree.

Solution

Divide the seconds by 60 to convert to decimals of a minute.

$$46°24'15'' = 46°24' + \frac{15''}{60}$$
$$= 46°24.25'$$

Divide the minutes by 60 to convert to decimals of a degree.

$$46°24.25' = 46° + \frac{24.25'}{60}$$
$$= 46.40°$$

Example 1.7

Convert a measurement of $36.12345°$ to degrees, minutes, and seconds.

Solution

Multiply the decimals of a degree by 60 to convert to minutes.

$$36.12345° = 36° + (0.12345°)(60)$$
$$= 36°07.407'$$

Multiply the decimals of a minute by 60 to convert to seconds.

$$36°07.407' = 36°07' + (0.407')(60)$$
$$= 36°07'24.42''$$

7. POLYGONS

A *polygon* is a closed figure bounded by straight lines (sides) lying in the same plane. Polygons have different names depending on the number of sides and the angles formed by those sides.

A polygon of three sides is known as a *triangle*. A *right triangle* is a triangle in which two sides form a right angle. If a triangle has two equal sides and two equal angles, it is an *isosceles triangle*; if all three sides and angles are equal, the triangle is an *equilateral triangle*. An *oblique triangle* is a triangle that has no right angle and no equal sides. (See Fig. 1.7.)

Figure 1.7 *Triangles*

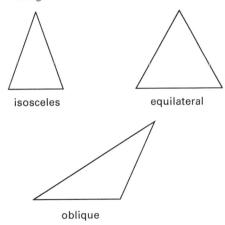

isosceles   equilateral

oblique

Two triangles are *similar* if their corresponding angles are equal and their corresponding sides are proportional. (See Fig. 1.8.) Two triangles are *congruent* if their corresponding sides (and consequently their corresponding angles as well) are equal. To show that two triangles are congruent, it is enough to show that two pairs of corresponding sides are equal and that the corresponding angles between the two sides are equal. Triangles can be congruent even if their equal sides and angles are not in the same positions.

Figure 1.8 *Similar Triangles*

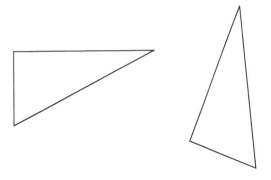

A *trapezoid* is a four-sided polygon that has two parallel sides and two nonparallel sides.

A *rectangle* is a trapezoid where all angles are right angles. A *square* is a rectangle where all sides are of equal length. Figure 1.9 shows different types of trapezoids, and Fig. 1.10 shows different types of rectangles.

The sum of the interior angles of a closed polygon can be found from Eq. 1.1 where n is the number of sides of the polygon. Polygons with the same number of sides will always have interior angles that sum to the same amount. For example, the sum of the interior angles of a triangle is always $180°$, the sum of the interior angles of a rectangle is always $360°$, the sum of the interior angles of a five-sided figure is always $540°$, and so on.

$$\text{sum of interior angles} = (n-2)(180°) \qquad 1.1$$

Figure 1.9 *Trapezoids*

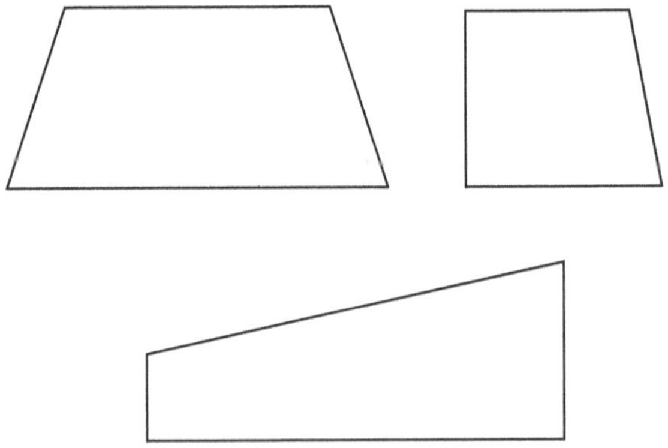

Figure 1.10 *Rectangles*

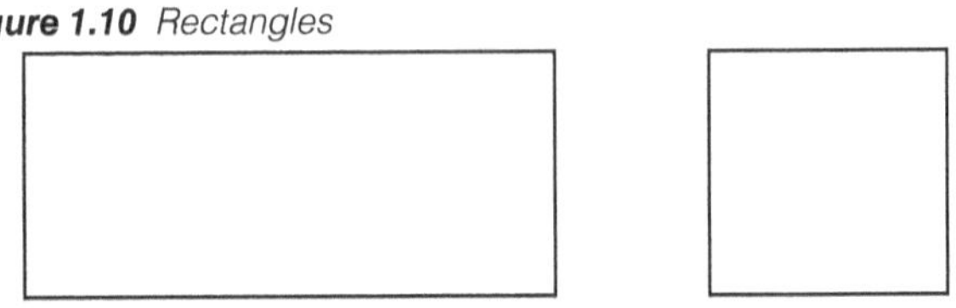

8. CIRCLES

A *circle* is a closed plane curve where all points on the curve are equidistant from the center of the enclosed space. The distance from the center of the circle to any point on the circle is called the *radius* of the circle. The distance across the circle through the center is called the *diameter* (i.e., one diameter is two radii).

Any portion of the curve of a circle is called an *arc*; an arc equal to one-half the circumference of a circle is called a *semicircle*. A *central angle* is an angle formed by two radii (often represented using the Greek letter Δ (delta) or θ (theta).

A central angle is measured by its intercepted arc; the central angle has the same number of degrees as the arc it intercepts (e.g., a $60°$ central angle intercepts a $60°$ arc). A semicircle and a central angle are shown in Fig. 1.11.

Figure 1.11 *Semicircle and Central Angle*

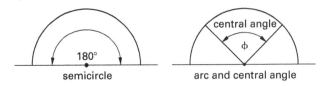

central angle

$180°$

ϕ

semicircle   arc and central angle

A straight line between any points on a circle is called a *chord* (designated LC in equations). The perpendicular *bisector* of a chord passes through the center of the circle. A line that touches the circle at only one point is called a *tangent,* and a line that intersects a circle at two points is called a *secant.* From any outside point, two tangents to a circle may be drawn (one on either side). The distances along the two tangents to the points of tangency are always equal, and the radius of a circle is always perpendicular to the tangents at the point of tangency.

The angle formed by two chords is equal to one-half its intercepted arc, and the angle formed by a tangent and a chord is equal to one-half its intercepted arc.

These elements of a circle are illustrated in Fig. 1.12 through Fig. 1.14.

Figure 1.12 *Chord, Secant, and Tangent*

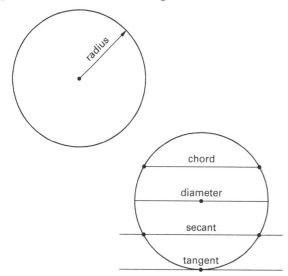

Figure 1.13 *Tangents and Chords*

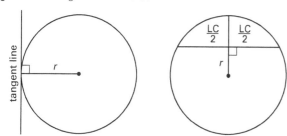

Figure 1.14 *Tangents to a Circle*

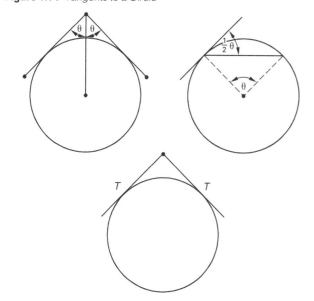

A figure bounded by an arc of a circle and two radii of the same circle is called a *sector* of the circle.

A figure bounded by a chord and an arc of a circle is called a *segment* of a circle.

Two circles of different radius but with the same center are called *concentric circles*. Concentric circles, sectors, and segments are illustrated in Fig. 1.15.

Figure 1.15 *Concentric Circles, Sectors, and Segments*

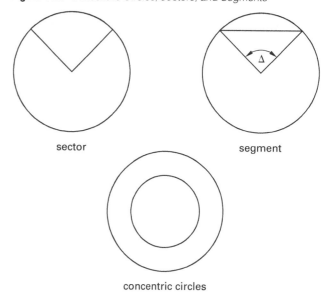

9. SOLID GEOMETRY

Polygons and circles are two-dimensional figures, and the study of them and their properties is called *plane geometry*. *Solid geometry* is the study of three-dimensional figures (*solids*), such as cubes, cones, pyramids, and spheres.

A *polyhedron* is any solid formed by plane surfaces. Like polygons, polyhedrons have different names depending on the number and placement of the edges and faces of the polyhedron.

A polyhedron with parallel edges and parallel faces is called a *prism*.

A prism with edges perpendicular to the bases is known as a *right prism*. (See Fig. 1.16.) A *cylinder* is a right prism with circular bases.

A *pyramid* is a polyhedron with a polygon for a base and faces that are all triangles with a common vertex. A *right pyramid* is a pyramid in which the base is a regular polygon, and a line from the vertex to the center of the base is perpendicular to the base.

The *altitude* of a pyramid is the perpendicular distance from the vertex to the base. The *slant height* of a right pyramid is the altitude of one of the lateral faces.

Figure 1.16 Right Prisms

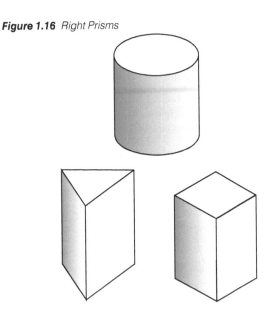

A *frustum* of a pyramid is a portion of a pyramid's base separated from the top of the pyramid by a vertical plane parallel to the base. If the pyramid that is truncated by the vertical plane is a right pyramid, the frustum is a *right frustum*. Frusta are illustrated in Fig. 1.17.

Figure 1.17 Pyramids and Cones

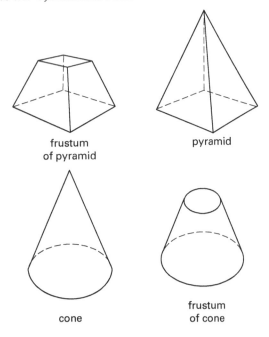

frustum
of pyramid

pyramid

cone

frustum
of cone

A *cone* is a polyhedron with a circular base and sides that taper evenly up to a vertex. A *right circular cone* is a cone in which a line from the vertex to the center of the base is perpendicular to the base. Like a pyramid, a cone has an altitude and a slant height.

The *altitude of a cone* is the perpendicular distance from the vertex to the base, and the *slant height* of a cone is the distance from the vertex to the base measured along the surface.

A vertical plane can also be inserted parallel to the base of a right circular cone to create a frustum.

10. DRAWING GEOMETRIC FIGURES

Geometric figures are drawn using a compass and straightedge.

To bisect a given angle, A, draw an arc intersecting the sides of the angle; then, at these intersections, B and C, draw arcs of equal radii. A line from the vertex to the intersection of these two arcs, D, bisects the angle A. (See Fig. 1.18.)

Figure 1.18 Bisecting an Angle

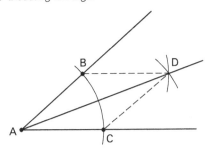

To draw a line perpendicular to a given line, w, through a given point on the line, P, draw an arc intersecting the line at points A and B. Use both A and B as radius points[3] to draw arcs with lengths more than half of AB. The arcs intersect at point D. A line through D and P is perpendicular to line w. (See Fig. 1.19.)

Figure 1.19 Drawing a Perpendicular Line

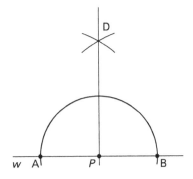

To draw the perpendicular bisector of a line AB, use A as a radius point, and draw an arc with a radius more than half the length of AB. Use B as a radius point, and construct an arc with a radius the same length as the

---

[3]The radius point is the point from which an arc is swung. The radius point is equidistant from any point on the arc.

first arc. A line through the intersections of these arcs, points C and D, is the perpendicular bisector of line AB. (See Fig. 1.20.)

Figure 1.20 *Drawing a Perpendicular Bisector*

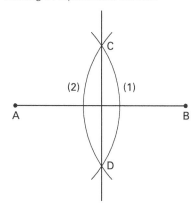

To circumscribe a circle about a triangle, draw the perpendicular bisectors of two sides of the triangle. Their intersection is at the center of the circle. The radius of the circle is the distance from the circle's center to any vertex of the triangle. (See Fig. 1.21.)

Figure 1.21 *Circumscribing a Circle about a Triangle*

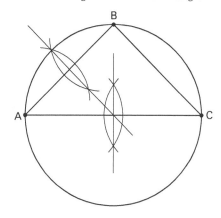

To inscribe a circle in a triangle, draw the bisectors of two of the angles of the triangle. Their intersection is at the center of the circle. (See Fig. 1.22.)

Figure 1.22 *Inscribing a Circle in a Triangle*

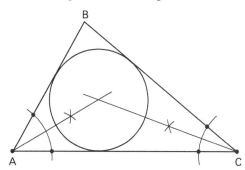

To locate the center of a circle that is not about a triangle, select three points on the circle, connect them with two chords, and then draw perpendicular bisectors to the chords. Their intersection is at the center of the circle. (See Fig. 1.23.)

Figure 1.23 *Locating a Circle's Center*

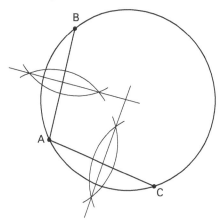

11. PRACTICE PROBLEMS

1. What is the sum of an angle measuring $46°19'22''$ and an angle measuring $35°51'40''$?

- (A) $81°21'25''$
- (B) $81°32'18''$
- (C) $82°11'02''$
- (D) $82°35'31''$

2. An angle is measured twice for an accumulated value of $237°27'17''$. The average measurement of the angle is most nearly

- (A) $118°43'39''$
- (B) $118°44'08''$
- (C) $119°14'09''$
- (D) $119°43'39''$

3. An angle is measured six times for an accumulated value of $390°13'24''$. The average measurement of the angle is most nearly

- (A) $65°02'14''$
- (B) $65°03'04''$
- (C) $66°04'04''$
- (D) $75°02'14''$

4. How is an angle of $223°37'48''$ expressed in degrees and decimals of a degree?

(A) $223.38°$

(B) $223.41°$

(C) $223.63°$

(D) $223.75°$

5. Most nearly, how is an angle of $303.107778°$ expressed in degrees, minutes, and seconds?

(A) $303°04'40''$

(B) $303°06'28''$

(C) $303°06'36''$

(D) $303°10'08''$

6. Which of the following statements about angles are true?

I.   An acute angle is an angle of less than $90°$.

II. An obtuse angle is an angle of $180°$.

III. Complementary angles are two angles with a sum of $90°$.

IV. Supplementary angles are two angles with a sum of $180°$.

(A) I only

(B) I and IV only

(C) I, II, and III

(D) I, III, and IV

SOLUTIONS

1. Sum the angles.

$$46°19'22'' + 35°51'40'' = 81°70'62'' = 82°11'02''$$

The answer is (C).

2. The average measurement of the angle is

$$\frac{237°27'17''}{2} = \frac{236°86'77''}{2}$$
$$= 118°43'38.5'' \quad (118°43'39'')$$

The answer is (A).

3. The average measurement of the angle is

$$\frac{390°13'24''}{6} = \frac{390°12'84''}{6} = 65°02'14''$$

The answer is (A).

4. Convert the seconds to decimals of a minute.

$$223°37'48'' = 223°37' + \frac{48''}{60} = 223°37.8'$$

Convert the minutes to decimals of a degree.

$$223°37.8' = 223° + \frac{37.8'}{60} = 223.63°$$

The answer is (C).

5. Multiply the decimals of a degree by 60 to convert to minutes.

$$303.107778° = 303° + (0.107778')(60)$$
$$= 303°6.46668'$$

Multiply the decimals of a minute by 60 to convert to seconds.

$$303°6.46668' = 303°6' + (0.46668')(60)$$
$$= 303°6'28.0008'' \quad (303°6'28'')$$

The answer is (B).

6. An acute angle is an angle of less than $90°$, so statement I is true. Complementary angles are two angles with a sum of $90°$, so statement III is true. Supplementary angles are two angles with a sum of $180°$, so statement IV is true.

An obtuse angle is any angle of more than $90°$ and less than $180°$. An angle of $180°$ is called a straight angle. Statement II is false.

The answer is (D).

2 Dimensional Equations

1. MEASUREMENT

A *measurement* consists of two things: a number that expresses quantity and a unit of measure. Surveyors and surveying technicians are intricately involved in measurements and in converting measurements expressed in one unit of measure to their equivalents in other units of measure.

In converting values from one unit of measure to another, it is just as important to find the correct unit of measure as it is to find the correct quantity.

2. DEFINITION OF DIMENSIONAL EQUATION

A *dimensional equation* is one that contains units of measure but does not contain the corresponding numerical values. For example, to express in cubic yards the volume of a dump truck bed with dimensions of 6 ft × 8 ft × 4 ft, multiply the dimensions to find the volume in cubic feet (192 ft^3). Since there are 27 ft^3 in a cubic yard, divide 192 by 27 to find that the bed has a volume of 7 yd^3. This operation is written as

$$\frac{(6\text{ ft})(8\text{ ft})(4\text{ ft})}{27\ \dfrac{\text{ft}^3}{\text{yd}^3}} = 7\text{ yd}^3$$

The dimensional equation that corresponds to this operation is

$$\frac{(\text{ft})(\text{ft})(\text{ft})}{\dfrac{\text{ft}^3}{\text{yd}^3}} = \frac{\dfrac{\text{ft}^3}{1}}{\dfrac{\text{ft}^3}{\text{yd}^3}} = \left(\frac{\text{ft}^3}{1}\right)\left(\frac{\text{yd}^3}{\text{ft}^3}\right) = \text{yd}^3$$

Inserting the numerical values, the equation reads

$$\frac{\left(\dfrac{6}{1}\text{ ft}\right)\left(\dfrac{8}{1}\text{ ft}\right)\left(\dfrac{4}{1}\text{ ft}\right)}{\dfrac{27}{1}\ \dfrac{\text{ft}^3}{\text{yd}^3}} = \left(\frac{192\text{ ft}^3}{1}\right)\left(\frac{1}{27}\ \frac{\text{yd}^3}{\text{ft}^3}\right)$$

$$= 7\text{ yd}^3$$

Example 2.1

A rectangular tract of land measures 300 ft by 200 ft. Write a dimensional equation, including the measured quantities, for finding the area of the tract in acres.

Solution

The dimensional equation is

$$A = \frac{(\text{ft})(\text{ft})}{\dfrac{\text{ft}^2}{\text{ac}}}$$

$$= \frac{\left(\dfrac{300}{1}\text{ ft}\right)\left(\dfrac{200}{1}\text{ ft}\right)}{\dfrac{43{,}560}{1}\ \dfrac{\text{ft}^2}{\text{ac}}}$$

$$= \left(\frac{60{,}000}{1}\text{ ft}^2\right)\left(\frac{1}{43{,}560}\ \frac{\text{ac}}{\text{ft}^2}\right)$$

$$= 1.38\text{ ac}$$

Example 2.2

A vehicle is traveling 36 mph. Write a dimensional equation, including the measured quantities in the equation, for finding the velocity of the vehicle in feet per second.

Solution

The dimensional equation is

$$v = \frac{\left(\dfrac{\text{mi}}{\text{hr}}\right)\left(\dfrac{\text{ft}}{\text{mi}}\right)}{\dfrac{\text{sec}}{\text{hr}}}$$

$$v = \frac{\left(\dfrac{36 \ \text{mi}}{1 \ \text{hr}}\right)\left(\dfrac{5280 \ \text{ft}}{1 \ \text{mi}}\right)}{\dfrac{3600 \ \text{sec}}{1 \ \text{hr}}}$$

$$= \left(\frac{36 \ \text{mi}}{1 \ \text{hr}}\right)\left(\frac{5280 \ \text{ft}}{1 \ \text{mi}}\right)\left(\frac{1 \ \text{hr}}{3600 \ \text{sec}}\right)$$

$$= 53 \ \text{ft/sec}$$

3. FORM FOR PROBLEM-SOLVING

The use of common conversion factors can make it unnecessary to set up a dimensional equation for solving problems involving several measured quantities. However, setting up a single equation that includes the numbers expressing quantity but not units of measure (similar to the dimensional equation) both facilitates cancellation and eases the process of checking the calculation for accuracy. Each number that expresses quantity should be shown in the equation.

For example, to find the area of a 10 in circle, do not simply write $A = (\pi/4)D^2 = 78.5 \ \text{in}^2$; write out $A = (\pi/4)(10 \ \text{in})^2 = 78.5 \ \text{in}^2$. For solutions that involve unfamiliar formulas, it is good practice to write the formula and then substitute the measured quantities.

Example 2.3

What is the cost of concrete, delivered to a site at $36 per cubic yard, for a parking lot 100 ft long, 54 ft wide, and 4 in thick?

Solution

Where length and width are measured in feet and thickness in inches, use a common fraction of a foot as the thickness measurement. Four inches is exactly one-third of a foot. The dimensional equation is

$$\$ = \frac{(\text{ft})(\text{ft})(\text{ft})\left(\dfrac{\$}{\text{yd}}\right)}{\dfrac{\text{ft}^3}{\text{yd}^3}}$$

$$= \frac{(100 \ \text{ft})(54 \ \text{ft})\left(\dfrac{4}{12} \ \text{ft}\right)\left(36 \ \dfrac{\$}{\text{yd}}\right)}{\dfrac{27 \ \text{ft}^3}{1 \ \text{yd}^3}}$$

$$= \$2400$$

Example 2.4

What is the cost of filling a rectangular tank 100 ft long, 40 ft wide, and 10 ft deep, with water at a cost of $0.06 per 1000 gallons?

Solution

$0.06 per 1000 gallons is equal to $0.00006 per gallon. The dimensional equation is

$$\$ = (\text{ft})(\text{ft})(\text{ft})\left(\frac{\text{gal}}{\text{ft}^3}\right)\left(\frac{\$}{\text{gal}}\right)$$

$$= (100 \ \text{ft})(40 \ \text{ft})(10 \ \text{ft})\left(7.5 \ \frac{\text{gal}}{\text{ft}^3}\right)(\$0.06)\left(0.00006 \ \frac{\$}{\text{gal}}\right)$$

$$= \$18$$

4. PRACTICE PROBLEMS

1. An excavation project will cost $0.30 per cubic yard of earth excavated. Most nearly, what is the cost of excavation of a ditch of rectangular cross section that is 3 ft wide, 4 ft deep, and 324 ft long?

(A) $4

(B) $11

(C) $14

(D) $43

2. A 6 ft by 3 ft concrete box culvert 54 ft long is to be constructed. The walls, footing, and deck must be 6 in thick. Wing walls can be disregarded. Most nearly, how many cubic yards of concrete are needed to construct the culvert?

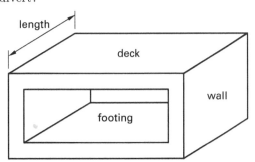

(A) 20 yd³

(B) 36 yd³

(C) 56 yd³

(D) 60 yd³

3. Most nearly, what is the mass of water in a full rectangular tank 8 ft long, 5 ft wide, and 5 ft deep?

(A) 1.3 tons

(B) 6.3 tons

(C) 8.4 tons

(D) 10.5 tons

4. A cross-sectional view of a concrete curb and gutter is shown.

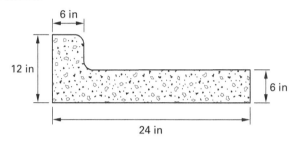

Most nearly, how many lineal feet of curb and gutter can be poured with 1 yd³ of concrete?

(A) 11 ft

(B) 18 ft

(C) 22 ft

(D) 27 ft

5. A water truck weighs 11,808 lbf when empty and 28,468 lbf when loaded with water. Most nearly, how many gallons of water does the truck hold?

(A) 1400 gal

(B) 2000 gal

(C) 3400 gal

(D) 4800 gal

6. A canal is to be excavated to a trapezoidal cross section 540 ft long, 30 ft at the top, 6 ft at the bottom, and 5 ft deep. Excavation costs $0.50 per cubic yard. What is the cost of excavation?

7. A drainage ditch on level ground is to be 162 ft long, 4 ft wide at the bottom, and 12 ft wide at the top, with an average depth of 2 ft. How many cubic yards of earth must be excavated to make the ditch?

8. How tall must a cylindrical tank 10 ft in diameter be in order to have a capacity of 3000 gal?

9. A parking lot is 100 ft by 81 ft. Pavement for the parking lot costs $9.00 per square yard. What is the cost of paving the parking lot?

10. A rectangular building lot 32 ft wide has an area of 3840 ft². How long is the lot?

11. An electric power line is to be built from city A to city B. City A is 16 mi to the north and 12 mi to the west of city B.

(a) What length of wire is needed to connect the two cities?

(b) If the wire weighs 50 lbf per 100 ft, what is the total weight of wire needed?

12. A right triangular plot of land measures 320 yards west to east and ¼ mile south to north. A second rectangular plot of land measures 250 yards west to east and has the same total acreage. Which plot will require the most perimeter fencing?

13. A shaft with a diameter of 24 in is drilled 54 ft deep and filled with concrete as part of a bridge pier. How many cubic yards of concrete are needed to fill the shaft?

14. A new swimming pool is 100 ft long, 50 ft wide, 2 ft deep at the shallow end, and 10 ft deep at the deep end. Water for the pool costs $0.20 per 1000 gallons. What will it cost to fill the pool with water?

15. A 24 lbm wheel of cheese is 8 in thick with a 16 in diameter. The cheese is priced at $1 per pound. A 30° sector is cut from the wheel for sale.

(a) What is the cost of the sector?

(b) What is the volume of the sector?

16. An open rectangular concrete tank is 10 ft long, 5 ft wide, and 4 ft 6 in tall on the outside. The walls and floor of the tank are 6 in thick. How many gallons of water will the tank contain when three-quarters full?

17. A piece of property to be purchased for a highway right-of-way is bounded by an arc of a circle with a 500 ft radius and a chord of that circle with a central angle of 90°. What is the area of the piece of property?

18. A cylindrical water tank contains 60,000 gal of water when the water is 5 ft deep. What is the diameter of the tank?

19. A lot 150 ft long and 100 ft wide is to be leveled for building construction. The fill at the front of the lot is 1.4 ft, and the fill at the rear of the lot is 2.2 ft. Shrinkage of soil can be disregarded. How many cubic yards of dirt will be needed to make the fill?

20. Write a dimensional equation to convert the given quantities to an equivalent quantity in the unit of measure indicated.

(a) Convert 48 in to feet.

(b) Convert 3 ft to inches.

(c) Convert 72 ft^2 to square yards.

(d) Convert 588 ft to yards.

(e) Convert 121 yd to feet.

(f) Convert 2 mi to feet.

(g) Convert 4 ft^2 to square inches.

(h) Convert 432 in^2 to square feet.

(i) Convert 5 yd^2 to square feet.

(j) Convert 81 ft^2 to square yards.

(k) Convert 2 ac to square feet.

(l) Convert 21,780 ft^2 to acres.

(m) Convert 3 ft^3 to cubic inches.

(n) Convert 3456 in^3 to cubic feet.

(o) Convert 5 yd^3 to cubic feet.

(p) Convert 135 ft^3 to cubic yards.

(q) Convert 3 gal to cubic inches.

(r) Convert 693 in^3 to gallons.

(s) Convert 4 gal of water to pounds.

(t) Convert 25 lbm of water to gallons.

(u) Convert 87,120 ft^2 to acres.

(v) Convert 1320 ft to miles.

(w) Convert 7 yd^2 to square feet.

21. Find the needed quantities by including the given quantities within a dimensional equation.

(a) Find the number of square yards in a driveway 20 ft wide and 54 ft long.

(b) Find the cost of a concrete sidewalk 4 ft wide, 81 ft long, and 4 in thick if concrete costs $30 per cubic yard.

(c) How many acres are in a rectangular plot 545 ft long and 400 ft wide?

(d) What is the weight in tons of 2000 gal of water?

(e) Find the velocity in feet per second of a vehicle traveling 72 miles per hour.

(f) What is the weight in tons of the water in a full cylindrical tank 10 ft tall with a 10 ft diameter?

(g) What is the cost of excavating a ditch of rectangular cross section 3 ft wide, 4 ft deep, and 324 ft long if excavation costs $0.30 per cubic yard?

SOLUTIONS

1. The cost of excavation is

$$\frac{(3 \text{ ft})(4 \text{ ft})(324 \text{ ft})\left(\dfrac{\$0.30}{\text{yd}^3}\right)}{27 \ \dfrac{\text{ft}^3}{\text{yd}^3}} = \$43.20 \quad (\$43)$$

The answer is (D).

2. The volume is

$$V = \frac{((7 \text{ ft})(4 \text{ ft}) - (6 \text{ ft})(3 \text{ ft}))(54 \text{ ft})}{27 \ \dfrac{\text{ft}^3}{\text{yd}^3}}$$

$$= 20 \text{ yd}^3$$

The answer is (A).

3. The mass of water in the tank is

$$m = \frac{(8 \text{ ft})(5 \text{ ft})(5 \text{ ft})\left(62.5 \ \dfrac{\text{lbm}}{\text{ft}^3}\right)}{2000 \ \dfrac{\text{lbm}}{\text{ton}}}$$

$$= 6.25 \text{ tons} \quad (6.3 \text{ tons})$$

The answer is (B).

4. The length of curb and gutter that can be poured is

$$L = \frac{27 \ \dfrac{\text{ft}^3}{\text{yd}^3}}{(1.0 \text{ ft})(0.5 \text{ ft}) + (1.5 \text{ ft})(0.5 \text{ ft})}$$

$$= 21.6 \text{ ft} \quad (22 \text{ ft})$$

The answer is (C).

5. Water weighs 8.33 lbf per gallon. The capacity of the truck in gallons is

$$\frac{28{,}468 \text{ lbf} - 11{,}808 \text{ lbf}}{8.33 \ \dfrac{\text{lbf}}{\text{gal}}} = 2000 \text{ gal}$$

The answer is (B).

6. Solve by writing a dimensional equation. The cost is

$$\text{cost} = \frac{\left(\dfrac{30 \text{ ft} + 6 \text{ ft}}{2}\right)(5 \text{ ft})(540 \text{ ft})\left(\dfrac{\$0.50}{1 \text{ yd}^3}\right)}{27 \ \dfrac{\text{ft}^3}{\text{yd}^3}}$$

$$= \$900.00$$

7. Solve by writing a dimensional equation. The volume is

$$V = \frac{\left(\dfrac{12 \text{ ft} + 4 \text{ ft}}{2}\right)(2 \text{ ft})(162 \text{ ft})}{27 \ \dfrac{\text{ft}^3}{\text{yd}^3}} = 96 \text{ yd}^3$$

8. Solve by writing a dimensional equation. The needed height of the tank is

$$h = \frac{3000 \text{ gal}}{\left(7.5 \ \dfrac{\text{gal}}{\text{ft}^3}\right)\left(\dfrac{\pi}{4}\right)(10 \text{ ft})^2} = 5.1 \text{ ft}$$

9. Solve by writing a dimensional equation. The cost is

$$\text{cost} = \frac{(100 \text{ ft})(81 \text{ ft})\left(\dfrac{\$9.00}{1 \text{ yd}^2}\right)}{9 \ \dfrac{\text{ft}^2}{\text{yd}^2}} = \$8100.00$$

10. Solve by writing a dimensional equation. The length is

$$d = \frac{3840 \text{ ft}^2}{32 \text{ ft}} = 120 \text{ ft}$$

11.

(a) Solve by writing a dimensional equation. The length is

$$L = \sqrt{(16 \text{ mi})^2 + (12 \text{ mi})^2} = 20 \text{ mi}$$

(b) Solve by writing a dimensional equation. The weight is

$$W = \frac{(20 \text{ mi})\left(5280 \ \dfrac{\text{ft}}{\text{mi}}\right)(50 \text{ lbf})}{(100 \text{ ft})\left(2000 \ \dfrac{\text{lbf}}{\text{ton}}\right)} = 26.4 \text{ ton}$$

12. Solve by writing dimensional equations.

The area of the triangle is

$$\left(\frac{1}{2}\right)(320 \text{ yd})\left(3 \frac{\text{ft}}{\text{yd}}\right)\left(\frac{1}{4} \text{ mi}\right)\left(5280 \frac{\text{ft}}{\text{mi}}\right) = 633,600 \text{ ft}^2$$

The perimeter of the triangle is

$$\sqrt{(960)^2 + (1320 \text{ ft})^2} + 960 \text{ ft} + 1320 \text{ ft} = 3912 \text{ ft}$$

The length of the sides of the rectangle perpendicular to the 250 yd sides is

$$\frac{633,600 \text{ ft}^2}{(250 \text{ yd})\left(3 \frac{\text{ft}}{\text{yd}}\right)} = 845 \text{ ft}$$

The perimeter of the rectangle is

$$(2)\left(3 \frac{\text{ft}}{\text{yd}}\right)(250 \text{ yd}) + (2)(845 \text{ ft}) = 3190 \text{ ft}$$

The triangular plot of land will need the most fencing.

13. Solve by writing a dimensional equation. The volume of the shaft is

$$V = \frac{\pi(1 \text{ ft})^2(54 \text{ ft})}{27 \frac{\text{ft}^3}{\text{yd}^3}} = 6.3 \text{ yd}^3$$

14. Solve by writing a dimensional equation. A cost of $0.20 per 1000 gallons is $0.00020 per gallon. The cost is

$$\text{cost} = \left(\frac{2 \text{ ft} + 10 \text{ ft}}{2}\right)(50 \text{ ft})(100 \text{ ft})\left(7.5 \frac{\text{gal}}{\text{ft}^3}\right)\left(\frac{\$0.00020}{\text{gal}}\right)$$
$$= \$45.00$$

15.

(a) Solve by writing a dimensional equation. The cost is

$$\left(\frac{30°}{360°}\right)(24 \text{ lbm})\left(\frac{\$1.00}{1 \text{ lbm}}\right) = \$2.00$$

(b) Solve by writing a dimensional equation. The volume is

$$V = \left(\frac{30°}{360°}\right)\pi(8 \text{ in})^2(8 \text{ in}) = 134 \text{ in}^3$$

16. Solve by writing a dimensional equation. The volume is

$$V = \left(\frac{3}{4}\right)(10 \text{ ft} - (2 \times 0.5 \text{ ft}))(5 \text{ ft} - (2 \times 0.5 \text{ ft}))(4.5 \text{ ft} - 0.5 \text{ ft})\left(7.48 \frac{\text{gal}}{\text{ft}^3}\right)$$
$$= 807.84 \text{ gal} \quad (810 \text{ gal})$$

17. Solve by writing a dimensional equation. The area is

$$A = \left(\frac{90°}{360°}\right)\left(\frac{\pi}{4}\right)(1000 \text{ ft})^2 - \left(\frac{1}{2}\right)(500 \text{ ft})(500 \text{ ft})$$
$$= 71,350 \text{ ft}^2$$

18. Solve by writing a dimensional equation. The diameter is

$$D = 2\sqrt{\frac{60,000 \text{ gal}}{\left(7.5 \frac{\text{gal}}{\text{ft}^3}\right)(5 \text{ ft})\pi}} = 45 \text{ ft}$$

19. Solve by writing a dimensional equation. The volume is

$$V = \frac{\left(\frac{1.4 \text{ ft} + 2.2 \text{ ft}}{2}\right)(150 \text{ ft})(100 \text{ ft})}{27 \frac{\text{ft}^3}{\text{yd}^3}} = 1000 \text{ yd}^3$$

20.

(a)

$$\frac{48 \text{ in}}{12 \frac{\text{in}}{\text{ft}}} = (48 \text{ in})\left(\frac{1 \text{ ft}}{12 \text{ in}}\right) = 4 \text{ ft}$$

(b)

$$(3 \text{ ft})\left(12 \frac{\text{in}}{\text{ft}}\right) = 36 \text{ in}$$

(c)

$$\frac{72 \text{ ft}^2}{9 \frac{\text{ft}^2}{\text{yd}^2}} = (72 \text{ ft}^2)\left(\frac{1 \text{ yd}^2}{9 \text{ ft}^2}\right) = 8 \text{ yd}^2$$

(d)

$$\frac{588 \text{ ft}}{3 \frac{\text{ft}}{\text{yd}}} = 196 \text{ yd}$$

(e)

$$(121 \text{ yd})\left(3 \ \frac{\text{ft}}{\text{yd}}\right) = 363 \text{ ft}$$

(f)

$$(2 \text{ mi})\left(5280 \ \frac{\text{ft}}{\text{mi}}\right) = 10{,}560 \text{ ft}$$

(g)

$$(4 \text{ ft}^2)\left(144 \ \frac{\text{in}^2}{\text{ft}^2}\right) = 576 \text{ in}^2$$

(h)

$$\frac{432 \text{ in}^2}{144 \ \frac{\text{in}^2}{\text{ft}^2}} = 3 \text{ ft}^2$$

(i)

$$(5 \text{ yd}^2)\left(9 \ \frac{\text{ft}^2}{\text{yd}^2}\right) = 45 \text{ ft}^2$$

(j)

$$\frac{81 \text{ ft}^2}{9 \ \frac{\text{ft}^2}{\text{yd}^2}} = 9 \text{ yd}^2$$

(k)

$$(2 \text{ ac})\left(43{,}560 \ \frac{\text{ft}^2}{\text{ac}}\right) = 87{,}120 \text{ ft}^2$$

(l)

$$\frac{21{,}780 \text{ ft}^2}{43{,}560 \ \frac{\text{ft}^2}{\text{ac}}} = 0.5 \text{ ac}$$

(m)

$$(3 \text{ ft}^3)\left(1728 \ \frac{\text{in}^3}{\text{ft}^3}\right) = 5184 \text{ in}^3$$

(n)

$$\frac{3456 \text{ in}^3}{1728 \ \frac{\text{in}^3}{\text{ft}^3}} = 2 \text{ ft}^3$$

(o)

$$(5 \text{ yd}^3)\left(27 \ \frac{\text{ft}^3}{\text{yd}^3}\right) = 135 \text{ ft}^3$$

(p)

$$\frac{135 \text{ ft}^3}{27 \ \frac{\text{ft}^3}{\text{yd}^3}} = 5 \text{ yd}^3$$

(q)

$$(3 \text{ gal})\left(231 \ \frac{\text{in}^3}{\text{gal}}\right) = 693 \text{ in}^3$$

(r)

$$\frac{693 \text{ in}^3}{231 \ \frac{\text{in}^3}{\text{gal}}} = 3 \text{ gal}$$

(s)

$$(4 \text{ gal})\left(8\frac{1}{3} \ \frac{\text{lbm}}{\text{gal}}\right) = 33\frac{1}{3} \text{ lbm}$$

(t)

$$\frac{25 \text{ lbm}}{8\frac{1}{3} \ \frac{\text{lbm}}{\text{gal}}} = 3 \text{ gal}$$

(u)

$$\frac{87{,}120 \text{ ft}^2}{43{,}560 \ \frac{\text{ft}^2}{\text{ac}}} = 2 \text{ ac}$$

(v)

$$\frac{1320 \text{ ft}}{5280 \frac{\text{ft}}{\text{mi}}} = \frac{1}{4} \text{ mi}$$

(w)

$$(7 \text{ yd}^2)\left(9 \frac{\text{ft}^2}{\text{yd}^2}\right) = 63 \text{ ft}^2$$

21.

(a)

$$A = \frac{(20 \text{ ft})(54 \text{ ft})}{9 \frac{\text{ft}^2}{\text{yd}^2}}$$

$$= (20 \text{ ft})(54 \text{ ft})\left(\frac{1 \text{ yd}^2}{9 \text{ ft}^2}\right)$$

$$= 120 \text{ yd}^2$$

(b)

$$\text{cost} = \frac{(4 \text{ ft})(81 \text{ ft})\left(\dfrac{4 \text{ in}}{12 \frac{\text{in}}{\text{ft}}}\right)\left(\dfrac{\$30}{1 \text{ yd}^3}\right)}{\left(27 \dfrac{\text{ft}^3}{\text{yd}^3}\right)}$$

$$= (4 \text{ ft})(81 \text{ ft})\left(\frac{1}{3} \text{ ft}\right)\left(\frac{1 \text{ yd}^3}{27 \text{ ft}^3}\right)\left(\frac{\$30}{1 \text{ yd}^3}\right)$$

$$= \$120$$

(c)

$$A = \frac{(545 \text{ ft})(400 \text{ ft})}{43{,}560 \frac{\text{ft}^2}{\text{ac}}} = 5 \text{ ac}$$

(d)

$$W = \frac{(2000 \text{ gal})\left(8.33 \frac{\text{lbm}}{\text{gal}}\right)}{2000 \frac{\text{lbm}}{\text{ton}}} = 8.33 \text{ tons}$$

(e)

$$v = \frac{\left(72 \frac{\text{mi}}{\text{hr}}\right)\left(5280 \frac{\text{ft}}{\text{mi}}\right)}{3600 \frac{\text{sec}}{\text{hr}}}$$

$$= 106 \text{ ft/sec}$$

(f)

$$W = \frac{\left(\dfrac{\pi}{4}\right)(10 \text{ ft})^2(10 \text{ ft})\left(62.5 \dfrac{\text{lbf}}{\text{ft}^3}\right)}{2000 \dfrac{\text{lbf}}{\text{ton}}}$$

$$= 24.5 \text{ ton}$$

(g)

$$\text{cost} = \frac{(3 \text{ ft})(4 \text{ ft})(324 \text{ ft})\left(\dfrac{\$0.30}{\text{yd}^3}\right)}{27 \dfrac{\text{ft}^3}{\text{yd}^3}} = \$43.20$$

3 Systems of Units

1. THE ENGLISH SYSTEM

American colonists from England brought with them the system of weights and measures in use at the time in England, called the *English system.*

The English system was originally based on parts of the body, such as the foot, the hand, and the thumb. In the fourteenth century, the English king proclaimed the length of the English inch to be the length of three barleycorns laid end to end. Later, more sophisticated methods were used for standardization, but the system had no uniform conversion factors. The units used in the English system are summarized in Table 3.1.

Table 3.1 *English System Weights and Measures*

linear measure	square measure	cubic measure
12 in = 1 ft	$144 \text{ in}^2 = 1 \text{ ft}^2$	$1728 \text{ in}^3 = 1 \text{ ft}^3$
3 ft = 1 yd	$9 \text{ ft}^2 = 1 \text{ yd}^2$	$27 \text{ ft}^3 = 1 \text{ yd}^3$
$5\frac{1}{2}$ yd = 1 rod	$43{,}560 \text{ ft}^2 = 1 \text{ ac}$	$231 \text{ in}^3 = 1 \text{ gal}$
5280 ft = 1 mi	$640 \text{ ac} = 1 \text{ mi}^2$	$7.5 \text{ gal} = 1 \text{ ft}^3$

During the seventeenth and eighteenth centuries, the English system was well standardized (12 inches per foot, 3 feet per yard, $5\frac{1}{2}$ yards per rod, and 16 ounces per pound), but such was not the case in the rest of Europe. There was such a wide variety of weights and measures in use that commerce was difficult.

2. THE METRIC SYSTEM

The variety of weights and measurements in Europe prompted the French National Assembly to enact a decree in 1790 that directed the French Academy of Sciences to find standards for all weights and measures. The French Academy was supposed to work with the Royal Society of London, but the English did not

participate, so the French proceeded alone. The result was the *metric system*, which uses base ten in converting units of measure.

The metric system spread rapidly in the nineteenth century. In 1872, France called an international meeting that was attended by 26 nations, including the United States, to further refine the system. The meeting resulted in the establishment of the *International Bureau of Weights and Measures*, and in 1960 an extensive revision and simplification resulted in the *International System of Units* (officially abbreviated as *SI units*), which is in use in most countries today.

3. THE SI SYSTEM

The SI system uses base ten to express multiples and submultiples, just as the metric system has always done. The seven base units of measurement are given in Table 3.2.

Table 3.2 *Seven Base Units of SI Measurement*

value measured	unit	abbreviation
length	meter	m
mass	kilogram	kg
time	second	s
temperature	kelvin	K
electric current	ampere	A
amount of substance	mole	mol
luminous intensity	candela	cd

The *meter* is defined as the distance traveled by light in a vacuum in $1/299\,792\,458$ of a second. The SI unit of measure is the square meter, m^2, and the SI unit of volume is the cubic meter, m^3. Land is measured by the *hectare*, equal to $10\,000 \text{ m}^2$. Fluid volume is measured by the *liter*, equal to 0.001 m^3.

The *second* is defined as the duration of 9,192,631,770 oscillations between the two hyperfine levels of the ground state of a cesium 133 atom.

To convert multiples and submultiples of units in the SI system, the decimal point is moved to the right or left, just as in working with decimal fractions. To facilitate use of the system, names are given to the various powers of ten. For example, a centimeter is one-hundredth of a meter, and a kilometer is one thousand meters. To express meters in centimeters, move the decimal two places to the right, and to express meters in kilometers, move the decimal three places to the left. The prefixes and symbols for all SI unit prefixes are shown in Table 3.3.

Table 3.3 *Prefixes for SI Units*

multiple or submultiple	prefix	symbol
$1,000,000,000,000 = 10^{12}$	tera	T
$1,000,000,000 = 10^{9}$	giga	G
$1,000,000 = 10^{6}$	mega	M
$1000 = 10^{3}$	kilo	k
$100 = 10^{2}$	hecto	h
$10 = 10$	deka	da
$0.1 = 10^{-1}$	deci	d
$0.01 = 10^{-2}$	centi	c
$0.001 = 10^{-3}$	milli	m
$0.000\,001 = 10^{-6}$	micro	μ
$0.000\,000\,001 = 10^{-9}$	nano	n
$0.000\,000\,000\,001 = 10^{-12}$	pico	p
$0.000\,000\,000\,000\,001 = 10^{-15}$	femto	f
$0.000\,000\,000\,000\,000\,000\,001 = 10^{-18}$	atto	a

4. CONVERSION OF INCHES TO DECIMALS OF A FOOT

Engineering plans usually show dimensions of structures in feet and inches, while elevations are established in feet and decimals of a foot. Therefore, it is often necessary to convert between inches and decimals of a foot to establish finished elevations.

The key to conversion is shown in Table 3.4. The values of 1 in and $\frac{1}{8}$ in are important parts of the key.

Conversions can be made without Table 3.4 by memorizing the following values.

$3\text{ in} = 0.25\text{ ft}$

$6\text{ in} = 0.50\text{ ft}$

$9\text{ in} = 0.75\text{ ft}$

$4\text{ in} = 0.33\text{ ft}$

$8\text{ in} = 0.67\text{ ft}$

$1\text{ in} = 0.08\text{ ft}$

$\frac{1}{8}\text{ in} = 0.01\text{ ft}$

In converting measurements expressed in feet, inches, and fractions of an inch to feet and decimals of a foot, convert the inches and fractions of an inch separately to decimals of a foot, then add the three parts. In some cases, it is more efficient to convert a value to the nearest greater value that is easy to remember, then subtract an appropriate value to arrive at the final value.

Table 3.4 *Conversion of Inches to Decimals of a Foot*

inch measurement	decimal measurement	conversion
$\frac{1}{8}$ in	0.01 ft	$\frac{1}{8}$ in = 1 in ÷ 8 ÷ 12 = 0.0104 ft
1 in	0.08 ft	1 in = $\frac{1}{12}$ ft = 1 ÷ 12 = 0.0833 ft
2 in	0.17 ft	1 in + 1 in = $\frac{2}{12}$ ft = $\frac{1}{6}$ ft = 0.166 ft
3 in	0.25 ft	$\frac{3}{12}$ ft = $\frac{1}{4}$ ft = 0.250 ft
4 in	0.33 ft	$\frac{4}{12}$ ft = $\frac{1}{3}$ ft = 0.333 ft
5 in	0.42 ft	4 in + 1 in = 0.333 + 0.083 = 0.416 ft
6 in	0.50 ft	$\frac{6}{12}$ ft = $\frac{1}{2}$ ft = 0.500 ft
7 in	0.58 ft	6 in + 1 in = 0.500 + 0.083 = 0.583 ft
8 in	0.67 ft	$\frac{8}{12}$ ft = $\frac{2}{3}$ ft = 0.666 ft
9 in	0.75 ft	$\frac{9}{12}$ ft = $\frac{3}{4}$ ft = 0.750 ft
10 in	0.83 ft	9 in + 1 in = 0.750 + 0.083 = 0.833 ft
11 in	0.92 ft	10 in + 1 in = 0.833 + 0.083 = 0.916 ft
12 in	1.00 ft	—

Example 3.1

What is 11 ft $9\frac{1}{8}$ in expressed in feet and decimals of a foot?

Solution

Convert the values using the conversions in Table 3.4. 9 in is equivalent to 0.75 ft, and $\frac{1}{8}$ in is equivalent to 0.01 ft. Summing the values, the measurement in feet and decimals of a foot is

$$11.00\text{ ft} + 0.75\text{ ft} + 0.01\text{ ft} = 11.76\text{ ft}$$

5. CONVERSION OF DECIMALS OF A FOOT TO INCHES

In converting measurements expressed in feet and decimals of a foot to feet, inches, and fractions of an inch, either find or recall the decimal of a foot value for the full inch nearest to and less than the given measurement. Then convert the remainder, which should be in hundredths of a foot, to a fraction expressed in eighths of an inch. Or, recall the decimal of a foot value for the full inch nearest to and more than the given measurement, and subtract to arrive at the given measurement. Conversions are made only to the nearest $\frac{1}{8}$ inch in this procedure.

6. OTHER SYSTEMS OF UNITS

Some measures used in the Public Land Survey System (also known as the Rectangular Survey System) are listed in Table 3.5.

Table 3.5 *Measures of the Public Land Survey System*

1 m = 39.37 in exactly
1 m = 39.37/12 = 3.2808333 U.S. survey ft
1 U.S. survey ft = 12/39.37 = 0.3048006 m
1 international ft = 0.3048 m
1 Gunter's chain = 100 links = 66 ft
1 Gunter's link = 7.92 in
1 Gunter's chain = 4 rods = 4 poles
1 mi = 80 chains
10 chains^2 = 435,600 ft^2 = 10 ac

7. PRACTICE PROBLEMS

1. A pit is to be filled with soil as part of a construction project. The volume of the pit is 54 ft^3. The construction firm purchases soil in cubic yards. How many cubic yards of soil are needed to fill the pit?

(A)   2.0 yd^3

(B)   7.5 yd^3

(C)   9.0 yd^3

(D)   18 yd^3

2.

(a) Convert 4 yd to inches.

(b) Convert 288 in^2 to square feet.

(c) Convert 3 ac to square feet.

(d) Convert 0.5 ft^2 to square inches.

(e) Convert 2 gal to cubic inches.

(f) Convert 15 gal to cubic feet.

3. What is 10 ft 11$^1\!/_4$ in expressed in feet and decimals of a foot?

(A)   10.79 ft

(B)   10.88 ft

(C)   10.94 ft

(D)   10.97 ft

4. What is the measurement of 8.72 ft expressed in feet and inches?

(A)   8 ft 4$^3\!/_4$ in

(B)   8 ft 6$^5\!/_8$ in

(C)   8 ft 7$^1\!/_2$ in

(D)   8 ft 8$^5\!/_8$ in

5. Convert each value from feet and decimals to feet and inches.

(a) 0.36 ft

(b) 1.35 ft

(c) 2.69 ft

(d) 2.94 ft

(e) 3.52 ft

(f) 3.87 ft

(g) 4.76 ft

(h) 4.79 ft

(i) 4.83 ft

(j) 5.06 ft

(k) 5.60 ft

(l) 6.08 ft

(m) 6.16 ft

(n) 6.25 ft

(o) 6.67 ft

(p) 7.81 ft

(q) 8.21 ft

(r) 9.23 ft

(s) 9.27 ft

SOLUTIONS

1. Convert the volume of the pit to cubic yards.

$$\frac{54 \text{ ft}^3}{27 \ \dfrac{\text{ft}^3}{\text{yd}^3}} = 2 \text{ yd}^3$$

The answer is (A).

2.

(a)   144 in

(b)   2 ft^2

(c)   130,680 ft^2

(d)   72 in^2

(e)   462 in^3

(f)   2 ft^3

3. From Table 3.4, 11 in = 0.92 ft. $\frac{1}{4}$ in is double the length of $\frac{1}{8}$ in, so

$$\frac{1}{4} \text{ in} = (2)\left(\frac{1}{8} \text{ in}\right)$$

$$\frac{1}{8} \text{ in} = 0.01 \text{ ft}$$

$$\frac{1}{4} \text{ in} = (2)(0.01 \text{ ft})$$

$$= 0.02 \text{ ft}$$

Summing the values, the measurement in feet and decimals of a foot is

$$10 \text{ ft} + 0.92 \text{ ft} + 0.02 \text{ ft} = 10.94 \text{ ft}$$

The answer is (C).

4. Using Table 3.4, convert 0.72 ft to inches. The nearest decimal value less than 0.72 ft is 8 in = 0.67 ft, which leaves a remainder of 0.05 ft. $\frac{1}{8}$ in = 0.01 ft, so

$$0.05 \text{ ft} = (5)(0.01 \text{ ft})$$

$$0.01 \text{ ft} = \frac{1}{8} \text{ in}$$

$$0.05 \text{ ft} = (5)\left(\frac{1}{8} \text{ in}\right)$$

$$= \frac{5}{8} \text{ in}$$

Summing the values, the measurement in feet and inches is

$$8 \text{ ft} + 8 \text{ in} + \frac{5}{8} \text{ in} = 8 \text{ ft } 8\frac{5}{8} \text{ in}$$

The answer is (D).

5.

(a) $4\frac{1}{3}$ in

(b)   1 ft $4\frac{1}{5}$ in

(c)   2 ft $8\frac{1}{4}$ in

(d)   2 ft $11\frac{1}{4}$ in

(e)   3 ft $6\frac{1}{4}$ in

(f)   3 ft $10\frac{1}{2}$ in

(g)   4 ft $9\frac{1}{8}$ in

(h)   4 ft $9\frac{1}{2}$ in

(i)   4 ft 10 in

(j)   5 ft $\frac{3}{4}$ in

(k)   5 ft $7\frac{1}{4}$ in

(l)   6 ft 1 in

(m)   6 ft $1\frac{7}{8}$ in

(n)   6 ft 3 in

(o)   6 ft 8 in

(p)   7 ft $9\frac{3}{4}$ in

(q)   8 ft $2\frac{1}{2}$ in

(r)   9 ft $2\frac{3}{4}$ in

(s)   9 ft $3\frac{1}{4}$ in

Perimeter and Circumference

Nomenclature

C	circumference	ft	m
d	diameter	ft	m
r	radius	ft	m

1. DEFINITION

The sum of the lengths of the sides of a polygon is called the *perimeter* of the polygon.

Example 4.1

Find the perimeter of the right triangle shown.

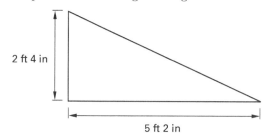

Solution

Convert the lengths to feet and decimals. From Table 3.4, 2 in = 0.17 ft, and 4 in = 0.33 ft. From the Pythagorean theorem, the length of the hypotenuse is

$$a^2 + b^2 = c^2$$
$$c = \sqrt{(5.17 \text{ ft})^2 + (2.33 \text{ ft})^2}$$
$$= 5.67 \text{ ft}$$

The perimeter of the triangle is

$$\text{perimeter} = 5.17 \text{ ft} + 2.33 \text{ ft} + 5.67 \text{ ft}$$
$$= 13.17 \text{ ft}$$

Example 4.2

Find the perimeter of the isosceles trapezoid shown.

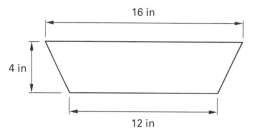

Solution

Find the actual length of one side of the trapezoid by treating the side as the hypotenuse of a triangle. The altitude of the triangle is 4 in. The base of the triangle is one-half the difference between the length of the top of the trapezoid and the length of the bottom of the trapezoid, or (16 in – 12 in)/2 = 2 in. From the Pythagorean theorem, the length of the side of the trapezoid is

$$\text{side} = \sqrt{(4 \text{ in})^2 + (2 \text{ in})^2} = 4.5 \text{ in}$$

The perimeter of the triangle is

$$\text{perimeter} = 12 \text{ in} + 4.5 \text{ in} + 16 \text{ in} + 4.5 \text{ in} = 37 \text{ in}$$

Example 4.3

The altitude and base of the right triangle shown are 12 in and 9 in, respectively. Find the triangle's perimeter.

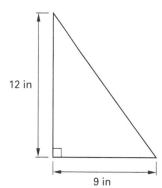

Solution

From the Pythagorean theorem, the length of the hypotenuse is

$$\text{hypotenuse} = \sqrt{(9 \text{ in})^2 + (12 \text{ in})^2} = 15 \text{ in}$$

The perimeter of the triangle is

$$\text{perimeter} = 9 \text{ in} + 12 \text{ in} + 15 \text{ in} = 36 \text{ in}$$

2. CIRCUMFERENCE OF A CIRCLE

The *circumference* of a circle is the distance around the circle. A circle contains 360°. Regardless of the size, the circumference of any circle is always approximately 3.14 times the length of the diameter. The exact ratio of the circumference to the diameter is the number π. It is an irrational number, but its value is usually considered to be 3.1416 or 3.14, depending on the accuracy desired, based on the accuracy of the measurement of the diameter. Thus, the circumference of any circle is

$$C = \pi d \qquad 4.1$$

Because the diameter is twice the radius,

$$C = 2\pi r \qquad 4.2$$

See Sec. 1.8 for more on the geometry of circles.

Example 4.4

Find the circumference of a 10 ft diameter circle.

Solution

From Eq. 4.1,

$$C = \pi d = \pi(10 \text{ ft}) = 31.42 \text{ ft}$$

Example 4.5

Find the circumference of a circle with a radius of 21 ft 9 in.

Solution

From Eq. 4.2,

$$C = 2\pi r = 2\pi(21.75 \text{ ft}) = 136.66 \text{ ft}$$

Example 4.6

Find the outside circumference of a concrete pipe with an inside diameter of 36 in and a wall thickness of 3 in.

Solution

The outside circumference will be based on the diameter of the entire pipe, including the wall. With a wall thickness of 3 in, the diameter is 36 in + (2)(3 in) = 42 in. From Eq. 4.1,

$$C = \pi d = \pi(42 \text{ in}) = 132 \text{ in}$$

Example 4.7

Find the diameter of a tank with a circumference of 62 ft 10 in.

Solution

From Table 3.4, 10 in is equal to 0.83 ft. From Eq. 4.1,

$$C = \pi d$$
$$d = \frac{C}{\pi} = \frac{62.83 \text{ ft}}{\pi} = 20 \text{ ft}$$

3. LENGTH OF AN ARC OF A CIRCLE

The *length of an arc* of a circle is proportional to its central angle. For example, a central angle of 90° (one-fourth of 360°) subtends an arc that is one-fourth the circumference in length, and an arc whose central angle is 45° is $(45°/360°)C$ or one-eighth the circumference in length.

Example 4.8

An arc of a 100 ft diameter circle has a central angle of 36°. What is the length of the arc?

Solution

The length of the arc is

$$L = \frac{\phi d \pi}{360°} = \frac{(36°)(100 \text{ ft})\pi}{360°} = 31 \text{ ft}$$

4. PRACTICE PROBLEMS

1. A right triangle has a base of 12 in and an altitude of 5 in. Most nearly, what is the perimeter of the triangle?

(A) 21 in

(B) 23 in

(C) 29 in

(D) 30 in

2. The floor of a room is 18 ft by 22 ft. Most nearly, what is the perimeter of the floor?

(A) 40 ft

(B) 50 ft

(C) 60 ft

(D) 80 ft

3. An isosceles triangle has a base of 12 in and an altitude of 8 in. Most nearly, what is the perimeter of the triangle?

(A) 23 in

(B) 28 in

(C) 32 in

(D) 35 in

4. Most nearly, what is the circumference of a 10 in diameter circle?

(A) 31 in

(B) 45 in

(C) 57 in

(D) 63 in

5. Most nearly, what is the circumference of a circle with a radius of 7 in?

(A) 22 in

(B) 34 in

(C) 44 in

(D) 58 in

6. A circle of 24 in radius has an arc with a central angle of 60°. Most nearly, what is the length of the arc?

(A) 9.5 in

(B) 19 in

(C) 25 in

(D) 30 in

7. A cylindrical tank has an outside circumference of 47.10 ft. Most nearly, what is the diameter of the tank?

(A) 13 ft

(B) 15 ft

(C) 23 ft

(D) 27 ft

8. A circle of 16 in diameter has an arc with a central angle of 45°. Most nearly, how long is the arc?

(A) 2.0 in

(B) 5.2 in

(C) 6.3 in

(D) 13 in

9. A tree trunk has a circumference of 3 ft 1¾ in. Most nearly, what is the diameter of the trunk?

(A) 0.9 ft

(B) 1.0 ft

(C) 1.2 ft

(D) 1.5 ft

10. What is the perimeter of the figure shown?

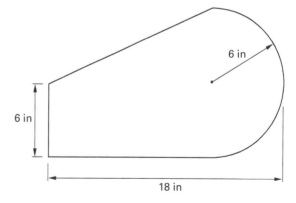

SOLUTIONS

1. From the Pythagorean theorem, the length of the hypotenuse is

$$a^2 + b^2 = c^2$$
$$c = \sqrt{(5 \text{ in})^2 + (12 \text{ in})^2}$$
$$= 13 \text{ in}$$

The perimeter of the triangle is

$$\text{perimeter} = 5 \text{ in} + 12 \text{ in} + 13 \text{ in} = 30 \text{ in}$$

The answer is (D).

2. The perimeter of the floor is equal to the length of all four sides, so double the length of one long side and double the length of one short side.

$$(2)(22 \text{ ft}) + (2)(18 \text{ ft}) = 80 \text{ ft}$$

The answer is (D).

3. For an isosceles triangle, a line from the vertex perpendicular to the base bisects the base. The length of a side of the triangle may be calculated by solving the right triangle formed by the line from the vertex to the base and half the base (6 in).

$$\text{side} = \sqrt{(8 \text{ in})^2 + (6 \text{ in})^2} = 10 \text{ in}$$

The perimeter is

$$\text{perimeter} = 10 \text{ in} + 10 \text{ in} + 12 \text{ in} = 32 \text{ in}$$

The answer is (C).

4. From Eq. 4.1,

$$C = \pi d$$
$$= \pi(10 \text{ in})$$
$$= 31.4 \text{ in} \quad (31 \text{ in})$$

The answer is (A).

5. From Eq. 4.2,

$$C = 2\pi r$$
$$= 2\pi(7 \text{ in})$$
$$= 43.98 \text{ in} \quad (44 \text{ in})$$

The answer is (C).

6. The length of the arc is

$$L = \frac{\phi d\pi}{360°}$$
$$= \frac{(60°)(48 \text{ in})\pi}{360°}$$
$$= 25.13 \text{ in} \quad (25 \text{ in})$$

The answer is (C).

7. From Eq. 4.1,

$$C = \pi d$$
$$d = \frac{C}{\pi}$$
$$= \frac{47.10 \text{ ft}}{\pi}$$
$$= 14.992 \text{ ft} \quad (15 \text{ ft})$$

The answer is (B).

8. The length of the arc is

$$L = \frac{\phi d\pi}{360°}$$
$$= \frac{(45°)(16 \text{ in})\pi}{360°}$$
$$= 6.28 \text{ in} \quad (6.3 \text{ in})$$

The answer is (C).

9. From Table 3.4, $1\tfrac{3}{4}$ in is equivalent to 0.14 ft. From Eq. 4.1,

$$C = \pi d$$
$$d = \frac{C}{\pi}$$
$$= \frac{3.14 \text{ ft}}{\pi}$$
$$= 1.0 \text{ ft}$$

The answer is (B).

10.

$$\text{arc} = \frac{2\pi r}{2} = \frac{2\pi(6 \text{ in})}{2} = 19 \text{ in}$$
$$\text{base} = 18 \text{ in} - 6 \text{ in} = 12 \text{ in}$$
$$\text{left side} = 6 \text{ in}$$
$$\text{top} = \sqrt{(6 \text{ in})^2 + (12 \text{ in})^2} = 13.4 \text{ in}$$
$$\text{total perimeter} = 50 \text{ in}$$

5 Area

Nomenclature

a, b, c	sides of a triangle or other shape	ft	m
A	area	ft^2	m^2
b	base	ft	m
d	diameter	ft	m
h	altitude	ft	m
l	length	ft	m
r	radius	ft	m
s	half the perimeter	ft	m
w	width	ft	m

Symbols

ϕ	angle	deg	deg

1. DEFINITION

An *area* is the surface within a set of lines. The area of a triangle is the surface within the three sides, the area of a circle is the surface within the circumference, and so on.

Area is measured in square units of distance: square inches (in^2), square feet (ft^2), square miles (mi^2), and so on.

A square inch is a square, each side of which is 1 in long. The rectangle shown in Fig. 5.1 has an area of 6 in^2: it could be exactly covered by six squares, each 1 in on a side.

The area formulas given in this chapter can all be found in Table 5.1 at the end of the chapter.

2. AREA OF A RECTANGLE

The area of a rectangle is equal to the product of the length and the width.

$$A = lw \qquad 5.1$$

Figure 5.1 *Square Inch*

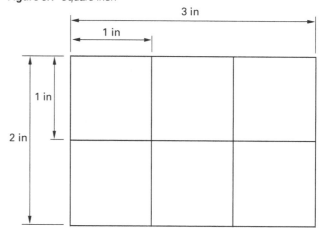

Example 5.1

Find the area of the floor of a room that is 20.25 ft long and 16.33 ft wide.

Solution

From Eq. 5.1,

$$\begin{aligned} A &= lw \\ &= (20.25 \text{ ft})(16.33 \text{ ft}) \\ &= 330.7 \text{ ft}^2 \end{aligned}$$

Example 5.2

Find the area of the walls of a room that is 8.0 ft high, 20.0 ft long, and 15.0 ft wide.

Solution

Find the area of one wall running the length of the room. From Eq. 5.1,

$$\begin{aligned} A_1 &= lw \\ &= (20 \text{ ft})(8 \text{ ft}) \\ &= 160 \text{ ft}^2 \end{aligned}$$

The total area of the walls running the length of the room is

$$(2)(160 \text{ ft}^2) = 320 \text{ ft}^2$$

Find the area of one wall running the width of the room. From Eq. 5.1,

$$A_2 = lw$$
$$= (15 \text{ ft})(8 \text{ ft})$$
$$= 120 \text{ ft}^2$$

The total area of the walls running the width of the room is

$$(2)(120 \text{ ft}^2) = 240 \text{ ft}^2$$

The total area of the walls is

$$320 \text{ ft}^2 + 240 \text{ ft}^2 = 560 \text{ ft}^2$$

3. AREA OF A TRIANGLE

The area of a triangle is expressed in terms of its base and its altitude. Any side of a triangle can be called the base. The *vertex* of a triangle is the point of the angle opposite the base, and the *altitude* of a triangle is the perpendicular distance from the vertex to the base.

The area of any triangle is equal to one-half the product of the base and the altitude, as shown in Fig. 5.2.

Figure 5.2 *Area of a Triangle*

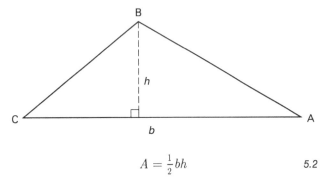

$$A = \tfrac{1}{2}bh \qquad\qquad 5.2$$

If the lengths of the three sides of a triangle, a, b, and c, are known, the area of the triangle can be found from Eq. 5.3.

$$A = \sqrt{s(s-a)(s-b)(s-c)}$$
$$s = \frac{(a+b+c)}{2} \qquad\qquad 5.3$$

Example 5.3

Find the area of a triangle with a base of 12 in and an altitude of 4 in.

Solution

$$A = \tfrac{1}{2}bh$$
$$= \left(\frac{1}{2}\right)(12 \text{ in})(4 \text{ in})$$
$$= 24 \text{ in}^2$$

Equation 5.2 also applies to a right triangle, in which case the altitude may be one of the sides, as illustrated in Ex. 5.4.

Example 5.4

Find the area of the right triangle shown.

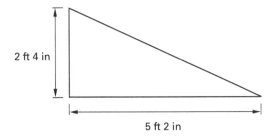

Solution

$$\text{base} = 5 \text{ ft } 2 \text{ in} = 5.17 \text{ ft}$$
$$\text{altitude} = 2 \text{ ft } 4 \text{ in} = 2.33 \text{ ft}$$
$$\text{area} = \frac{bh}{2} = \frac{(5.17 \text{ ft})(2.33 \text{ ft})}{2} = 6.02 \text{ ft}^2$$

Example 5.5

Find the area of a triangle with sides 32 ft, 46 ft, and 68 ft.

Solution

$$s = \frac{a+b+c}{2} = \frac{32 \text{ ft} + 46 \text{ ft} + 68 \text{ ft}}{2}$$
$$= 73 \text{ ft}$$

$$A = \sqrt{s(s-a)(s-b)(s-c)}$$
$$= \sqrt{(73 \text{ ft})(73 \text{ ft} - 32 \text{ ft})(73 \text{ ft} - 46 \text{ ft})(73 \text{ ft} - 68 \text{ ft})}$$
$$= 635.7 \text{ ft}^2$$

4. AREA OF A TRAPEZOID

The area of a trapezoid is equal to the product of the average width and the altitude. This may be expressed in another way: The area of a trapezoid is equal to one-half the sum of the two bases, a and b, times the altitude. Figure 5.3 illustrates these dimensions.

$$A = \tfrac{1}{2}h(a+b) \qquad\qquad 5.4$$

Figure 5.3 *Area of a Trapezoid*

Example 5.6

Find the area of the trapezoid shown.

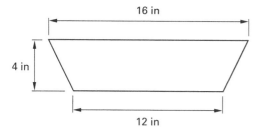

Solution

From Eq. 5.4,

$$A = \tfrac{1}{2}h(a+b) = \left(\tfrac{1}{2}\right)(4\text{ in})(16\text{ in} + 12\text{ in})$$
$$= 56\text{ in}^2$$

Example 5.7

A 100 ft long swimming pool is 4 ft deep at one end and 8 ft deep at the other end. Find the area of the trapezoidal section through the long axis of the pool.

Solution

From Eq. 5.4,

$$A = \tfrac{1}{2}h(a+b) = \left(\tfrac{1}{2}\right)(100\text{ ft})(4\text{ ft} + 8\text{ ft})$$
$$= 600\text{ ft}^2$$

Example 5.8

A drainage ditch has a trapezoidal cross section with a bottom width of 6 ft, a top width of 24 ft, and a depth of 4 ft. Find the area of the cross section.

Solution

From Eq. 5.4, using the depth of the ditch as the altitude,

$$A = \tfrac{1}{2}h(a+b) = \left(\tfrac{1}{2}\right)(4\text{ ft})(6\text{ ft} + 24\text{ ft})$$
$$= 60\text{ ft}^2$$

5. AREA OF A CIRCLE

The circumference of a circle is always 2π times the radius of the circle. Similarly, the area of a circle is always π times the square of its radius, or

$$A = \pi r^2 \qquad 5.5$$

Since the radius of a circle is equal to half the diameter, the area of a circle can also be expressed as

$$A = \pi\left(\frac{d}{2}\right)^2 = \left(\frac{\pi}{4}\right)d^2 \qquad 5.6$$

Example 5.9

Find the area of a 12 in circle.

Solution

From Eq. 5.5,

$$A = \pi r^2 = \pi(6\text{ in})^2$$
$$= 113\text{ in}^2$$

Example 5.10

Find the area of a 10 in circle.

Solution

From Eq. 5.6,

$$A = \left(\frac{\pi}{4}\right)d^2 = \left(\frac{\pi}{4}\right)(10\text{ in})^2$$
$$= 78.539\text{ in}^2 \quad (78.5\text{ in}^2)$$

Example 5.11

Find the area of an 11 in circle.

Solution

From Eq. 5.6,

$$A = \left(\frac{\pi}{4}\right)d^2 = \left(\frac{\pi}{4}\right)(11\text{ in})^2$$
$$= 95.033\text{ in}^2 \quad (95\text{ in}^2)$$

6. AREA OF A SECTOR OF A CIRCLE

The area of a sector of a circle is a fractional part of the area of the circle. The central angle of the sector, ϕ, is a measure of the fraction. As an example, a sector whose central angle is 90° is one-fourth of a circle because 90° is one-fourth of 360°. The area of a sector of a circle is illustrated in Fig. 5.4.

Figure 5.4 *Area of a Sector of a Circle*

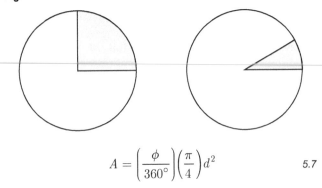

$$A = \left(\frac{\phi}{360°}\right)\left(\frac{\pi}{4}\right)d^2 \qquad 5.7$$

Example 5.12

Find the area of a sector in a 6 in diameter circle with a central angle of 30°.

Solution

$$A = \left(\frac{\phi}{360°}\right)\left(\frac{\pi}{4}\right)d^2 = \left(\frac{30°}{360°}\right)\left(\frac{\pi}{4}\right)(6\text{ in})^2$$

$$= 2.4\text{ in}^2$$

7. AREA OF A SEGMENT OF A CIRCLE

A segment of a circle is bounded by an arc and a straight line that connects the ends of the arc. The area of a segment is found by subtracting the area of the triangle formed by the chord and the two radii to its end points from the area of the sector formed by the two radii and the arc. The area of a segment of a circle is illustrated in Fig. 5.5.

$$A = \left(\frac{\phi}{360°}\right)\pi r^2 - \left(\frac{r^2}{2}\right)\sin\phi \qquad 5.8$$

Figure 5.5 *Area of a Segment of a Circle*

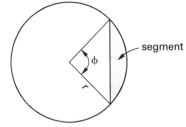

segment

Example 5.13

A segment of a circle with an 8 in radius has an arc subtending an angle of 90°. What is the area of the segment?

Solution

From Eq. 5.8,

$$A = \left(\frac{\phi}{360°}\right)\pi r^2 - \left(\frac{r^2}{2}\right)\sin\phi$$

$$= \left(\frac{90°}{360°}\right)\pi(8\text{ in})^2 - \left(\frac{(8\text{ in})^2}{2}\right)\sin 90°$$

$$= 18\text{ in}^2$$

8. COMPOSITE AREAS

Irregularly shaped areas can sometimes be divided into components that consist of geometric figures, the areas of which can be found. Total area can be found by adding the areas of the components. In some cases it may be appropriate to subtract the areas of geometric figures in order to find the net area desired.

Example 5.14

Find the area of the following figure.

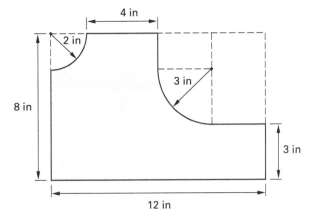

Solution

Treat the figure as a rectangle 12 in long and 8 in wide from which areas have been removed. From Eq. 5.1, the area of this rectangle is

$$A_{\text{rectangle}} = lw$$

$$= (12\text{ in})(8\text{ in})$$

$$= 96\text{ in}^2$$

To account for the curve in the upper-left corner, treat the area described by the curve as a sector of a circle with a radius of 2 in (and therefore a 4 in diameter). From Eq. 5.7,

$$A_1 = \left(\frac{\phi}{360°}\right)\left(\frac{\pi}{4}\right)d^2 = \left(\frac{90°}{360°}\right)\left(\frac{\pi}{4}\right)(4\text{ in})^2$$

$$= 3.14\text{ in}^2$$

To account for the curve in the upper-right corner, treat the area described by the curve as two rectangles and a sector of a circle with a radius of 3 in (and therefore a 6 in diameter).

The dimensions of the rectangular area directly above the curve with the 3 in radius are 2 in and 3 in. The area can be found using Eq. 5.1.

$$A_2 = lw = (2 \text{ in})(3 \text{ in}) = 6 \text{ in}^2$$

The dimensions of the rectangular area to the right of the curve with the 3 in radius are 3 in and 5 in. From Eq. 5.1,

$$A_3 = lw = (3 \text{ in})(5 \text{ in}) = 15 \text{ in}^2$$

The dimensions of the sector described by the curve with the 3 in radius can be found using Eq. 5.7.

$$A_4 = \left(\frac{\phi}{360°}\right)\left(\frac{\pi}{4}\right)d^2 = \left(\frac{90°}{360°}\right)\left(\frac{\pi}{4}\right)(6 \text{ in})^2$$
$$= 7.07 \text{ in}^2$$

The total area of the figure is

$$A_{\text{total}} = A_{\text{rectangle}} - A_1 - A_2 - A_3 - A_4$$
$$= 96 \text{ in}^2 - 3.14 \text{ in}^2 - 6 \text{ in}^2 - 15 \text{ in}^2 - 7.07 \text{ in}^2$$
$$= 64.79 \text{ in}^2$$

Table 5.1 *Area Formulas*

shape	figure	formula
circle		$A = \pi r^2$ $A = \pi\left(\frac{d}{2}\right)^2$ $= \left(\frac{\pi}{4}\right)d^2$
sector of a circle		$A = \left(\frac{\phi}{360°}\right) \times \left(\frac{\pi}{4}\right)d^2$
segment of a circle		$A = \left(\frac{\phi}{360°}\right)\pi r^2 - \left(\frac{r^2}{2}\right)\sin\phi$
rectangle		$A = lw$
triangle		$A = \frac{1}{2}bh$
right triangle		$A = \frac{1}{2}bh$ $A = \sqrt{\begin{array}{l}s(s-a)\\ \times(s-b)\\ \times(s-c)\end{array}}$
trapezoid		$A = \frac{1}{2}h(a+b)$

9. PRACTICE PROBLEMS

1. A right triangle has a base of 12 in and an altitude of 8 in. Most nearly, what is the area of the triangle?

(A) 24 in²

(B) 30 in²

(C) 48 in²

(D) 52 in²

2. Wallboard is needed to cover the walls and ceiling of a room 24 ft long, 16 ft wide, and 8 ft high (making no subtractions for doors and windows). Wallboard is only available in 4 ft by 8 ft sheets. How many sheets of wallboard are needed to cover the room?

(A) 10

(B) 12

(C) 22

(D) 32

3. Find the cross-sectional area of a ditch with a trapezoidal cross section 28 ft wide at the top, 4 ft wide at the bottom, and 6 ft deep.

(A) 64 ft²

(B) 68 ft²

(C) 96 ft²

(D) 140 ft²

4. Most nearly, what is the cross-sectional area of a highway fill with a trapezoidal cross section 44 ft wide at the top, 92 ft wide at the bottom, and 8 ft high?

(A) 350 ft²

(B) 400 ft²

(C) 540 ft²

(D) 740 ft²

5. Most nearly, what is the area of a circle with a 20 ft radius?

(A) 300 ft²

(B) 700 ft²

(C) 1300 ft²

(D) 3000 ft²

6. Most nearly, what is the area of a 10 ft diameter circle?

(A) 20 ft²

(B) 79 ft²

(C) 160 ft²

(D) 310 ft²

7. Most nearly, what is the area of a 60° sector of a 6 in circle?

(A) 4.7 in²

(B) 7.1 in²

(C) 14 in²

(D) 19 in²

8. A sector of a 12 ft diameter circle has a central angle of 90°. Most nearly, what is the area of the sector?

(A) 8.4 ft²

(B) 12 ft²

(C) 20 ft²

(D) 28 ft²

9. Find the area of a triangle with sides of 18 ft, 12 ft, and 10 ft.

(A) 20 ft²

(B) 57 ft²

(C) 90 ft²

(D) 96 ft²

10. A plan view of an excavation site is shown.

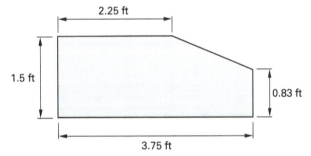

Most nearly, what is the total area of the site?

(A) 3.3 ft²

(B) 3.8 ft²

(C) 5.1 ft²

(D) 5.6 ft²

Mathematics
Basics

11. Find the area of each of the following figures.

(a)

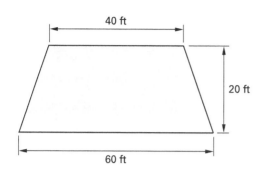

(b)

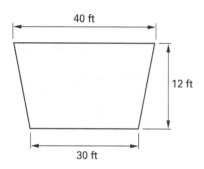

(c)

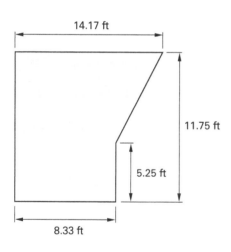

(d)

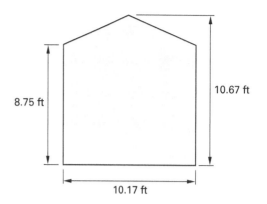

(e)

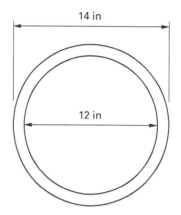

(f)

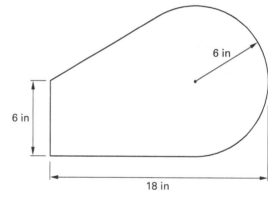

(g)

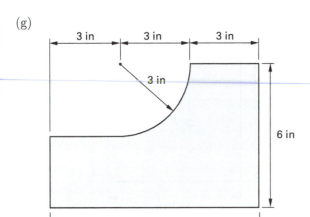

(h)

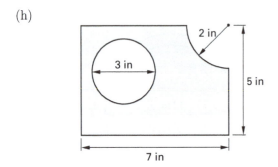

(i)

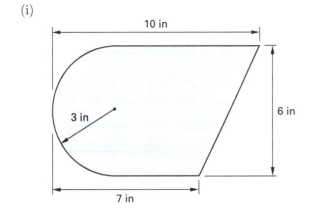

SOLUTIONS

1. From Eq. 5.2, the area is

$$A = \tfrac{1}{2}bh = \left(\tfrac{1}{2}\right)(12 \text{ in})(8 \text{ in}) = 48 \text{ in}^2$$

The answer is (C).

2. Find the area of one of the long walls. From Eq. 5.1, the area is

$$A_{\text{long wall}} = lw$$
$$= (8 \text{ ft})(24 \text{ ft})$$
$$= 192 \text{ ft}^2$$

Find the area of one of the short walls.

$$A_{\text{short wall}} = lw$$
$$= (8 \text{ ft})(16 \text{ ft})$$
$$= 128 \text{ ft}^2$$

Find the area of the ceiling.

$$A_{\text{ceiling}} = lw$$
$$= (16 \text{ ft})(24 \text{ ft})$$
$$= 384 \text{ ft}^2$$

The total area that must be covered is

$$2A_{\text{long wall}}$$
$$+2A_{\text{short wall}} + A_{\text{ceiling}} = (2)(192 \text{ ft}^2)$$
$$+(2)(128 \text{ ft}^2) + 384 \text{ ft}^2$$
$$= 1024 \text{ ft}^2$$

The quantity of sheets of wallboard needed is

$$\frac{1024 \text{ ft}^2}{32 \dfrac{\text{ft}^2}{\text{sheet}}} = 32 \text{ sheets}$$

The answer is (D).

3. From Eq. 5.4, the area is

$$A = \tfrac{1}{2}h(a+b)$$
$$= \left(\tfrac{1}{2}\right)(6 \text{ ft})(28 \text{ ft} + 4 \text{ ft})$$
$$= 96 \text{ ft}^2$$

The answer is (C).

4. From Eq. 5.4, the area is

$$A = \frac{1}{2}h(a+b)$$
$$= \left(\frac{1}{2}\right)(8 \text{ ft})(44 \text{ ft} + 92 \text{ ft})$$
$$= 544 \text{ ft}^2 \quad (540 \text{ ft}^2)$$

The answer is (C).

5. From Eq. 5.5, the area is

$$A = \pi r^2$$
$$= \pi(20 \text{ ft})^2$$
$$= 1256 \text{ ft}^2 \quad (1300 \text{ ft}^2)$$

The answer is (C).

6. From Eq. 5.6, the area is

$$A = \left(\frac{\pi}{4}\right)d^2$$
$$= \left(\frac{\pi}{4}\right)(10 \text{ ft})^2$$
$$= 78.5 \text{ ft}^2 \quad (79 \text{ ft}^2)$$

The answer is (B).

7. From Eq. 5.7, the area of the sector is

$$A = \left(\frac{\phi}{360°}\right)\left(\frac{\pi}{4}\right)d^2$$
$$= \left(\frac{60°}{360°}\right)\left(\frac{\pi}{4}\right)(6 \text{ in})^2$$
$$= 4.712 \text{ in}^2 \quad (4.7 \text{ in}^2)$$

The answer is (A).

8. From Eq. 5.7, the area is

$$A = \left(\frac{\phi}{360°}\right)\left(\frac{\pi}{4}\right)d^2$$
$$= \left(\frac{90°}{360°}\right)\left(\frac{\pi}{4}\right)(12 \text{ ft})^2$$
$$= 28.3 \text{ ft}^2 \quad (28 \text{ ft}^2)$$

The answer is (D).

9. From Eq. 5.3, the area is

$$A = \sqrt{s(s-a)(s-b)(s-c)}$$
$$= \sqrt{(20 \text{ ft})(20 \text{ ft} - 18 \text{ ft})(20 \text{ ft} - 12 \text{ ft})(20 \text{ ft} - 10 \text{ ft})}$$
$$= 57 \text{ ft}^2$$

The answer is (B).

10. Treat the site as a 1.5 ft $\times$ 3.75 ft rectangle, then subtract the area of the "missing" upper-right corner. From Eq. 5.1, the area of the "whole" site is

$$A = lw$$
$$= (1.5 \text{ ft})(3.75 \text{ ft})$$
$$= 5.625 \text{ ft}^2$$

Treat the "missing" upper-right corner as a triangle and find its area. The length of the base of the triangle is equal to the difference between the lower length and the upper length of the rectangle, 3.75 ft – 2.25 ft = 1.5 ft. The length of the altitude of the triangle is equal to the difference between the left side length and right side length of the rectangle, 1.5 ft – 0.83 ft = 0.67 ft. From Eq. 5.2,

$$A = \frac{1}{2}bh$$
$$= \left(\frac{1}{2}\right)(1.5 \text{ ft})(0.67 \text{ ft})$$
$$= 0.502 \text{ ft}^2$$

The total area of the site is

$$5.625 \text{ ft}^2 - 0.502 \text{ ft}^2 = 5.123 \text{ ft}^2 \quad (5.1 \text{ ft}^2)$$

The answer is (C).

11.

(a)

$$A = \left(\frac{40 \text{ ft} + 60 \text{ ft}}{2}\right)(20 \text{ ft}) = 1000 \text{ ft}^2$$

(b)

$$A = \left(\frac{40 \text{ ft} + 30 \text{ ft}}{2}\right)(12 \text{ ft}) = 420 \text{ ft}^2$$

(c)

$$A = (8.33 \text{ ft})(11.75 \text{ ft}) + \left(\frac{1}{2}\right)(5.84 \text{ ft})(6.50 \text{ ft})$$
$$= 116.9 \text{ ft}^2$$

Mathematics
Basics

(d)

$$A = (10.17 \text{ ft})(8.75 \text{ ft}) + \left(\frac{1}{2}\right)(10.17 \text{ ft})(1.92 \text{ ft})$$

$$= 99 \text{ ft}^2$$

(e)

$$A = \left(\frac{\pi}{4}\right)\left((14 \text{ in})^2 - (12 \text{ in})^2\right)$$

$$= 41 \text{ in}^2$$

(f)

$$A = (6 \text{ in})(12 \text{ in}) + \left(\frac{1}{2}\right)(6 \text{ in})(12 \text{ in}) + \left(\frac{1}{2}\right)\pi(6 \text{ in})^2$$

$$= 165 \text{ in}^2$$

(g)

$$A = (9 \text{ in})(6 \text{ in}) - (3 \text{ in})(3 \text{ in}) - \left(\frac{1}{4}\right)\pi(3 \text{ in})^2$$

$$= 38 \text{ in}^2$$

(h)

$$A = (7 \text{ in})(5 \text{ in}) - \left(\frac{\pi}{4}\right)(3 \text{ in})^2 - \left(\frac{1}{4}\right)\pi(2 \text{ in})^2$$

$$= 25 \text{ in}^2$$

(i)

$$A = (6 \text{ in})(4 \text{ in}) + \left(\frac{1}{2}\right)\pi(3 \text{ in})^2 + \left(\frac{1}{2}\right)(3 \text{ in})(6 \text{ in})$$

$$= 47 \text{ in}^2$$

6 Volume

Nomenclature

a	smaller base of a trapezoid	ft	m
A	area	ft^2	m^2
b	base	ft	m
d	diameter	ft	m
h	altitude or height	ft	m
l	length	ft	m
r	radius	ft	m
V	volume	ft^3	m^3
w	width	ft	m

Subscripts

b	base
i	inside
o	outside
p	prism

1. DEFINITION

Volume is the amount of space that a substance occupies or a surface encloses. Volume is measured in cubic units of length: cubic feet, cubic inches, cubic meters, and so on. The block shown in Fig. 6.1 has a volume of 6 cubic inches (6 in^3). One cubic inch is the volume of a cube that measures 1 in on each edge.

The volume formulas given in this chapter can all be found in Table 6.1 at the end of the chapter.

2. VOLUME OF A CUBE

A cube is a three-dimensional shape that has equal length, width, and height dimensions. The six sides of a cube all have equal dimensions and all meet at right angles. As a result, the volume may be defined as the product of the length, l, width, w, and height, h. Since

Figure 6.1 *Volume of a Block*

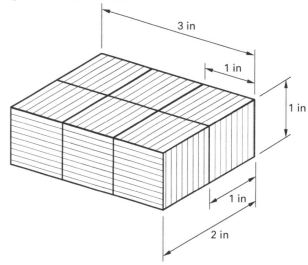

the length, width, and height are all equal, the volume may also be defined as the cube of any of those dimensions.

$$V = lwh \qquad\qquad 6.1$$

3. VOLUME OF RIGHT PRISMS AND CYLINDERS

The volume of a right prism or cylinder is the product of the area of the base and the altitude.

$$V = Ah_p \qquad\qquad 6.2$$

Example 6.1

Find the volume of a rectangular prism with an 8 in × 6 in base and an altitude of 10 in.

Solution

From Eq. 5.1,

$$\begin{aligned} A &= lw \\ &= (8\ \text{in})(6\ \text{in}) \\ &= 48\ \text{in}^2 \end{aligned}$$

From Eq. 6.2,

$$V = Ah_p$$
$$= (48 \text{ in}^2)(10 \text{ in})$$
$$= 480 \text{ in}^3$$

Example 6.2

Find the volume of a triangular prism with an 8 in altitude, and with a triangular base that has sides 3 in, 4 in, and 5 in long.

Solution

The base is a 3-4-5 right triangle. From Eq. 5.2,

$$A = \tfrac{1}{2}bh_b$$
$$= \left(\frac{1}{2}\right)(3 \text{ in})(4 \text{ in})$$
$$= 6 \text{ in}^2$$

From Eq. 6.2,

$$V = Ah_p$$
$$= (6 \text{ in}^2)(8 \text{ in})$$
$$= 48 \text{ in}^3$$

Example 6.3

How many cubic yards of dirt are needed for the highway fill shown?

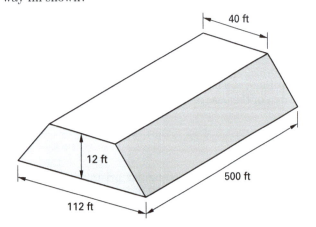

Solution

The highway fill is trapezoidal in shape. From Eq. 5.4,

$$A = \tfrac{1}{2}h_b(a+b)$$
$$= \left(\frac{1}{2}\right)(12 \text{ ft})(112 \text{ ft} + 40 \text{ ft})$$
$$= 912 \text{ ft}^2$$

From Eq. 6.2, the volume of fill needed is

$$V = Ah_p$$
$$= \frac{(912 \text{ ft}^2)(500 \text{ ft})}{27 \dfrac{\text{ft}^3}{\text{yd}^3}}$$
$$= 16{,}889 \text{ yd}^3$$

Example 6.4

The inside and outside diameters of the hollow cylinder shown are 6 in and 8 in, respectively. The cylinder is 5 in long. What is the volume of the hollow cylinder?

Solution

From Eq. 5.6, the cross-sectional area of the cylinder, including the hollow portion, is

$$A_{\text{outside}} = \left(\frac{\pi}{4}\right)d^2$$
$$= \left(\frac{\pi}{4}\right)(8 \text{ in})^2$$
$$= 50.265 \text{ in}^2$$

The cross-sectional area of the hollow portion is

$$A_{\text{inside}} = \left(\frac{\pi}{4}\right)d^2$$
$$= \left(\frac{\pi}{4}\right)(6 \text{ in})^2$$
$$= 28.274 \text{ in}^2$$

The area of the base of the cylinder is $50.265 \text{ in}^2 - 28.274 \text{ in}^2 = 21.991 \text{ in}^2$. From Eq. 6.2, the volume is

$$V = Ah_p$$
$$= (21.991 \text{ in}^2)(5 \text{ in})$$
$$= 109.955 \text{ in}^3$$

4. VOLUME OF A CONE

The volume of a right circular cone is equal to one-third the product of the area of its base and its altitude.

$$V = \frac{1}{3}\pi r^2 h \qquad \qquad 6.3$$

Example 6.5

Find the volume of a cone that is 6 in high with a 4 in diameter base.

Solution

From Eq. 6.3,

$$V = \frac{1}{3}\pi r^2 h$$
$$= \left(\frac{1}{3}\right)\pi(2 \text{ in})^2(6 \text{ in})$$
$$= 25 \text{ in}^3$$

5. VOLUME OF A PYRAMID

The volume of a pyramid is equal to one-third the product of the area of its base (found from Eq. 5.1) and its altitude.

$$V = \frac{1}{3}Ah \qquad \qquad 6.4$$

6. VOLUME OF A SPHERE

The volume of a sphere is based on its radius and is found from Eq. 6.5.

$$V = \frac{4}{3}\pi r^3 \qquad \qquad 6.5$$

Example 6.6

Find the volume of liquid that could be held in a spherical tank with a radius of 50 ft.

Solution

From Eq. 6.5,

$$V = \frac{4}{3}\pi r^3$$
$$= \left(\frac{4}{3}\right)\pi(50 \text{ ft})^3$$
$$= 523,599 \text{ ft}^3$$

Table 6.1 *Volume Formulas*

shape	figure	formula
cube		$V = lwh$
cylinder or prism		$V = Ah$
cone		$V = \frac{1}{3}\pi r^2 h$
pyramid		$V = \frac{1}{3}Ah$
sphere		$V = \frac{4}{3}\pi r^3$

7. PRACTICE PROBLEMS

1. A rectangular right prism has a base 3 ft wide and 4 ft long and an altitude of 6 ft. Most nearly, what is the volume of the prism?

(A) 48 ft³

(B) 72 ft³

(C) 96 ft³

(D) 108 ft³

2. A triangular right prism has a 9 in × 12 in × 15 in base and an altitude of 10 in. Most nearly, what is the volume of the prism?

(A) 270 in³

(B) 450 in³

(C) 540 in³

(D) 810 in³

3. Find the number of cubic feet of concrete (to the nearest tenth) in a pipe with an 8 in inside diameter, 2 in wall thickness, and 30 in length.

(A) 0.9 ft³

(B) 1.1 ft³

(C) 1.9 ft³

(D) 2.1 ft³

4. A cylindrical tank 6 ft in diameter must accommodate a volume of 223 ft³. Most nearly, how tall must the tank be?

(A) 5.8 ft

(B) 7.9 ft

(C) 9.0 ft

(D) 10.2 ft

5. Find the volume of a right prism whose altitude is 10 in and whose base is an isosceles triangle with a base of 8 in and an altitude of 3 in.

(A) 90 in³

(B) 110 in³

(C) 120 in³

(D) 140 in³

6. How many cubic yards of concrete are needed to pour a rectangular parking area 30 ft long, 27 ft wide, and 4 in thick?

(A) 8 yd³

(B) 10 yd³

(C) 22 yd³

(D) 40 yd³

7. A highway fill 810 ft long has a trapezoidal cross section 120 ft wide at the bottom, 80 ft wide at the top, and 8 ft deep. How many cubic yards of dirt are in the fill?

8. Find the volume of a cylinder with a diameter of 10 in and an altitude of 8 in.

SOLUTIONS

1. From Eq. 6.1, the area of the rectangle is

$$A = lw$$
$$= (3 \text{ ft})(4 \text{ ft})$$
$$= 12 \text{ ft}^2$$

From Eq. 5.1,

$$V = Ah_p$$
$$= (12 \text{ ft}^2)(6 \text{ ft})$$
$$= 72 \text{ ft}^3$$

The answer is (B).

2. The base is a 3-4-5 right triangle. From Eq. 5.2, the area of the base is

$$A = \tfrac{1}{2}bh$$
$$= \left(\tfrac{1}{2}\right)(9 \text{ in})(12 \text{ in})$$
$$= 54 \text{ in}^2$$

From Eq. 6.2, the volume is

$$V = Ah_p$$
$$= (54 \text{ in}^2)(10 \text{ in})$$
$$= 540 \text{ in}^3$$

The answer is (C).

3. Find the area of the base of the pipe. The total area is the difference between the outside area of the base and the inside area of the base. The outside diameter of the pipe, including the wall, is 8 in + (2)(2 in) = 12 in. From Eq. 5.6, the outside area is

$$A_{\text{outside}} = \left(\tfrac{\pi}{4}\right)d^2$$
$$= \left(\tfrac{\pi}{4}\right)(12 \text{ in})^2$$
$$= 113.097 \text{ in}^2$$

The inside area is

$$A_{\text{inside}} = \left(\tfrac{\pi}{4}\right)d^2$$
$$= \left(\tfrac{\pi}{4}\right)(8 \text{ in})^2$$
$$= 50.265 \text{ in}^2$$

The total area of the pipe is 113.097 in² – 50.265 in² = 62.832 in². From Eq. 6.2, the volume is

$$V = Ah_p$$
$$= \frac{(62.832 \text{ in}^2)(30 \text{ in})}{1728 \frac{\text{in}^3}{\text{ft}^3}}$$
$$= 1.0908 \text{ ft}^3 \quad (1.1 \text{ ft}^3)$$

The answer is (B).

4. Find the area of the tank's base. From Eq. 5.6,

$$A = \left(\tfrac{\pi}{4}\right)d^2$$
$$= \left(\tfrac{\pi}{4}\right)(6 \text{ ft})^2$$
$$= 28.27 \text{ ft}^2$$

From Eq. 6.2, the required height is

$$V = Ah_p$$
$$h = \frac{V}{A} = \frac{223 \text{ ft}^3}{28.27 \text{ ft}^2}$$
$$= 7.888 \text{ ft} \quad (7.9 \text{ ft})$$

The answer is (B).

5. Find the area of the base. From Eq. 5.2,

$$A = \tfrac{1}{2}bh$$
$$= \left(\tfrac{1}{2}\right)(8 \text{ in})(3 \text{ in})$$
$$= 12 \text{ in}^2$$

From Eq. 6.2, the volume is

$$V = Ah_p$$
$$= (12 \text{ in}^2)(10 \text{ in})$$
$$= 120 \text{ in}^3$$

The answer is (C).

6. Find the area of the lot. From Eq. 5.1,

$$A = lw$$
$$= (30 \text{ ft})(27 \text{ ft})$$
$$= 810 \text{ ft}^2$$

From Eq. 6.2, the volume is

$$V = Ah_p$$

$$= \frac{(810 \text{ ft}^2)(4 \text{ in})}{\left(12 \frac{\text{in}}{\text{ft}}\right)\left(27 \frac{\text{ft}^3}{\text{yd}^3}\right)}$$

$$= 10 \text{ yd}^3$$

The answer is (B).

7. The fill is a prism with a trapezoidal base. The area of the base is

$$A = \tfrac{1}{2}h(a+b) = \left(\frac{1}{2}\right)(8 \text{ ft})(80 \text{ ft} + 120 \text{ ft})$$

$$= 800 \text{ ft}^2$$

The volume in cubic yards is

$$V = Ah = \frac{(800 \text{ ft}^2)(810 \text{ ft})}{27 \frac{\text{ft}^3}{\text{yd}^3}}$$

$$= 24,000 \text{ yd}^3$$

8. The area of the cylinder's base is

$$A = \left(\frac{\pi}{4}\right)d^2 = \left(\frac{\pi}{4}\right)(10 \text{ in})^2$$

$$= 78.54 \text{ in}^2$$

The volume is

$$V = Ah = (78.54 \text{ in}^2)(8 \text{ in})$$

$$= 628 \text{ in}^3$$

7 Trigonometry

1. ANGLES

In trigonometry, an *angle* is the measure of the rotation of a *ray* (a half-line with its endpoint at the vertex of the angle) from one position to another in a counterclockwise direction.

Standard Position of an Angle

An angle is in *standard position* when its vertex is at the origin of a rectangular coordinate system and its initial side lies on the positive *x*-axis. (See Fig. 7.1.)

Figure 7.1 *Angle Terminology*

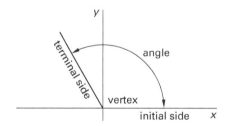

Quadrants

The coordinate axes of a rectangular coordinate system divide the plane into four *quadrants* designated I, II, III, and IV, as shown in Fig. 7.2. An angle in standard position is considered to be in a quadrant when its *terminal side* is in that quadrant.

Figure 7.2 *Angle Quadrants*

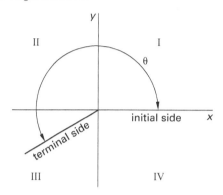

2. TRIGONOMETRIC FUNCTIONS OF ANY ANGLE

For any angle θ in standard position, the six most common trigonometric functions are given by Eq. 7.1 through Eq. 7.6. r is the hypotenuse, x is the side of the triangle adjacent to angle θ, and y is the side of the triangle opposite angle θ. (See Fig. 7.3 and Fig. 7.4.)

$$\sin \theta = \frac{y}{r} \qquad \text{7.1}$$

$$\cos \theta = \frac{x}{r} \qquad \text{7.2}$$

$$\tan \theta = \frac{y}{x} \qquad \text{7.3}$$

$$\csc \theta = \frac{r}{y} \qquad \text{7.4}$$

$$\sec \theta = \frac{r}{x} \qquad \text{7.5}$$

$$\cot \theta = \frac{x}{y} \qquad \text{7.6}$$

Equation 7.1, Eq. 7.2, and Eq. 7.3 (sine, cosine, and tangent) are the most frequently used in surveying calculations. The equations for these functions are easily remembered by a traditional memorization tool, "SOH-CAH-TOA": Sine = Opposite side over Hypotenuse, Cosine = Adjacent side over Hypotenuse, and Tangent = Opposite side over Adjacent side.

Figure 7.3 Functions of an Angle

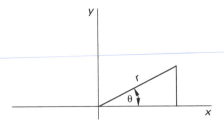

Figure 7.4 Sides of a Triangle

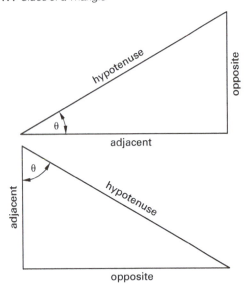

3. TRIGONOMETRIC FUNCTIONS OF AN ACUTE ANGLE

All angles in the northeastern quadrant of a rectangular coordinate system (quadrant I in Fig. 7.2) are *acute angles* and have positive values. In dealing with an acute angle, it is more convenient to consider it as part of a right triangle and express the trigonometric functions in terms of the sides of a triangle using Eq. 7.1 through Eq. 7.3.

4. RECIPROCAL OF A TRIGONOMETRIC FUNCTION

The *reciprocal* of a number is one divided by that number. The reciprocal of 3 is 1/3, the reciprocal of 2/3 is 3/2, and so on. Trigonometric functions are ratios of numbers and can be treated as such. Therefore, the reciprocal of $\sin \theta = 1/\sin \theta$. It follows that if $\sin \theta = y/r$, the reciprocal of $\sin \theta$ is r/y. Equation 7.7 through Eq. 7.12 show the reciprocal relationships between trigonometric functions.

$$\sin \theta = \frac{1}{\csc \theta} \qquad 7.7$$

$$\cos \theta = \frac{1}{\sec \theta} \qquad 7.8$$

$$\tan \theta = \frac{1}{\cot \theta} \qquad 7.9$$

$$\csc \theta = \frac{1}{\sin \theta} \qquad 7.10$$

$$\sec \theta = \frac{1}{\cos \theta} \qquad 7.11$$

$$\cot \theta = \frac{1}{\tan \theta} \qquad 7.12$$

Therefore, $\sin \theta$ and $\csc \theta$ are reciprocals, $\cos \theta$ and $\sec \theta$ are reciprocals, and $\tan \theta$ and $\cot \theta$ are reciprocals. From the relation between a function and its reciprocal,

$$\sin \theta \csc \theta = \left(\frac{y}{r}\right)\left(\frac{r}{y}\right) = 1 \qquad 7.13$$

$$\cos \theta \sec \theta = \left(\frac{x}{r}\right)\left(\frac{r}{x}\right) = 1 \qquad 7.14$$

$$\tan \theta \cot \theta = \left(\frac{y}{x}\right)\left(\frac{x}{y}\right) = 1 \qquad 7.15$$

5. COFUNCTIONS

Any function of an acute angle is equal to the *cofunction* of its *complementary angle*. (Complementary angles are two angles whose total is 90°.) The sine and cosine are cofunctions, as are the tangent and the cotangent (see Eq. 7.16 and Eq. 7.17).

$$\sin x = \cos(90° - x) \qquad 7.16$$

$$\tan x = \cot(90° - x) \qquad 7.17$$

For example, $\sin 30° = \cos 60°$, and $\tan 20° = \cot 70°$.

6. TRIGONOMETRIC FUNCTIONS OF 30°, 45°, AND 60°

Consider an equilateral triangle with sides of length 2. The bisector of any angle will bisect the opposite side and form a right triangle with acute angles of 30° and 60°, as shown in Fig. 7.5.

Figure 7.5 *30-60-90 Triangle*

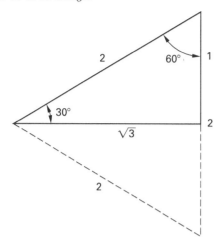

In Fig. 7.5,

$$\sin 30° = \frac{1}{2} = \cos 60° = 0.5$$

$$\sin 60° = \frac{\sqrt{3}}{2} = \cos 30° = 0.8660$$

$$\tan 30° = \frac{1}{\sqrt{3}} = \cot 60° = 0.5774$$

Consider an isosceles right triangle with two of its sides equal to 1, as shown in Fig. 7.6. The angles opposite these two sides will be 45°. Then,

$$\sin 45° = \frac{1}{\sqrt{2}} = 0.7071$$

$$\cos 45° = \frac{1}{\sqrt{2}} = 0.7071$$

$$\tan 45° = \frac{1}{1} = 1.0$$

Figure 7.6 *45-45-90 Triangle*

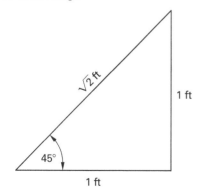

7. SLOPE ANGLES

There are several ways to express slope.

1. As the *angle of inclination* with the horizon. This is the vertical angle between the horizontal and a line rotated upward or downward. If the angle of inclination is positive (upward), it is called the *angle of elevation*; if the angle of inclination is negative (downward), it is called the *angle of depression*.

2. As a percentage value consisting of the rise or drop over a given distance, divided by the run, multiplied by 100. This may also be expressed as the tangent of the angle of inclination times 100. In the United States, this percentage "grade" is the most commonly used method for communicating slopes in transportation for streets, roads, highways, and rail tracks, as well as for surveying, construction, and civil engineering. (See Fig. 7.7.)

3. As a ratio, which is usually expressed as one part rise to so many parts run. For example, a slope ratio of 1 in 20 has a rise of 5 feet for every 100 feet of run. The word "in" is normally used, rather than the mathematical ratio notation of "1:20."

Figure 7.7 *Slope Expressed as a Percentage (grade)*

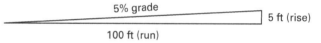

5% grade
5 ft (rise)
100 ft (run)

8. SOLVING RIGHT TRIANGLES

Solving a right triangle means finding the values of its three angles and the length of each of its sides. To solve a right triangle, two values must be known, either the lengths of two sides or the length of one side and the measure of one acute angle. If one acute angle of a right triangle is known, the other acute angle is the complement of it, since the sum of the interior angles of a triangle equals 180°.

Example 7.1

Solve the right triangle shown.

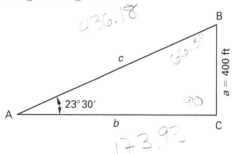

Solution

Angle B is the complement of angle A, so the measure of angle B is

$$B = 90° - 23°30' = 66°30'$$

Find the length of the hypotenuse. The length of the side opposite angle A is known, as is angle A, so from Eq. 7.1, the length of side c is

$$\sin\theta = \frac{y}{r}$$

$$\sin 23°30' = \frac{400 \text{ ft}}{c}$$

$$c = \frac{400 \text{ ft}}{\sin 23°30'} = 1000 \text{ ft}$$

b is the side adjacent to the known angle A, so from Eq. 7.3,

$$\tan\theta = \frac{y}{x}$$

$$\tan 23°30' = \frac{400 \text{ ft}}{b}$$

$$b = \frac{400 \text{ ft}}{\tan 23°30'} = 920 \text{ ft}$$

Example 7.2

Solve the right triangle shown.

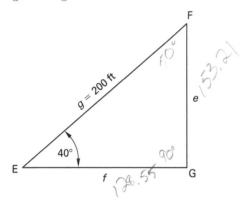

Solution

Angle F is the complement of angle E, so

$$F = 90° - 40° = 50°$$

Find the length of e using Eq. 7.1.

$$\sin\theta = \frac{y}{r}$$

$$\sin 40° = \frac{e}{200 \text{ ft}}$$

$$e = (200 \text{ ft})(\sin 40°) = 130 \text{ ft}$$

Find the length of f using Eq. 7.2.

$$\cos\theta = \frac{x}{r}$$

$$\cos 40° = \frac{f}{200 \text{ ft}}$$

$$f = (200 \text{ ft})(\cos 40°) = 150 \text{ ft}$$

9. ALTERNATE SOLUTION METHODS FOR RIGHT ANGLES

Since there are many different relationships between the sides and angles of right triangles, it is not surprising that triangle problems may have more than a single correct solution method. Some solutions are simpler than others. Practice is needed to be able to select the simplest solution procedure.

Example 7.3

Solve the right triangle shown.

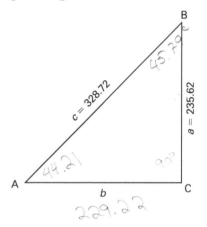

Solution

Find the solution from the inverse sine (arcsin or $\sin^{-1}$) function on a calculator.

$$\sin A = \frac{235.62}{328.72} = 0.7167802$$

$$A = \arcsin 0.7167802 = 45°47'21''$$

$$B = 90°00'00'' - 45°47'21'' = 44°12'39''$$

$$b = \sqrt{(328.72)^2 - (235.62)^2} = 229.22$$

Example 7.4

To measure the width of a river, points A and B are established on the west bank, and point C is established on the east bank. The distance between point A and point B, and the angle at A between AB and AC, are as shown.

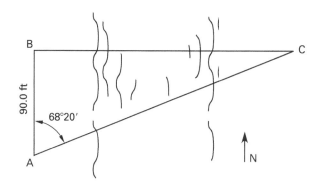

How wide is the river?

Solution

Using the definition of tangent as being opposite over adjacent sides (Eq. 7.3),

$$BC = (90.0 \text{ ft})(\tan 68°20') = (90.0 \text{ ft})(2.517)$$
$$= 226.5 \text{ ft}$$

Example 7.5

San Angelo, Texas, is due west of Waco, and Arlington is 100 mi due north of Waco. The bearing to Arlington from San Angelo is N 66°30′E, as shown.

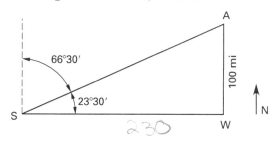

How far is San Angelo from Arlington?

Solution

The triangle formed by the three cities is a right triangle, and the angle at S is the complement of the bearing angle. Use Eq. 7.1 to solve for the length of SA.

$$\sin \theta = \frac{y}{r}$$
$$\sin 23°30' = \frac{100 \text{ mi}}{c}$$
$$c = \frac{100 \text{ mi}}{\sin 23°30'} = 250 \text{ mi}$$

Find the length of SW using Eq. 7.3.

$$\tan \theta = \frac{y}{x}$$
$$\tan 23°30' = \frac{100 \text{ mi}}{b}$$
$$b = \frac{100 \text{ mi}}{\tan 23°30'} = 230 \text{ mi}$$

10. GENERAL TRIANGLES

A *general triangle* (or *oblique triangle*) is a triangle that does not contain a right angle. All the angles in an oblique triangle may be acute, or there may be one obtuse angle and two acute angles. (See Fig. 7.8.)

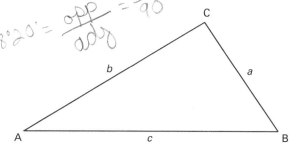

Figure 7.8 *General Triangle*

An oblique triangle can be solved if three of its parts, at least one of which is a side, are known. However, the solution is not as simple as the solution of a right triangle. An oblique triangle can be solved by forming two right triangles within the oblique triangle, but the task is made easier by the use of formulas. The most important formulas are the *law of sines* and the *law of cosines.*

Law of Sines

In any triangle, the sides are proportional to the sines of the opposite angles, as shown in Eq. 7.18 and Eq. 7.19.

$$\frac{a}{\sin A} = \frac{b}{\sin B} = \frac{c}{\sin C} \qquad 7.18$$

$$\frac{\sin A}{a} = \frac{\sin B}{b} = \frac{\sin C}{c} \qquad 7.19$$

Law of Cosines

Equation 7.20 through Eq. 7.25 are different forms of the law of cosines.

$$a^2 = b^2 + c^2 - 2bc \cos A \qquad 7.20$$
$$b^2 = c^2 + a^2 - 2ca \cos B \qquad 7.21$$
$$c^2 = a^2 + b^2 - 2ab \cos C \qquad 7.22$$

$$\cos A = \frac{b^2 + c^2 - a^2}{2bc} \qquad 7.23$$

$$\cos B = \frac{c^2 + a^2 - b^2}{2ca} \qquad 7.24$$

$$\cos C = \frac{a^2 + b^2 - c^2}{2ab} \qquad 7.25$$

The cosine of an angle greater than 90° and less than 180° has a negative algebraic sign. In substituting the cosines of angles in that range, the negative sign must be included. For example, in Eq. 7.20, a negative value for $2bc \cos A$ in Eq. 7.20 must be added to $b^2 + c^2$.

11. SOLVING OBLIQUE TRIANGLES

The choice of which of the two laws to use to solve a particular oblique triangle depends on which three parts of the triangle are known. If one side and two angles of a triangle are known (abbreviated SAA for side-angle-angle), the triangle can be solved using the law of sines. If one angle and two sides of a triangle are known (SSA, side-side-angle), the triangle can be solved using the law of sines.

If three sides of a triangle are known (SSS, side-side-side), the triangle can be solved using the law of cosines. If two sides and the angle included between the two sides are known (SAS, side-angle-side), the triangle can be solved using the law of cosines and the law of sines together.

Table 7.1 provides a summary of equations for solving the four cases of oblique triangles.

$$C = 180° - (A + B) \qquad 7.26$$

$$b = \frac{a \sin B}{\sin A} \qquad 7.27$$

$$c = \frac{a \sin C}{\sin A} \qquad 7.28$$

$$\text{area} = \tfrac{1}{2} ab \sin C \qquad 7.29$$

$$B = \arcsin \frac{b \sin A}{a} \qquad 7.30$$

$$c = \sqrt{a^2 + b^2 - 2ab \cos C} \qquad 7.31$$

$$A = \arcsin \frac{a \sin C}{c} \qquad 7.32$$

$$B = \arcsin \frac{b \sin C}{c} \qquad 7.33$$

$$A = \arccos \frac{b^2 + c^2 - a^2}{2bc} \qquad 7.34$$

Table 7.1 Oblique Triangle Equations

type	given	equations	basis
SAA	a, A, B	Equation 7.26	
		Equation 7.27	law of sines
		Equation 7.28	law of sines
		Equation 7.29	
SSA	a, b, A	Equation 7.30	law of sines
		Use Eq. 7.26 to find C.	law of sines
		Use Eq. 7.28 to find c.	law of sines
		Use Eq. 7.29 to find the area.	
SAS	a, C, b	Equation 7.31	law of cosines
		Equation 7.32	law of sines
		Equation 7.33	law of sines
		Use Eq. 7.29 to find the area.	
SSS	a, b, c	Equation 7.34	law of cosines
		Equation 7.35	law of cosines
		Use Eq. 7.26 to find C.	
		Use Eq. 7.29 to find the area.	

$$B = \arccos \frac{a^2 + c^2 - b^2}{2ac} \qquad 7.35$$

When two sides and the angle opposite one of them are given (SAS), it is often possible to construct two different triangles from the provided measurements. Example 7.6 illustrates such a scenario.

Example 7.6

Solve the general triangle shown.

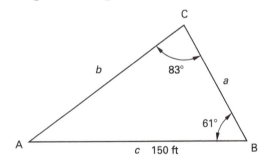

Solution

This general triangle is an SAA case. Use Eq. 7.26 to find A.

$$
\begin{aligned}
A &= 180° - (C + B) \\
&= 180° - (83° + 61°) \\
&= 36°
\end{aligned}
$$

With A, B, and C known, use Eq. 7.18 to find a and b.

$$
\frac{a}{\sin A} = \frac{c}{\sin C}
$$

$$
\begin{aligned}
a &= (\sin A)\left(\frac{c}{\sin C}\right) = (\sin 36°)\left(\frac{150 \text{ ft}}{\sin C}\right) \\
&= 89 \text{ ft}
\end{aligned}
$$

$$
\frac{b}{\sin B} = \frac{c}{\sin C}
$$

$$
\begin{aligned}
b &= (\sin B)\left(\frac{c}{\sin C}\right) = (\sin 61°)\left(\frac{150 \text{ ft}}{\sin 83°}\right) \\
&= 132 \text{ ft}
\end{aligned}
$$

Example 7.7

Solve both triangle ABC and triangle AB′C.

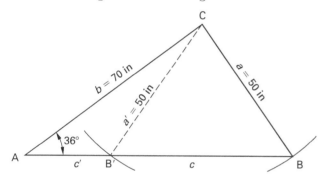

Solution

For triangle ABC:

From Eq. 7.19,

$$
\frac{\sin A}{a} = \frac{\sin B}{b}
$$

$$
\begin{aligned}
\sin B &= \frac{b \sin A}{a} = \frac{(70 \text{ in})\sin 36°}{50 \text{ in}} \\
&= 0.8229
\end{aligned}
$$

The angle B is

$$
B = \arcsin 0.8229 = 55°
$$

From Eq. 7.26,

$$
\begin{aligned}
C &= 180° - (A + B) \\
&= 180° - (36° + 55°) \\
&= 89°
\end{aligned}
$$

From Eq. 7.18,

$$
\frac{c}{\sin C} = \frac{a}{\sin A}
$$

$$
\begin{aligned}
c &= \frac{a \sin C}{\sin A} = \frac{(50 \text{ in})\sin 89°}{\sin 36°} \\
&= 85 \text{ in}
\end{aligned}
$$

For triangle AB′C:

Since sides a and a' are equal, angle $CB'B$ = angle B. Therefore, angle $CB'A = 180° - 55° = 125°$. From Eq. 7.26,

$$
\begin{aligned}
C &= 180° - (A + B') \\
&= 180° - (36° + 125°) \\
&= 19°
\end{aligned}
$$

From Eq. 7.18,

$$
\frac{c'}{\sin C} = \frac{a'}{\sin A}
$$

$$
\begin{aligned}
c' &= \frac{a' \sin C}{\sin A} = \frac{(50 \text{ in})\sin 19°}{\sin 36°} \\
&= 28 \text{ in}
\end{aligned}
$$

Example 7.8

Solve the triangle shown.

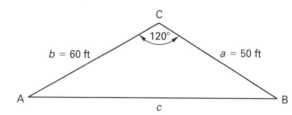

Solution

From Eq. 7.31,

$$
\begin{aligned}
c &= \sqrt{a^2 + b^2 - 2ab \cos C} \\
&= \sqrt{(50 \text{ ft})^2 + (60 \text{ ft})^2 - 2(50 \text{ ft})(60 \text{ ft})\cos 120°} \\
&= 95 \text{ ft}
\end{aligned}
$$

From Eq. 7.32 and Eq. 7.33,

$$A = \arcsin \frac{a \sin C}{c}$$

$$= \arcsin \frac{(50 \text{ ft}) \sin 120°}{95 \text{ ft}}$$

$$= 27°$$

$$B = \arcsin \frac{b \sin C}{c}$$

$$= \arcsin \frac{(60 \text{ ft}) \sin 120°}{95 \text{ ft}}$$

$$= 33°$$

Example 7.9

Solve a triangle ABC with sides $a = 3.0$, $b = 5.0$, and $c = 6.0$.

Solution

$$\cos A = \frac{b^2 + c^2 - a^2}{2bc} = \frac{(5.0)^2 + (6.0)^2 - (3.0)^2}{(2)(5.0)(6.0)}$$

$$= 0.8667$$
$$A = \arccos 0.8667 = 30°$$

$$\cos B = \frac{a^2 + c^2 - b^2}{2ac} = \frac{(3.0)^2 + (6.0)^2 - (5.0)^2}{(2)(3.0)(6.0)}$$

$$= 0.8667$$
$$B = \arccos 0.5556 = 56°$$

$$\cos C = \frac{a^2 + b^2 - c^2}{2ab} = \frac{(3.0)^2 + (5.0)^2 - (6.0)^2}{(2)(3.0)(5.0)}$$

$$= -0.0667$$
$$C = \arccos(-0.0667) = 94°$$

12. PRACTICE PROBLEMS

1. Most nearly, what is the sine of the angle θ in the graph shown?

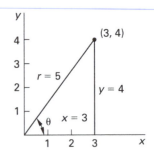

(A) 3/4

(B) 3/5

(C) 4/5

(D) 4/3

2. For the illustration shown, determine the sine, cosine, tangent, cotangent, secant, and cosecant of angle θ.

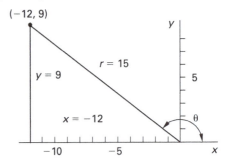

3. For the illustration shown, determine the sine, cosine, tangent, cotangent, secant, and cosecant of angle θ.

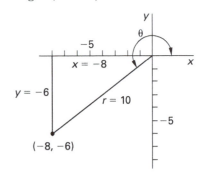

4. For the illustration shown, determine the sine, cosine, tangent, cotangent, secant, and cosecant of angle θ.

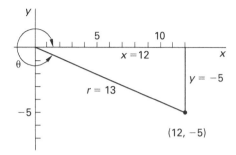

5. In the right triangle shown, given that angle A and side a are known, which trigonometric function—sine, cosine, or tangent—can be used to find the length of b?

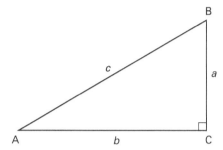

6. For (a) through (g), the two given values are known. Which trigonometric function—sine, cosine, or tangent—can be used to find each of the other two values?

(a) $A = 46°44'$, and $b = 156$ ft. Find a and c.

(b) $B = 53°21'$, and $c = 300$ ft. Find a and b.

(c) $A = 38°19'$, and $c = 700$ ft. Find a and b.

(d) $a = 600$ ft, and $c = 1000$ ft. Find A and B.

(e) $b = 400$ ft, and $c = 500$ ft. Find A and B.

(f) $B = 55°10'$, and $b = 378$ ft. Find a and c.

(g) $A = 33°40'$, and $a = 250$ ft. Find b and c.

7. For (a) through (f), completely solve the right triangle ABC with the given values as the only values known.

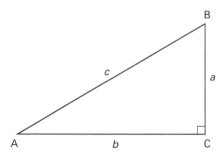

(a) $A = 28°41'$, $b = 540$ ft

(b) $B = 55°13'$, $a = 371$ ft

(c) $B = 61°29'$, $b = 466$ ft

(d) $A = 33°15'$, $c = 263$ ft

(e) $B = 58°55'$, $c = 562$ ft

(f) $a = 300$ ft, $b = 400$ ft

8. A person casts a shadow 10 ft long when the angle of elevation of the sun is $31°18'$. Most nearly, how tall is the person?

(A) 5 ft 1 in

(B) 5 ft 9 in

(C) 6 ft 1 in

(D) 6 ft 4 in

9. The angle of elevation from a point 160 ft from the foot of a flagpole to the top of the flagpole is $50°40'$. Most nearly, how tall is the flagpole?

(A) 180 ft

(B) 200 ft

(C) 210 ft

(D) 230 ft

10. A 500 ft tall tower is erected near a road intersection. The angle of depression to the intersection is $30°$. Most nearly, how far from the tower is the road intersection?

(A) 540 ft

(B) 860 ft

(C) 900 ft

(D) 930 ft

11. A 20 ft ladder leans against the top of a building. The angle of elevation of the ladder is $60°$. Most nearly, how tall is the building?

(A) 11 ft

(B) 13 ft

(C) 17 ft

(D) 20 ft

12. A general triangle MNO has two known angles: M = 38°48′45″ and O = 82°23′56″, and one known length, MN = 298.34 ft. Most nearly, what is the length of side OM?

(A) 190 ft

(B) 240 ft

(C) 260 ft

(D) 300 ft

13. For (a) through (c), solve the general triangle with the values given.

(a) triangle ABC: angle A = 34°18′24″, angle C = 62°12′55″, line AB = 1347.77 ft

(b) triangle PQR: angle P = 118°34′24″, angle Q = 23°06′54″, line QR = 526.30 ft

(c) triangle XYZ: angle X = 82°46′58″, angle Y = 58°54′20″, line YZ = 345.43 ft

14. The three sides of general triangle EFG are EF = 125.83 ft, FG = 171.25 ft, and GE = 155.13. What are angles E, F, and G?

15. For (a) through (j), if the indicated values are known, which law can be used to solve a general triangle?

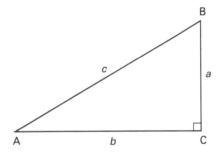

(a) a, b, A

(b) b, c, A

(c) b, A, C

(d) A, B, a

(e) a, c, B

(f) b, c, B

(g) A, a, b

(h) a, b, c

(i) A, C, a

(j) c, b, a

SOLUTIONS

1.

$$\sin\theta = \frac{y}{r} = 4/5$$

The answer is (C).

2.

$$\sin\theta = \frac{9}{15}$$
$$\cos\theta = -\frac{12}{15}$$
$$\tan\theta = -\frac{9}{12}$$
$$\cot\theta = -\frac{12}{9}$$
$$\sec\theta = -\frac{15}{12}$$
$$\csc\theta = \frac{15}{9}$$

3.

$$\sin\theta = -\frac{6}{10}$$
$$\cos\theta = -\frac{8}{10}$$
$$\tan\theta = \frac{6}{8}$$
$$\cot\theta = \frac{8}{6}$$
$$\sec\theta = -\frac{10}{8}$$
$$\csc\theta = -\frac{10}{6}$$

4.

$$\sin\theta = -\frac{5}{13}$$
$$\cos\theta = \frac{12}{13}$$
$$\tan\theta = -\frac{5}{12}$$
$$\cot\theta = -\frac{12}{5}$$
$$\sec\theta = \frac{13}{12}$$
$$\csc\theta = -\frac{13}{5}$$

5. The tangent function can be used to find the length b.

6.

(a) The tangent function can be used to find the length a. The cosine function can be used to find the length c.

(b) The cosine function can be used to find the length a. The sine function can be used to find the length b.

(c) The sine function can be used to find the length a. The cosine function can be used to find the length b.

(d) The sine function can be used to find the angle A. The cosine function can be used to find the angle B.

(e) The cosine function can be used to find the angle A. The sine function can be used to find the angle B.

(f) The tangent function can be used to find the length a. The sine function can be used to find the length c.

(g) The tangent function can be used to find the length b. The sine function can be used to find the length c.

7.

(a)

$$B = 90° - 28°41' = 61°19'$$
$$a = (540 \text{ ft}) \tan 28°41' = 295 \text{ ft}$$
$$c = \frac{540 \text{ ft}}{\cos 28°41'} = 616 \text{ ft}$$

(b)

$$A = 90° - 55°13' = 34°47'$$
$$b = (371 \text{ ft}) \tan 55°13' = 534 \text{ ft}$$
$$c = \frac{371 \text{ ft}}{\cos 55°13'} = 650 \text{ ft}$$

(c)

$$A = 90° - 61°29' = 28°31'$$
$$a = \frac{466 \text{ ft}}{\tan 61°29'} = 253 \text{ ft}$$
$$c = \frac{466 \text{ ft}}{\sin 61°29'} = 530 \text{ ft}$$

(d)

$$B = 90° - 33°15' = 56°45'$$
$$a = (263 \text{ ft}) \sin 33°15' = 144 \text{ ft}$$
$$b = (263 \text{ ft}) \cos 33°15' = 220 \text{ ft}$$

(e)

$$A = 90° - 58°55' = 31°05'$$
$$a = (562 \text{ ft}) \cos 58°55' = 290 \text{ ft}$$
$$b = (562 \text{ ft}) \sin 58°55' = 481 \text{ ft}$$

(f)

$$A = \arctan \frac{300 \text{ ft}}{400 \text{ ft}} = 36°52'$$
$$B = \arctan \frac{400 \text{ ft}}{300 \text{ ft}} = 53°08'$$
$$c = \sqrt{(300 \text{ ft})^2 + (400 \text{ ft})^2} = 500 \text{ ft}$$

8. From Eq. 7.3

$$\tan \theta = \frac{y}{x}$$
$$y = x \tan \theta = (10 \text{ ft}) \tan 31°18'$$
$$= 6 \text{ ft } 1 \text{ in}$$

The answer is (C).

9. From Eq. 7.3,

$$\tan \theta = \frac{y}{x}$$
$$y = x \tan \theta = (160 \text{ ft}) \tan 50°40'$$
$$= 195 \text{ ft} \quad (200 \text{ ft})$$

The answer is (B).

10. From Eq. 7.3,

$$\tan \theta = \frac{y}{x}$$
$$x = \frac{y}{\tan \theta} = \frac{500 \text{ ft}}{\tan 30°}$$
$$= 866 \text{ ft} \quad (860 \text{ ft})$$

The answer is (B).

11. From Eq. 7.1,

$$\sin \theta = \frac{y}{r}$$
$$y = r \sin \theta = (20 \text{ ft}) \sin 60°$$
$$= 17 \text{ ft}$$

The answer is (C).

12. From Eq. 7.26,

$$N = 180° - (38°48'45'' + 82°23'56'') = 58°47'19''$$

From the law of sines,

$$\frac{NO}{\sin M} = \frac{OM}{\sin N} = \frac{MN}{\sin O}$$

$$OM = \frac{MN \sin N}{\sin O} = \frac{(298.34 \text{ ft}) \sin 58°47'19''}{\sin 82°23'56''}$$

$$= 257.42 \text{ ft} \quad (260 \text{ ft})$$

The answer is (C).

13.

(a)

$$B = 180° - (34°18'24'' + 62°12'55'')$$

$$= 83°28'41''$$

$$BC = \frac{(1347.77 \text{ ft}) \sin 34°18'24''}{\sin 62°12'55''}$$

$$= 858.63 \text{ ft}$$

$$CA = \frac{(1347.77 \text{ ft}) \sin 83°28'41''}{\sin 62°12'55''}$$

$$= 1513.55 \text{ ft}$$

(b)

$$R = 180° - (118°34'24'' + 23°06'54'')$$

$$= 38°18'42''$$

$$PQ = \frac{(526.30 \text{ ft}) \sin 38°18'42''}{\sin 118°34'24''}$$

$$= 371.52 \text{ ft}$$

$$RP = \frac{(526.30 \text{ ft}) \sin 23°06'54''}{\sin 118°34'24''}$$

$$= 235.27 \text{ ft}$$

(c)

$$Z = 180° - (82°46'58'' + 58°54'20'')$$

$$= 38°18'42''$$

$$XY = \frac{(345.43 \text{ ft}) \sin 38°18'42''}{\sin 82°46'58''}$$

$$= 215.86 \text{ ft}$$

$$ZX = \frac{(345.43 \text{ ft}) \sin 58°54'20''}{\sin 82°46'58''}$$

$$= 298.16 \text{ ft}$$

14. From the law of cosines, Eq. 7.23 through Eq. 7.25,

$$\cos E = \frac{(EF)^2 + (GE)^2 - (FG)^2}{2(EF)(GE)}$$

$$= \frac{(125.83)^2 + (155.13)^2 - (171.25)^2}{(2)(125.83)(155.13)}$$

$$E = 74°17'18''$$

$$\cos F = \frac{(FG)^2 + (EF)^2 - (GE)^2}{2(FG)(EF)}$$

$$= \frac{(171.25)^2 + (125.83)^2 - (155.13)^2}{(2)(171.25)(125.83)}$$

$$F = 60°41'40''$$

$$\cos G = \frac{(GE)^2 + (FG)^2 - (EF)^2}{2(GE)(FG)}$$

$$= \frac{(155.13)^2 + (171.25)^2 - (125.83)^2}{(2)(155.13)(171.25)}$$

$$G = 45°01'02''$$

As a check, $E + F + G = 180°$.

15.

(a) law of sines

(b) law of cosines

(c) law of sines

(d) law of sines

(e) law of cosines

(f) law of sines

(g) law of sines

(h) law of cosines

(i) law of sines

(j) law of cosines

8 Rectangular Coordinate System

1. DIRECTED LINE

A *directed line* is a line or line segment with an explicit direction indicated. For example, suppose a surveyor establishes a west-east line through a point on a monument. This point is called the *origin*. Moving east from the origin, the surveyor marks off points 1 ft apart, labeling them $+1, +2, +3$, and so on. Then, moving west from the origin, the surveyor marks off points 1 ft apart, labeling them $-1, -2, -3$, and so on. The surveyor then establishes the directed line shown in Fig. 8.1. From left to right (west to east), the direction is positive, and the numbers increase in value. From right to left (east to west), the direction is negative, and the numbers decrease in value.

Figure 8.1 *Directed Line*

A directed line on a plane can be called an *axis*. The horizontal (or west-east) line is called the *x-axis*. A point on the axis associated with a particular number is called the *graph* of the number, and the number is called the *coordinate* of the point. This coordinate is the directed distance from the origin to the point. Point A in Fig. 8.1 has the *x*-coordinate -4, which is the distance from 0 to A (not from A to 0). Point B has the *x*-coordinate $+3$, which is the distance from 0 to B (not B to 0). The distances from point A to point B and vice versa are

$$\text{distance AB} = 3 - (-4) = 7$$
$$\text{distance BA} = -4 - (+3) = -7$$

The word "distance" is used loosely in this chapter; a more appropriate term for distance on a directed line would be "the measure of travel in a specified direction." The actual length from A to B, or B to A, is the absolute value of the length, 7.

In general, if P_1 and P_2 are any two points on the *x*-axis with coordinates x_1 and x_2, respectively, then

$$P_1P_2 = x_2 - x_1 \qquad 8.1$$
$$P_2P_1 = x_1 - x_2 \qquad 8.2$$

A directed line that is positive from south to north and negative from north to south is called the *y-axis*. In the example with the monument, the surveyor could establish a south-north line through the same point on the monument and mark points on the line at 1 ft intervals from the origin in a northerly direction, labeling them $+1, +2, +3$, and so on, and draw a south-north line with points at 1 ft intervals in a southerly direction labeled $-1, -2, -3$, and so on.

Figure 8.2 shows a vertical line through the origin representing this south-north line.

Figure 8.2 *x-axis and y-axis*

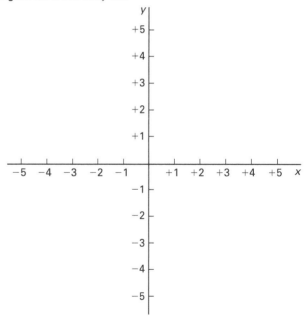

If P_1 and P_2 are any two points on the *y*-axis with coordinates y_1 and y_2, respectively, then

$$P_1P_2 = y_2 - y_1 \qquad 8.3$$
$$P_2P_1 = y_1 - y_2 \qquad 8.4$$

2. THE RECTANGULAR COORDINATE SYSTEM

The French mathematician René Descartes (1596-1650) devised the *rectangular coordinate system*, sometimes called the *Cartesian plane*. The rectangular coordinate system consists of an x-axis (horizontal axis) and a y-axis (vertical axis), which are directed lines as shown in Fig. 8.3. This system uses an ordered pair of coordinates to locate a point. The ordered pair consists of the x-coordinate and the y-coordinate enclosed in parentheses, with the x-coordinate always written first, followed by a comma and the y-coordinate.

Figure 8.3 Rectangular Coordinate System

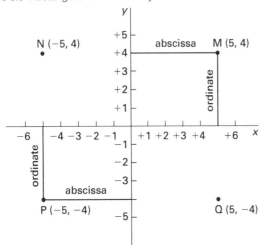

In Fig. 8.3, point M has the coordinates $(5, 4)$. The horizontal distance from the y-axis to M (5, in this case) is known as the *abscissa*. The vertical distance from the x-axis to M (4, in this case) is known as the *ordinate*. The abscissa and ordinate are measured from the axis to a point, not from a point to the axis, in accordance with distances on a directed line. In Fig. 8.3, point N has an abscissa of -5 and an ordinate of $+4$; point P has an abscissa of -5 and an ordinate of -4; and point Q has an abscissa of $+5$ and an ordinate of -4.

The x- and y-axes divide the Cartesian plane into four parts, called *quadrants*, numbered in a counterclockwise direction as shown in Fig. 8.4. Signs of the coordinates of points in each quadrant are also shown in Fig. 8.4. In quadrant I, x is positive and y is positive; in quadrant II, x is negative and y is positive; in quadrant III, x is negative and y is negative; and in quadrant IV, x is positive and y is negative.

Figure 8.4 Signs of the Quadrants

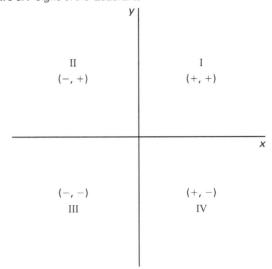

Example 8.1

Determine the coordinates of the points on the Cartesian plane shown.

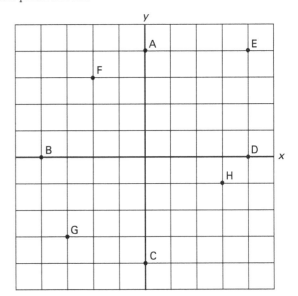

Solution

A: $(0, 4)$

B: $(-4, 0)$

C: $(0, -4)$

D: $(4, 0)$

E: $(4, 4)$

F: $(-2, 3)$

G: $(-3, -3)$

H: $(3, -1)$

3. DISTANCE FORMULA

A formula for finding the distance between any two points in a rectangular coordinate system can be derived from the *Pythagorean theorem*: In a right triangle, the square of the length of the hypotenuse equals the sum of the squares of the lengths of the other two sides.

If c represents the length of the hypotenuse and a and b are the lengths of the other two sides,

$$c^2 = a^2 + b^2 \qquad 8.5$$

Taking the square root of both sides of the equation,

$$c = \sqrt{a^2 + b^2} \qquad 8.6$$

In Fig. 8.5, P_1 and P_2 represent any two points in a rectangular coordinate system with the coordinates $(x_1, \ y_1)$ and $(x_2, \ y_2)$. If a horizontal line passes through P_1 and a vertical line passes through P_2, they will intersect at Q, forming a right triangle with the line P_1P_2 as the hypotenuse. The x-coordinate of Q will be the same as the x-coordinate of P_2, x_2; and the y-coordinate of Q will be the same as the y-coordinate of P_1, y_1.

Figure 8.5 *Distance Between Points*

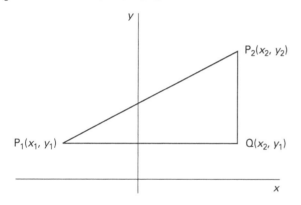

By the Pythagorean theorem,

$$(P_1P_2)^2 = (P_1Q)^2 + (QP_2)^2 \qquad 8.7$$

$$P_1P_2 = \sqrt{(x_2 - x_1)^2 + (y_2 - y_1)^2} \qquad 8.8$$

The distance r from the origin to any point is called the *radius*. This distance is always considered to be positive, regardless of its direction or quadrant. (See Fig. 8.6.)

From the Pythagorean theorem,

$$r = \sqrt{x^2 + y^2} \qquad 8.9$$

Example 8.2

Find the distance between points $P_1(-3, -1)$ and $P_2(5, 5)$.

Figure 8.6 *Radius*

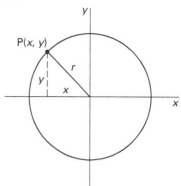

Solution

From Eq. 8.8,

$$
\begin{aligned}
P_1P_2 &= \sqrt{(x_2 - x_1)^2 + (y_2 - y_1)^2} \\
&= \sqrt{\left(5 - (-3)\right)^2 + \left(5 - (-1)\right)^2} \\
&= 10
\end{aligned}
$$

Example 8.3

Find the distance from the origin to the point $(-3, 4)$.

Solution

From Eq. 8.9,

$$
\begin{aligned}
r &= \sqrt{x^2 + y^2} \\
&= \sqrt{(-3)^2 + (4)^2} \\
&= 5
\end{aligned}
$$

4. MIDPOINT OF A LINE

In Fig. 8.7, point M is the midpoint of the line PQ.

The x-coordinate of M is the distance from the y-axis to M. This is equal to the distance from the y-axis to L, which is the average of the x-coordinates of P and Q.

$$\frac{6 + (-2)}{2} = 2$$

Figure 8.7 *Midpoint of a Line*

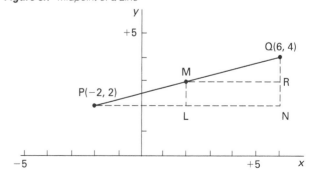

The y-coordinate of M is the distance from the x-axis to M. This is equal to the distance from the x-axis to R, which is the average of the y-coordinates of P and Q.

$$\frac{4+2}{2} = 3$$

The midpoint M of a line P_1P_2, then, has the coordinates

$$M\left(\frac{x_1+x_2}{2}, \frac{y_1+y_2}{2}\right) \qquad 8.10$$

Example 8.4

Find the midpoint of a line with endpoints $P_1(-2, 3)$ and $P_2(-8, -3)$.

Solution

From Eq. 8.10,

$$M\left(\frac{x_1+x_2}{2}, \frac{y_1+y_2}{2}\right) = M\left(\frac{-2+(-8)}{2}, \frac{3+(-3)}{2}\right)$$
$$= M(-5, 0)$$

5. INTERSECTION OF TWO LINES

The intersection of two straight lines is a point. The location of the point of intersection can be determined by solving the equations of the lines simultaneously[1] for x and y.

Example 8.5

Find the point of intersection of two lines with the equations $x + y - 6 = 0$ and $x - 2y + 6 = 0$.

Solution

Rearrange the two equations to isolate x.

$$x + y - 6 = 0$$
$$x = 6 - y$$
$$x - 2y + 6 = 0$$
$$x = 2y - 6$$

At the point of intersection, the values of x and y are the same in both equations. Set x in equation 1 equal to x in equation 2 and solve for y.

$$6 - y = 2y - 6$$
$$y = 4$$

Substitute 4 for y in the first equation.

$$x + y - 6 = 0$$
$$x + 4 - 6 = 0$$
$$x = 2$$

The point of intersection of the two lines is at $x = 2$, $y = 4$.

6. PRACTICE PROBLEMS

1. (a) In the rectangular coordinate system, plot the points given and draw lines PQ, QR, RS, ST, and TP.

P(12, 16)

Q(−14, 18)

R(−12, 1)

S(−14, −17)

T(14, −14)

(b) Find the length of each line and the perimeter of the shape they describe.

2. What is the radius of a line from the origin to point P(3,4)?

   (A)   4.2

   (B)   5.0

   (C)   5.7

   (D)   7.0

3. For (a) through (c), find the radii for the points given.

(a) (−3, 3)

(b) (−5, −12)

(c) (5, −5)

---

[1] Simultaneous equations are those with common values for the variables.

SOLUTIONS

1.

(a)

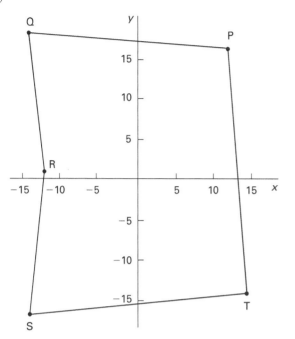

(b) From Eq. 8.8,

$$P_1P_2 = \sqrt{(x_2 - x_1)^2 + (y_2 - y_1)^2}$$
$$PQ = \sqrt{(-14 - 12)^2 + (18 - 16)^2} = 26$$
$$QR = \sqrt{\left(-12 - (-14)\right)^2 + (1 - 18)^2} = 17$$
$$RS = \sqrt{\left(-14 - (-12)\right)^2 + (-17 - 1)^2} = 18$$
$$ST = \sqrt{\left(14 - (-14)\right)^2 + \left(-14 - (-17)\right)^2} = 28$$
$$TP = \sqrt{(12 - 14)^2 + \left(16 - (-14)\right)^2} = 30$$

The perimeter is

$$P = 26 + 17 + 18 + 28 + 30 = 119$$

2. From Eq. 8.9,

$$r = \sqrt{x^2 + y^2}$$
$$= \sqrt{(3)^2 + (4)^2}$$
$$= 5.0$$

The answer is (B).

3.

(a)

$$r = \sqrt{x^2 + y^2}$$
$$= \sqrt{(-3)^2 + (3)^2}$$
$$= 4.242$$

(b)

$$r = \sqrt{x^2 + y^2}$$
$$= \sqrt{(-5)^2 + (-12)^2}$$
$$= 13$$

(c)

$$r = \sqrt{x^2 + y^2}$$
$$= \sqrt{(5)^2 + (-5)^2}$$
$$= 7.071$$

Mathematics Basics

Directions

Nomenclature

A	azimuth	deg	deg
B	bearing	deg	deg

1. DIRECTION OF LINES

The *direction of a line* is the angle between the line and a reference line, known as a *meridian*. The meridian may be a *true meridian* (*geodetic meridian*), a true circle of the earth passing through the true geographic poles; a *magnetic meridian*, a meridian parallel to the earth's magnetic field, the direction of which is defined by a compass needle; or a *grid meridian*, a line of a rectangular grid drawn over a map, established for a plane coordinate system.

In addition to the different meridians from which direction can be measured, the direction of a line may be expressed as an azimuth or a bearing (see Sec. 9.2 and Sec. 9.3), and can be measured in degrees, radians, or grads.[1] Familiarity with all these variables is essential to surveying practice. Direction of a line is discussed in Sec. 8.1.

2. AZIMUTH

The *azimuth* of a line is the horizontal angle of the line as measured clockwise from the meridian (i.e., usually measured from the north).[2] Azimuths vary from 0° to 360° and are not limited to 90°, so there is no need to divide the circle into quadrants. Figure 9.1 shows azimuths of 40°, 140°, 220°, and 320°.

Figure 9.1 *Representative Azimuths*

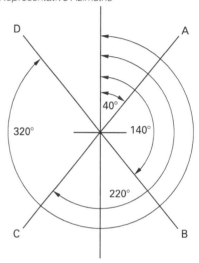

3. BEARING

The *bearing of a line* is the horizontal angle between the meridian and the line. The bearing of a line is always acute (i.e., cannot exceed 90°), so the full horizontal circle is divided into four *quadrants:* northeast, southeast, southwest, and northwest. Figure 9.2 shows a bearing of 40°, measured between the meridian and a line in each of the four quadrants. The quadrant in which a line is located is designated in the bearing by preceding the angle with N (for north) or S (for south) and following the angle with E (for east) or W (for west). For example, in Fig. 9.2, angle D is a 40° angle in the northwest quadrant.

4. CONVERTING BEARING TO AZIMUTH

The bearing of a line is used to give the direction of a course in most land surveys, while the azimuth is used in topographic surveys and some route surveys. A bearing can be converted to the equivalent azimuth, and vice versa.

In the northeast quadrant, the azimuth angle is equal to the bearing angle; for example, N76°30′E = 76°30′.

---

[1] A *grad* is 1/400 of a full circle (0.9° or $\pi/200$ rad) and is rarely used in surveying.
[2] Angles of a line are measured from either the north or south, but are never measured from the east or west.

Figure 9.2 *Bearing Quadrants*

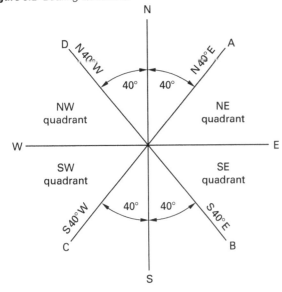

In the southeast quadrant, the azimuth angle is the difference between 180° and the bearing angle.

$$A_{SE} = 180° - B \qquad 9.1$$

In the southwest quadrant, the azimuth angle is the sum of 180° and the bearing angle.

$$A_{SW} = 180° + B \qquad 9.2$$

In the northwest quadrant, the azimuth angle is the difference between 360° and the bearing angle.

$$A_{NW} = 360° - B \qquad 9.3$$

5. CONVERTING AZIMUTH TO BEARING

In the northeast quadrant, the bearing angle equals the azimuth. The prefix N and the suffix E are added to the measurement to indicate it is an azimuth; for example, 52°73′ = N52°73′E.

In the southeast quadrant, the bearing angle is the difference between 180° and the azimuth angle. The prefix S and suffix E are added to the converted measurement.

$$B_{SE} = 180° - A \qquad 9.4$$

In the southwest quadrant, the bearing angle is the difference between the azimuth angle and 180°. The prefix S and suffix W are added to the measurement.

$$B_{SE} = A - 180° \qquad 9.5$$

In the northwest quadrant, the bearing angle is the difference between 360° and the azimuth. The prefix N and suffix W are added to the measurement.

$$B_{SE} = 360° - A \qquad 9.6$$

6. RADIAN MEASURE

A *radian* is an angle that, when situated as the central angle of a circle, is subtended by an arc with a length equal to the radius of the circle. Radians are shown in Fig. 9.3.

Figure 9.3 *Radians*

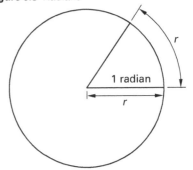

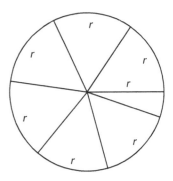

The circumference of a circle is 2π times the length of the radius, r. Therefore, the number of radians (i.e., arcs of length r) that can be applied to the circumference of a circle is 2π.

$$2\pi \text{ rad} = 360° \qquad 9.7$$

$$\pi \text{ rad} = 180° \qquad 9.8$$

Dividing both sides of Eq. 9.8 by π gives the equation for the equivalent angle of a measurement in radians.

$$1 \text{ rad} = \frac{180°}{\pi} \qquad 9.9$$

Dividing both sides of Eq. 9.8 by 180 gives the equation for the equivalent measurement in radians of a given angle.

$$1° = \frac{\pi}{180} \text{ rad} \qquad 9.10$$

For example, from Eq. 9.10, an angle of 60° measured in radians is

$$60° = (60)\left(\frac{\pi}{180} \text{ rad}\right) = \frac{\pi}{3} \text{ rad}$$

From Eq. 9.9, 2 rad measured in degrees is

$$2 \text{ rad} = (2)\left(\frac{180°}{\pi}\right) = \frac{360°}{\pi}$$

7. GRID DIRECTIONS VERSUS GEODETIC DIRECTIONS

When using a rectangular coordinate system such as the State Plane Coordinate System or the Universal Transverse Mercator (UTM) System, there are two meridians that must be considered, grid north and true (geodetic) north. In a plane coordinate system, the grid north and true north will not coincide except along the central meridian. This is because the true meridians converge toward the poles, while north-south grid lines in a plane coordinate system are parallel to the central meridian.

The *mapping angle* (also known as the *convergence angle*, *grid declination*, or *variation*) is the angular difference between grid north and true north. (See Fig. 9.4.) Mapping angles are either positive or negative depending on whether the location is east or west of the central meridian (regardless of the projection used).

Figure 9.4 *Mapping Angle (convergence) in a State Plane Coordinate System*

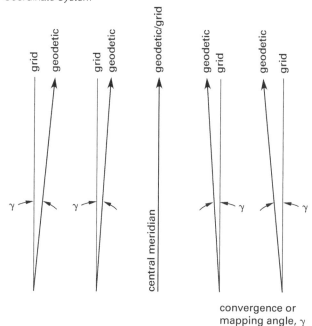

8. MAGNETIC DIRECTIONS

Two meridians must be considered when using magnetic directions, the magnetic meridian and the geodetic (true) meridian. The two differ because the magnetic

poles do not coincide with the earth's geographic poles; in some areas, the needle points east of true north, and in some areas it points west of true north.

The horizontal angle between the magnetic meridian and the true meridian is known as the *declination*. Zero declination is found along a line between areas. This line is known as the *agonic line*. The agonic line generally passes through Florida and the Great Lakes, but it is constantly changing its location. East of this line, declination is west (−); west of this line, declination is east (+). In the United States, declination varies from 0° to 23°.

Declination in any one point varies daily, annually, and secularly (over a long period of time such as a century). The declination for a particular location in a particular year can be obtained from the United States Geological Survey.

9. PRACTICE PROBLEMS

1. Most nearly, what is the measurement of a 30° angle expressed in radians?

(A) $\dfrac{\pi}{10}$ rad

(B) $\dfrac{\pi}{6}$ rad

(C) $\dfrac{\pi}{5}$ rad

(D) $\dfrac{\pi}{3}$ rad

2. For (a) through (e), find the measurement of the given angle in radians.

(a) 45°

(b) 120°

(c) 90°

(d) 15°

(e) 270°

3. Most nearly, what is the measurement in degrees of an angle of $\dfrac{\pi}{3}$ rad?

(A) 45°

(B) 60°

(C) 90°

(D) 120°

4. For (a) through (d), determine the measurement of the given angle in degrees.

(a) $\dfrac{\pi}{2}$ rad

(b) $\dfrac{\pi}{4}$ rad

(c) $\dfrac{\pi}{12}$ rad

(d) $\dfrac{3\pi}{2}$ rad

5. Convert the given bearings to azimuths.

(a) S85°13′16″W

(b) N74°24′01″W

(c) S08°19′19″E

(d) N84°28′13″E

(e) S83°03′28″E

(f) N07°26′33″W

(g) N27°57′45″E

(h) S05°17′25″W

(i) S04°18′12″E

(j) N57°08′02″W

6. Convert the given azimuths to bearings.

(a) 291°37′06″

(b) 12°13′47″

(c) 106°12′46″

(d) 232°31′18″

(e) 93°04′02″

(f) 65°11′37″

(g) 337°15′11″

(h) 267°25′51″

(i) 102°27′38″

(j) 317°40′53″

7. For a magnetic bearing of S56°10′W and declination of 7°30′W, draw the equivalent true bearing.

8. For a magnetic bearing of N48°30′W and declination of 4°20′E, draw the equivalent true bearing.

9. For a magnetic bearing of S68°10′E and declination of 5°40′W, draw the equivalent true bearing.

10. Convert the magnetic bearing of S89°50′E and declination of 6°30′W to the equivalent azimuth and determine the true bearing.

11. Convert the magnetic bearing of N85°15′W to its true bearing given the declination of 3°30′W.

SOLUTIONS

1. From Eq. 9.10,

$$1° = \frac{\pi}{180} \text{ rad}$$

$$30° = (30)\left(\frac{\pi}{180} \text{ rad}\right) = \frac{\pi}{6} \text{ rad}$$

The answer is (B).

2.

(a) $\frac{\pi}{4}$ rad

(b) $\frac{2\pi}{3}$ rad

(c) $\frac{\pi}{2}$ rad

(d) $\frac{\pi}{12}$ rad

(e) $\frac{3\pi}{2}$ rad

3. From Eq. 9.9,

$$1 \text{ rad} = \frac{180°}{\pi}$$

$$\frac{\pi}{3} \text{ rad} = \left(\frac{\pi}{3}\right)\left(\frac{180°}{\pi}\right) = 60°$$

The answer is (B).

4.

(a) 90°

(b) 45°

(c) 15°

(d) 270°

5.

(a) The bearing is in the southwest quadrant. From Eq. 9.2,

$$A_{\text{SW}} = 180° + B$$
$$= 180° + 85°13'16''$$
$$= 265°13'16''$$

(b) 285°35'59''

(c) 171°40'41''

(d) 84°28'13''

(e) 96°56'32''

(f) 352°33'27''

(g) 27°57'45''

(h) 185°17'25''

(i) 175°41'48''

(j) 302°51'58''

6.

(a) By definition, the azimuth is in the northwest quadrant. From Eq. 9.6,

$$B_{\text{SE}} = 360° - A$$
$$= 360° - 291°37'06''$$
$$= \text{N}68°22'54''\text{W}$$

(b) N12°13'47''E

(c) S73°47'14''E

(d) S52°31'18''W

(e) S86°55'58''E

(f) N65°11'37''E

(g) N22°44'49''W

(h) S87°25'51''W

(i) S77°32'22''E

(j) N42°19'07''W

7.

Both the geodetic (or grid) meridians as well as the magnetic meridians are drawn based on the provided data. This allows a graphic view of the equivalent true bearing as a guide to determining the true bearing.

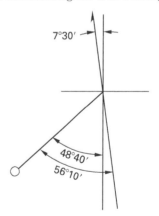

true bearing ———— S 48°40' W

8.

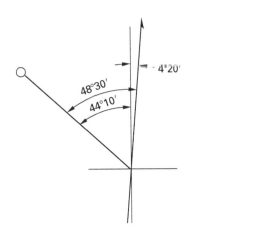

true bearing ——————— N 44°10' W

9.

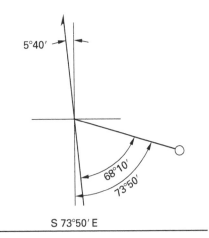

true bearing ——————— S 73°50' E

10. Both the geodetic (or grid) meridians and the magnetic meridians are drawn. Plotting the given magnetic bearing allows a graphic view of the equivalent azimuth and true bearing.

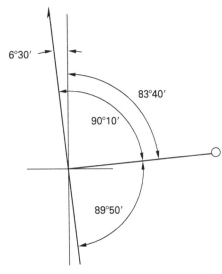

true bearing ——————— N 83°40' E

11.

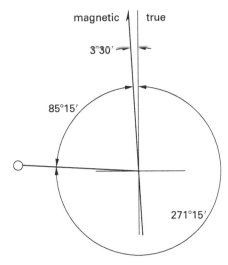

Since the magnetic bearing and declination both lie west of true north, summing them yields the true bearing:

$$N85° \, 15'W + 3° \, 30' = N88° \, 45'W$$

Alternatively, since the true azimuth (as measured clockwise from north) is also given, the true bearing can simply be determined by subtracting from 360°:

$$360° - 271.25° = 88.75° \text{ or } N88°45'W$$

10

Analytical Geometry

Nomenclature

a	x-intercept	–	–
A	coefficient of x	–	–
b	y-intercept	–	–
B	coefficient of y	–	–
C	constant	–	–
D	distance	–	–
h	x-coordinate of center of circle	–	–
k	y-coordinate of center of circle	–	–
m	slope	–	–
r	radius	ft	m
x	x-coordinate	–	–
y	y-coordinate	–	–

1. FIRST-DEGREE EQUATIONS

A *first-degree equation*, or *equation of the first degree*, is an equation where all unknown values (*unknowns* or *variables*) are of the first power (that is, not raised to any power larger than one). For example, the statement "the sum of two unknown numbers is eight" is expressed in algebraic terms as $x + y = 8$. In this equation there are two unknowns of the first power, represented by x and y.

More than one pair of numbers will make this statement true (e.g., if $x = 1$, $y = 7$; if $x = 2$, $y = 6$; if $x = 3$, $y = 5$; and so on). Each ordered pair is a solution to the equation; collectively, they form the *solution set* of the equation. If these pairs of numbers are always expressed in the order of x first and y second, they are called *ordered pairs* and are written symbolically in the form (x, y), for example, $(1, 7)$, $(2, 6)$, or $(3, 5)$.

The *roots* of an equation with a single unknown are the values of the unknown that satisfy the equation (make it true). For example, the statement "four times an unknown number minus three is equal to five" is expressed in algebraic terms as $4x - 3 = 5$. The letter x represents the unknown. A value of 2 is the only value that satisfies the equation, so 2 is the *root* of the equation. A first-degree equation with a single unknown will have exactly one root.

2. GRAPHS OF FIRST-DEGREE EQUATIONS WITH TWO VARIABLES

To graph a first-degree equation with two variables, consider each ordered pair that satisfies the equation to be the coordinates of a point. If several of these points are plotted on a rectangular coordinate system and connected with a line, the system will show the *graph of the equation*. When the points in the graph of an equation are connected, they lie in a straight line.

Consider the following equation.

$$2x - 3y = 12$$

To solve the equation—that is, to find ordered pairs that satisfy it—first rearrange the equation to isolate one of the unknowns.

$$-3y = -2x + 12$$
$$y = \tfrac{2}{3} x - 4$$

Next, assign various values to x and solve for the corresponding values of y (e.g., when $x = 0$, $y = -4$). (An infinite number of values could be given to x and y.)

$$x = -3, 0, 3, 6, 9$$
$$y = -6, -4, -2, 0, 2$$

Plotting each of these ordered pairs as the coordinates of a point creates the graph of the equation $2x - 3y = 12$, as shown in Fig. 10.1.

Figure 10.1 *Straight Line*

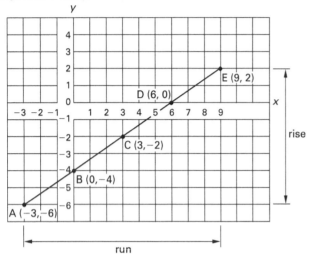

Example 10.1

Find three solutions of the equation $2x + y = 10$.

Solution

Rearrange the equation to solve for y.

$$2x + y = 10$$
$$y = 10 - 2x$$

Assign three different values to x, and find the corresponding values of y.

$$y = 10 - 2x = 10 - (2)(3) = 4 \quad [x = 3]$$
$$y = 10 - 2x = 10 - (2)(2) = 6 \quad [x = 2]$$
$$y = 10 - 2x = 10 - (2)(5) = 0 \quad [x = 5]$$

Writing the solutions as ordered pairs gives $(3, 4)$, $(2, 6)$, and $(5, 0)$.

Example 10.2

Find the coordinates of the points at which the equation $x - y + 5 = 0$ intersects the x- and y-axes.

Solution

The equation intersects the x-axis when y is zero, and it intersects the y-axis when x is zero. Set $y = 0$ and solve for x.

$$x - y + 5 = 0$$
$$x = y - 5$$
$$x = 0 - 5$$
$$x = -5$$

Set $x = 0$ and solve for y.

$$x - y + 5 = 0$$
$$y = 5 - x$$
$$= 5 - 0$$
$$= 5$$

The graph of the equation intersects the y-axis at point $(-5, 0)$ and the x-axis at point $(0, 5)$.

3. SLOPE OF A LINE

Slope is the ratio of the change in vertical distance to the change in horizontal distance. When a line rises from left to right, its slope is positive; when a line falls from left to right, its slope is negative. The slope, m, of a line connecting points (x_1, y_1) and (x_2, y_2) is found from Eq. 10.1.

$$m = \frac{y_2 - y_1}{x_2 - x_1} \qquad \text{10.1}$$

Example 10.3

What is the slope of the line that connects point $(1, 2)$ and point $(3, 6)$?

Solution

From Eq. 10.1,

$$m = \frac{y_2 - y_1}{x_2 - x_1} = \frac{6 - 2}{3 - 1} = 2$$

4. LINEAR EQUATIONS

A *linear equation* is a first-degree equation that describes a line. Equation 10.2 represents the *general form* of a linear equation, where A is the positive coefficient of x, B is the coefficient of y, and C is a constant.

$$Ax + By + C = 0 \qquad \text{10.2}$$

The slope is then

$$m = -\frac{A}{B} \qquad \text{10.3}$$

For example, the slope of the line represented by the linear equation $2x - 3y = 12$ (shown in Fig. 10.1) is $-(\frac{2}{-3})$ or $\frac{2}{3}$. The numerator 2 is the coefficient of x and the denominator -3 is the coefficient of y. When the coefficients A and B are opposite in sign, the slope is positive; when A and B are similar in sign, the slope is negative.

The equation can also be written as

$$y = \frac{2}{3}x - \frac{12}{3}$$

The coefficient of x is $\frac{2}{3}$, which is the slope of the line.

Example 10.4

Rearrange the first-degree equation $3x + 4y = 8$ into the general form, and determine the slope of the line.

Solution

From Eq. 10.2, the general form of the equation is

$$Ax + By + C = 0$$
$$3x + 4y = 8$$
$$3x + 4y - 8 = 0$$

From Eq. 10.3, the slope of the line is

$$m = -\frac{A}{B} = -\frac{3}{4}$$

5. EQUATIONS OF HORIZONTAL AND VERTICAL LINES

From Eq. 10.3, the slope of the line that contains the points $(-4, 5)$ and $(4, 5)$ is

$$m = -\frac{A}{B} = \frac{5 - 5}{4 - (-4)} = 0$$

This is a horizontal line with the equation $y = 5$, as shown in Fig. 10.2. In the general equation of a line, $Ax + By + C = 0$, the coefficient of x for a horizontal line is 0, so the general equation of a horizontal line is

$$By + C = 0 \qquad \text{10.4}$$

The slope of the line that contains the points $(4, 5)$ and $(4, -5)$ is

$$m = -\frac{A}{B} = \frac{-5 - 5}{4 - 4} = \frac{-10}{0}$$

This number is undefined, so the line is vertical and its slope is undefined (or infinite). The equation of the line is $x = 4$, as shown in Fig. 10.2. The general equation of a vertical line is

$$Ax + C = 0 \qquad \text{10.5}$$

6. *X*- AND *Y*-INTERCEPTS

If $x = 0$ in the equation $2x - 3y = 12$, then $y = -4$. The point $(0, -4)$ is where the graph of the equation crosses the y-axis (see Fig. 10.1). The distance from the x-axis to this point is known as the *y-intercept* of the equation, represented by the symbol b.

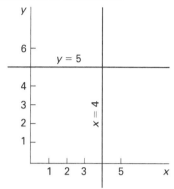

Figure 10.2 *Horizontal and Vertical Lines*

If $y = 0$ in the equation $2x - 3y = 12$, then $x = 6$. The point $(6, 0)$ is where the graph of the equation crosses the x-axis. The distance from the y-axis to this point is known as the *x-intercept* of the equation, represented by the symbol a.

The coordinates of the point where any line crosses the x-axis are $(a, 0)$ and the coordinates of the point where any line crosses the y-axis are $(0, b)$.

If the equation $2x - 3y = 12$ is rewritten as the equivalent equation $y = \frac{2}{3}x - 4$, the constant term -4 is the y-coordinate of the point where the line crosses the y-axis. It is, in fact, the y-intercept. For any equation $Ax + By + C = 0$, the y-intercept is

$$b = -\frac{C}{B} \qquad \text{10.6}$$

If the equation $2x - 3y = 12$ is rewritten as the equivalent equation $x = \frac{3}{2}y + 6$, the constant term 6 is the x-coordinate of the point where the line crosses the x-axis; it is the x-intercept. For any equation $Ax + By + C = 0$, the x-intercept is

$$a = -\frac{C}{A} \qquad \text{10.7}$$

In summary, for any linear equation $Ax + By + C = 0$,

$$m = \text{slope} = -\frac{A}{B}$$

$$a = x\text{-intercept} = -\frac{C}{A}$$

$$b = y\text{-intercept} = -\frac{C}{B}$$

Example 10.5

Find the slope, x-intercept, and y-intercept of the line $3x + 2y - 6 = 0$.

Solution

From Eq. 10.3, the slope is

$$m - -\frac{A}{B} = \frac{3}{2} = -1.5$$

From Eq. 10.7, the x-intercept is

$$a = -\frac{C}{A} = -\frac{-6}{3} = 2$$

From Eq. 10.6, the y-intercept is

$$b = -\frac{C}{B} = -\frac{-6}{2} = 3$$

7. PARALLEL LINES

Two different lines with the same slope are called *parallel lines*.

In Fig. 10.3, the lines $2x - 3y = -12$ and $2x - 3y = 12$ have the same slope, $\frac{2}{3}$. Writing the equations in equivalent form shows that the slope for each line is the same; only the y-intercepts differ.

$$y = \frac{2}{3}x + 4$$

$$y = \frac{2}{3}x - 4$$

Figure 10.3 *Parallel Lines*

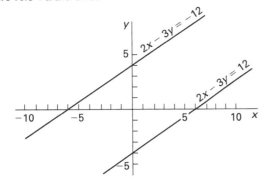

8. PERPENDICULAR LINES

Two lines that intersect at right angles are called *perpendicular lines*. For two lines to be perpendicular, the slope of one must be the negative reciprocal of the other, or

$$m_1 = -\frac{1}{m_2} \qquad \text{10.8}$$

In Fig. 10.4, the lines $2x - 3y = 12$ and $3x + 2y = 8$ are perpendicular lines, as can be seen when written in the form $y = mx + b$.

$$y = \frac{2}{3}x - 4$$

$$y = -\frac{3}{2}x + 4$$

Figure 10.4 *Perpendicular Lines*

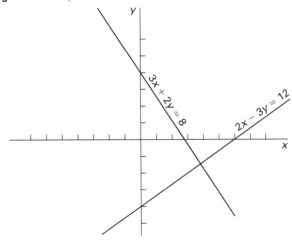

9. PERPENDICULAR DISTANCE FROM A POINT TO A LINE

Equation 10.9 can be used to find the perpendicular distance from a point of known coordinates (x, y) to a line of a known equation.

$$D = \frac{|Ax + By + C|}{\sqrt{A^2 + B^2}} \qquad \text{10.9}$$

Example 10.6

Find the perpendicular distance D from the point P $(-2, 4)$ to the line $4x - 3y - 16 = 0$.

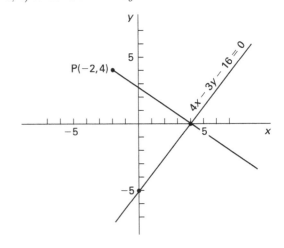

Solution

From Eq. 10.9,

$$D = \frac{|Ax + By + C|}{\sqrt{A^2 + B^2}} = \frac{|(4)(-2) - (3)(4) - 16|}{\sqrt{(4)^2 + (-3)^2}}$$
$$= 7.2$$

10. WRITING THE EQUATION OF A LINE

An equation may be written for any straight line if sufficient information is known. Any of the following combinations of information are sufficient.

- one point on the line and the slope

- two points on the line

- the x-intercept and y-intercept

- the slope of the line and the y-intercept

- the slope of the line and the x-intercept

11. POINT-SLOPE FORM OF THE EQUATION OF A LINE

If $P(x, y)$ is any point on a line with slope m through point $P_1(x_1, y_1)$, then the *point-slope form* of the line can be found by rearranging Eq. 10.1 into Eq. 10.10.

$$y - y_1 = m(x - x_1) \qquad 10.10$$

Example 10.7

Write the general form of the equation of a line through point $(4, -2)$ with a slope of 2.

Solution

From Eq. 10.10,

$$y - y_1 = m(x - x_1)$$
$$y - (-2) = (2)(x - 4)$$
$$y + 2 = 2x - 8$$
$$-2x + y + 10 = 0$$
$$2x - y - 10 = 0$$

12. TWO-POINT FORM OF THE EQUATION OF A LINE

If $P(x, y)$ is any point on a line that passes through points $P_1(x_1, y_1)$ and $P_2(x_2, y_2)$, then the *two-point form* of the line is found from Eq. 10.11 and Eq. 10.12.

$$y - y_1 = \left(\frac{y_2 - y_1}{x_2 - x_1}\right)(x - x_1) \qquad 10.11$$

$$\frac{y - y_1}{x - x_1} = \frac{y_2 - y_1}{x_2 - x_1} \qquad 10.12$$

When writing the two-point form of a line, either point may be designated as point 1.

Example 10.8

Write the equation of the line through points $(1, 4)$ and $(3, -2)$.

Solution

From Eq. 10.12,

$$\frac{y - y_1}{x - x_1} = \frac{y_2 - y_1}{x_2 - x_1}$$
$$\frac{y - 4}{x - 1} = \frac{-2 - 4}{3 - 1}$$
$$y - 4 = (-3)(x - 1)$$
$$3x + y - 7 = 0$$

13. INTERCEPT FORM OF THE EQUATION OF A LINE

The *intercept form* of the equation of a line, with x-intercept a and y-intercept b, neither equal to zero, is

$$\frac{x}{a} + \frac{y}{b} = 1 \qquad 10.13$$

Example 10.9

Write the intercept form of the line with x-intercept 3 and y-intercept -4.

Solution

From Eq. 10.13,

$$\frac{x}{a} + \frac{y}{b} = 1$$
$$\frac{x}{3} + \frac{y}{-4} = 1$$
$$-4x + 3y = -12$$
$$4x - 3y = 12$$

14. SLOPE-INTERCEPT FORM OF THE EQUATION OF A LINE

If the slope of a line and its y-intercept are known, the *slope-intercept form* of the line can be written using Eq. 10.14.

$$y = mx + b \qquad 10.14$$

Example 10.10

Write the slope-intercept form of the line of slope 2 and y-intercept -3.

Solution

From Eq. 10.14, the slope-intercept form is

$$y = 2x - 3$$

Example 10.11

Write the equation of a line through point $(3, -1)$ that is perpendicular to the line $2x + 3y = 6$.

Solution

The slope of the line $2x + 3y = 6$ is $m_1 = -2/3$.

The slope of the perpendicular line is $m_2 = 3/2$.

The equation of the perpendicular line is

$$y - y_1 = m_2(x - x_1)$$
$$y - (-1) = \left(\frac{3}{2}\right)(x - 3)$$
$$3x - 2y = 11$$

15. EQUATION OF A CIRCLE

A *circle* is a curve, all points on which are equidistant from a point called the *center*. The distance of all points from the center is known as the *radius*.

Figure 10.5 Circle Centered at (0,0)

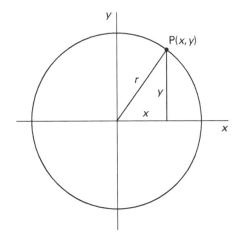

If the center of the circle is at the origin, as in Fig. 10.5, the equation of the circle is

$$x^2 + y^2 = r^2 \qquad \text{10.15}$$

If P is any point on the circle, its coordinates must satisfy the equation $x^2 + y^2 = r^2$.

If the center of the circle is at point $Q(h, k)$, as in Fig. 10.6, the equation becomes

$$(x - h)^2 + (y - k)^2 = r^2 \qquad \text{10.16}$$

Figure 10.6 Circle Centered at (h, k)

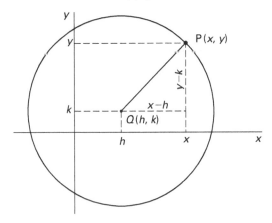

The *general form* for this equation is

$$x^2 + y^2 + Dx + Ey + F = 0 \qquad \text{10.17}$$

Example 10.12

Find the equation of the circle with center $(2, -1)$ and radius 3.

Solution

$$(x - 2)^2 + (y + 1)^2 = (3)^2$$
$$x^2 + y^2 - 4x + 2y - 4 = 0$$

Example 10.13

Write the equation $x^2 + 10x + y^2 - 6y + 18 = 0$ in the form $(x - h)^2 + (y - k)^2 = r^2$.

Solution

Complete the square.

$$x^2 + 10x + y^2 - 6y = -18$$
$$x^2 + 10x + 25 + y^2 - 6y + 9 = -18 + 25 + 9$$
$$(x + 5)^2 + (y - 3)^2 = 16$$

The center of the circle is at $(-5, 3)$, and the radius is 4.

16. TRANSLATION OF AXES

Solving simultaneous equations in which the coefficients of x and y are large numbers can be simplified by reducing the value of the coefficients. This can be done without changing the values of the equations by translating the axes.

In Fig. 10.7, let P be any point with coordinates (x, y) with respect to the axes OX and OY. Establish the new axes, O′X′ and O′Y′, respectively, parallel to the old axes, so that the new origin, O′, has the coordinates (h, k) with respect to the old axes. The coordinates of the point P will then be (x', y') with respect to the new axes.

$$x = x'+ h \qquad 10.18$$

$$x' = x - h \qquad 10.19$$

$$y = y'+ k \qquad 10.20$$

$$y' = y - k \qquad 10.21$$

Figure 10.7 *Transformation of Axes*

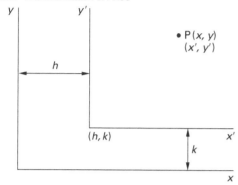

Example 10.14

Point P has the coordinates $(5, 3)$. Find the coordinates of P from the origin O′ $(3, 1)$.

Solution

$$x' = x - h = 5 - 3 = 2$$
$$y' = y - k = 3 - 1 = 2$$

The coordinates of P are $(2, 2)$.

17. PRACTICE PROBLEMS

1. Find two solutions for each equation.

Example: $2x+ y = 7$
$$(2, 3), (3, 1)$$

(a) $2x - 3y = 5$

(b) $x - 2y = 4$

(c) $3x + 2y = 6$

2. Find the coordinates of the points at which the graph of each equation intersects the x-axis and the y-axis.

Example: $3x - 2y = 12$
$$(4,0), (0,-6)$$

(a) $x - y = 0$

(b) $2x + 3y = 18$

(c) $x + y = 4$

3. Determine the slope of the line through each pair of points.

Example: $(-2, 4), (4, -3)$
$$m = \frac{-3 - 4}{4 + 2} = -7/6$$

(a) $(3, 2), (6, 8)$

(b) $(1, 3), (4, 5)$

(c) $(0, -2), (-3, 5)$

(d) $(6, -4), (2, -3)$

(e) $(-3, 4), (3, 4)$

(f) $(1, -5), (-1, 3)$

4. Rearrange the equation in the form $Ax+ By+ C = 0$ with A positive, and determine the slope m of each.

Example: $3y - 2x - 4 = 0$
$$2x - 3y + 4 = 0$$
$$m = -\frac{A}{B} = -\frac{+2}{-3} = \frac{2}{3}$$

(a) $3x+ 4y = 6$

(b) $y = -2x+ 5$

(c) $-4x+ 2y+ 8 = 0$

(d) $y = -5x$

5. Find the slope of each line. Express as a common fraction showing the algebraic sign.

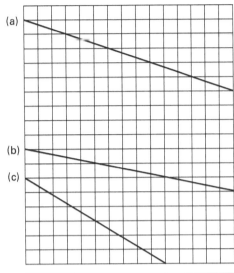

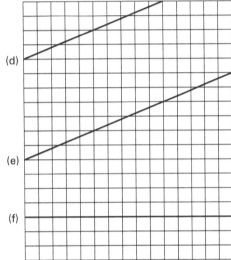

6. Graph each equation, plotting at least three points, and write the equation along the line.

(a) $3x - 2y = 0$

(b) $2x - 3y + 30 = 0$

(c) $3x + 2y + 6 = 0$

(d) $x + y = 0$

(e) $y + 11 = 0$

7. Write each equation in the form $y = mx + b$.

Example: $3x + 2y + 6 = 0$
$$y = -\frac{3}{2}x - 3$$

(a) $4x + 5y + 10 = 0$

(b) $2x - y - 10 = 0$

(c) $3x + 4y + 8 = 0$

(d) $x - y = 0$

(e) $2x - 3y - 12 = 0$

8. Write each equation in the form $Ax + By + C = 0$.

Example: $y = -\frac{2}{3}x - 2$
$$2x + 3y + 6 = 0$$

(a) $y = \frac{2}{5}x - 2$

(b) $y = 2x - 10$

(c) $y = \frac{3}{2}x + \frac{5}{2}$

(d) $y = -x + 5$

(e) $y = -\frac{2}{3}x + 4$

9. Write each equation in the form $y = mx + b$ and plot the graph. Write the equation in the form $Ax + By + C = 0$ along each graph.

Example: $2x - 3y + 30 = 0$
$$y = \frac{2}{3}x + 10$$

(a) $x + 5y - 60 = 0$

(b) $x - y = 0$

(c) $3x + 4y + 24 = 0$

(d) $x + y + 4 = 0$

(e) $2x - 5y - 50 = 0$

10. Write each equation in the form $Ax + By + C = 0$ and indicate the slope, the x-intercept, and the y-intercept of each. (Hint: $m = -A/B$, x-intercept $= -C/A$, and y-intercept $= -C/B$.)

$$\textit{Example:} \quad 4y - 3x = 10$$
$$3x - 4y + 10 = 0$$

$m = 3/4 \quad x\text{-intercept} = -10/3 \quad y\text{-intercept} = 5/2$

(a) $x - y + 5 = 0$

(b) $-2x + 3y = 12$

(c) $x - y = 0$

(d) $6y = 4x + 2$

(e) $2x - 4y = 2$

11. Find the slope of each line. Indicate which pairs of lines are parallel and which are perpendicular.

$$\textit{Example:} \quad 3x + 4y = 8$$
$$m = -\frac{3}{4}$$

(a) $2x - 3y - 3 = 0$

(b) $10x - 6y = 18$

(c) $y = -\frac{3}{2}x + 5$

(d) $5x - 3y = 9$

(e) $6x + 8y = 12$

(f) $4x - 3y = 7$

(g) $3x + 5y = -10$

(h) $y = -\frac{3}{4}x - 3$

12. Find the perpendicular distance from point P to the line indicated.

$\textit{Example:}$ $\text{P}(-8, -1), 2x - 3y - 6 = 0$

$$D = \frac{|(2)(-8) + (-3)(-1) + (-6)|}{\sqrt{(2)^2 + (-3)^2}} = 5.3$$

(a) $\text{P}(3, 3), 3x - 2y + 4 = 0$

(b) $\text{P}(-10, 8), x - y + 5 = 0$

(c) $\text{P}(9, 6), 5x - 2y + 10 = 0$

(d) $\text{P}(-8, -6), 3x + 2y = 0$

(e) $\text{P}(12, -6), 3x - 2y - 6 = 0$

13. Write the equation of the line through the given point with the given slope.

$\textit{Example:}$ $(1, 4), m = -1/2$

$$y - 4 = \frac{(-1)(x - 1)}{2}$$
$$2y - 8 = -x + 1$$
$$x + 2y - 9 = 0$$

(a) $(3, 1), m = -2$

(b) $(-4, 3), m = 2/3$

(c) $(-2, 5), m = 0$

(d) $(2, -3), m = -2/3$

14. Write the equation of the line through the two given points in the form $Ax + By + C = 0$.

$\textit{Example:}$ $(-4, 3), (0, -2)$

$$\frac{y - y_1}{x - x_1} = \frac{y_2 - y_1}{x_2 - x_1}$$
$$\frac{y - 3}{x - (-4)} = \frac{-2 - 3}{0 - (-4)}$$
$$(-5)(x + 4) = (4)(y - 3)$$
$$5x + 4y + 8 = 0$$

(a) $(-3, 2), (1, 4)$

(b) $(2, -3), (5, -2)$

(c) $(3, 4), (-3, 4)$

(d) $(-2, -6), (3, -4)$

15. Write the equation of the lines with the given x- and y-intercepts in the form $Ax + By + C = 0$.

$\textit{Example:}$ $a = -3, b = 4$

$$\frac{x}{a} + \frac{y}{b} = 1$$
$$\frac{x}{-3} + \frac{y}{4} = 1$$
$$4x + (-3y) = -12$$
$$4x - 3y + 12 = 0$$

(a) $a = -4, b = 3$

(b) $a = 1, b = -4$

16. Write the equation of the line that has a slope of $3/4$ and a y-intercept of -3.

17. Write the equation of the line whose y-intercept is 4 and that is perpendicular to the line $4x + 3y + 9 = 0$.

18. Write the equation of the line through the point $(0, 8)$ and parallel to the line whose equation is $y = -3x + 4$.

19. Write the equations of two lines through the point $(5, 5)$, one parallel and one perpendicular to the line $2x + y - 4 = 0$.

20. For each point, find the new coordinates if the axes are translated to a new origin located at $(4, 3)$.

(a) $(4, 6)$

(b) $(-7, 3)$

(c) $(-3, -2)$

(d) $(0, 0)$

(e) $(8, -2)$

(f) $(7, 7)$

SOLUTIONS

1.

(a) $(7, 3), (4, 1)$

(b) $(6, 1), (8, 2)$

(c) $(4, -3), (6, -6)$

Other solutions are possible.

2.

(a) $(0, 0), (0, 0)$

(b) $(9, 0), (0, 6)$

(c) $(4, 0), (0, 4)$

3.

(a) $m = \dfrac{8 - 2}{6 - 3} = 2$

(b) $m = \dfrac{5 - 3}{4 - 1} = 2/3$

(c) $m = \dfrac{5 + 2}{-3} = -7/3$

(d) $m = \dfrac{-3 + 4}{2 - 6} = -1/4$

(e) $m = \dfrac{4 - 4}{3 + 3} = 0$

(f) $m = \dfrac{3 + 5}{-1 - 1} = -4$

4.

(a)

$$3x + 4y = 6$$
$$3x + 4y - 6 = 0$$
$$m = -\frac{3}{4} = -3/4$$

(b)

$$y = -2x + 5$$
$$2x + y - 5 = 0$$
$$m = -\frac{2}{1} = -2$$

(c)

$$-4x + 2y + 8 = 0$$
$$4x - 2y - 8 = 0$$
$$m = -\frac{4}{-2} = 2$$

(d)

$$y = -5x$$
$$5x + y = 0$$
$$m = -\frac{5}{1}$$
$$= -5$$

5.

(a) $-\frac{1}{3}$

(b) $-\frac{1}{5}$

(c) $-\frac{3}{5}$

(d) $\frac{2}{5}$

(e) $\frac{2}{5}$

(f) 0

6.

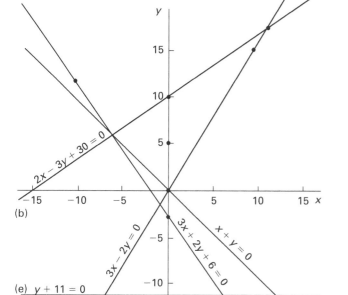

7.

(a)

$$4x + 5y + 10 = 0$$
$$5y = -4x - 10$$
$$y = -\frac{4}{5}x - 2$$

(b)

$$2x - y - 10 = 0$$
$$-y = -2x + 10$$
$$y = 2x - 10$$

(c)

$$3x + 4y + 8 = 0$$
$$4y = -3x - 8$$
$$y = -\frac{3}{4}x - 2$$

(d)

$$x - y = 0$$
$$-y = -x$$
$$y = x$$

(e)

$$2x - 3y - 12 = 0$$
$$-3y = -2x + 12$$
$$y = \frac{2}{3}x - 4$$

8.

(a)

$$y = \frac{2}{5}x - 2$$
$$5y = 2x - 10$$
$$2x - 5y - 10 = 0$$

(b)

$$y = 2x - 10$$
$$-2x + y + 10 = 0$$
$$2x - y - 10 = 0$$

(c)

$$y = \frac{3}{2}x + \frac{5}{2}$$
$$2y = 3x + 5$$
$$3x - 2y + 5 = 0$$

(d)

$$y = -x + 5$$
$$x + y - 5 = 0$$

(e)

$$y = -\frac{2}{3}x + 4$$
$$3y = -2x + 12$$
$$2x + 3y - 12 = 0$$

9.

(a)

$$x + 5y - 60 = 0$$
$$y = -\frac{1}{5}x + 12$$

(b)

$$x - y = 0$$
$$y = x$$

(c)

$$3x + 4y + 24 = 0$$
$$y = -\frac{3}{4}x - 6$$

(d)

$$x + y + 4 = 0$$
$$y = -x - 4$$

(e)

$$2x - 5y - 50 = 0$$
$$y = \frac{2}{5}x - 10$$

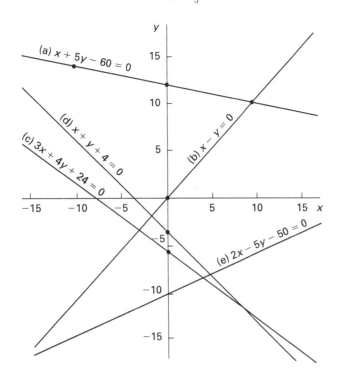

10.

(a) $x - y + 5 = 0$

$m = 1$

$x\text{-intercept} = -5$

$y\text{-intercept} = 5$

(b) $2x - 3y + 12 = 0$

$m = 2/3$

$x\text{-intercept} = -6$

$y\text{-intercept} = 4$

(c) $x - y = 0$

$m = 1$

$x\text{-intercept} = 0$

$y\text{-intercept} = 0$

(d) $4x - 6y + 2 = 0$

$m = 2/3$

$x\text{-intercept} = -1/2$

$y\text{-intercept} = 1/3$

(e) $2x - 4y - 2 = 0$

$m = 1/2$

$x\text{-intercept} = 1$

$y\text{-intercept} = -1/2$

11.

(a) slope $= \frac{2}{3}$

(b) slope $= \frac{5}{3}$

(c) slope $= -\frac{3}{2}$

(d) slope $= \frac{5}{3}$

(e) slope $= -\frac{3}{4}$

(f) slope $= \frac{4}{3}$

(g) slope $= -\frac{3}{5}$

(h) slope $= -\frac{3}{4}$

Lines b and d are parallel. Lines e and h are parallel. Lines a and c are perpendicular. Lines b and d are both perpendicular to line g. Lines e and h are both perpendicular to line f.

12.

(a)

$$D = \frac{|(3)(3) - (2)(3) + 4|}{\sqrt{(3)^2 + (2)^2}} = 1.9$$

(b)

$$D = \frac{|(1)(-10) + (-1)(8) + 5|}{\sqrt{(1)^2 + (1)^2}} = 9.2$$

(c)

$$D = \frac{|(5)(9) - (2)(6) + 10|}{\sqrt{(5)^2 + (-2)^2}} = 8.0$$

(d)

$$D = \frac{|(3)(-8) + (2)(-6)|}{\sqrt{(3)^2 + (2)^2}} = 10.0$$

(e)

$$D = \frac{|(3)(12) + (-2)(-6) - 6|}{\sqrt{(3)^2 + (-2)^2}} = 11.6$$

13.

(a)

$$y - 1 = (-2)(x - 3)$$
$$= -2x + 6$$
$$2x + y - 7 = 0$$

(b)

$$y - 3 = \frac{(2)(x + 4)}{3}$$
$$3y - 9 = 2x + 8$$
$$2x - 3y + 17 = 0$$

(c)

$$y - 5 = 0$$
$$y = 5$$

(d)

$$y + 3 = \frac{(-2)(x - 2)}{3}$$
$$3y + 9 = -2x + 4$$
$$2x + 3y + 5 = 0$$

14.

(a)

$$\frac{y - y_1}{x - x_1} = \frac{y_2 - y_1}{x_2 - x_1}$$
$$\frac{y - 2}{x - (-3)} = \frac{4 - 2}{1 - (-3)}$$
$$(2)(x + 3) = (4)(y - 2)$$
$$x - 2y + 7 = 0$$

(b)

$$\frac{y - y_1}{x - x_1} = \frac{y_2 - y_1}{x_2 - x_1}$$
$$\frac{y - (-3)}{x - 2} = \frac{-2 - (-3)}{5 - 2}$$
$$(1)(x - 2) = (3)(y + 3)$$
$$x - 3y - 11 = 0$$

(c)

$$\frac{y - y_1}{x - x_1} = \frac{y_2 - y_1}{x_2 - x_1}$$
$$\frac{y - 4}{x - 3} = \frac{4 - 4}{-3 - 3}$$
$$(0)(x - 3) = (-6)(y - 4)$$
$$y - 4 = 0$$

(d)

$$\frac{y - y_1}{x - x_1} = \frac{y_2 - y_1}{x_2 - x_1}$$
$$\frac{y - (-6)}{x - (-2)} = \frac{-4 - (-6)}{3 - (-2)}$$
$$(2)(x + 2) = (5)(y + 6)$$
$$2x - 5y - 26 = 0$$

15.

(a)

$$\frac{x}{a} + \frac{y}{b} = 1$$
$$\frac{x}{-4} + \frac{y}{3} = 1$$
$$3x + (-4y) = -12$$
$$3x - 4y + 12 = 0$$

(b)

$$\frac{x}{a} + \frac{y}{b} = 1$$
$$\frac{x}{1} + \frac{y}{-4} = 1$$
$$-4x + y = -4$$
$$4x - y - 4 = 0$$

16.

$$y = mx + b$$
$$= \frac{3x}{4} + (-3)$$
$$4y = 3x - 12$$
$$3x - 4y - 12 = 0$$

17. The slope of the line $4x + 3y + 9 = 0$ is $-4/3$. For the perpendicular line,

$$m = \frac{3}{4}$$
$$y = mx + b = \frac{3x}{4} + 4$$
$$4y = 3x + 16$$
$$3x - 4y + 16 = 0$$

18. For the parallel line,

$$m = -3$$
$$y = mx + b = -3x + 8$$
$$3x + y - 8 = 0$$

19. The slope of the line $2x + y - 4 = 0$ is -2. For the parallel line,

$$y - y_1 = m(x - x_1)$$
$$y - 5 = (-2)(x - 5)$$
$$2x + y - 15 = 0$$

The slope of the perpendicular line is 1/2. For the perpendicular line,

$$y - y_1 = m(x - x_1)$$
$$y - 5 = \left(\frac{1}{2}\right)(x - 5)$$
$$x - 2y + 5 = 0$$

20.

(a) $(0, 3)$

(b) $(-11, 0)$

(c) $(-7, -5)$

(d) $(-4, -3)$

(e) $(4, -5)$

(f) $(3, 4)$

11 Statistics of Measurements

Nomenclature

R	range	–	–
$\overline{x}$	average	–	–

Symbols

σ	standard deviation	–	–

1. CONCEPT OF MEASUREMENTS

In surveying, *measurements* are observations made to determine unknown quantities. Some typical modern surveying measurements include measurement of an angle using a digital theodolite, measurement of a distance between two points using an electronic distance measurement (EDM) instrument, and measurement of a geographic position using a Global Positioning System (GPS) receiver.

Modern surveying utilizes equipment capable of highly precise measurements. Despite this sophistication, all observational equipment and the practitioners who use it are imperfect, so all measurements are only estimates of true quantities. As a result, in order for a surveyor's measurements to approach as nearly as possible the actual value of unknown quantities, redundant observations must be made.

2. SIGNIFICANT DECIMAL PLACES

The number of *significant decimal places* refers to the number of digits that have meaning based on the precision of the measurements.[1] When measurements with different levels of precision are combined, the resulting measurement cannot be more precise than the least

precise measurement. It is important to determine how many digits in a measurement have meaning and to use that many digits for recorded values. Recording too few digits can result in *round-off errors* in subsequent calculations, but recording too many digits misrepresents the precision of the measurement. The number of decimal places refers to the number of digits to the right of the decimal. This should not be confused with the term "significant figures," which refers to digits that provide useful information. As an example, the measurement 1000.0 has one decimal place, while it has five significant figures, since the ".0" suggests the presumed precision of the measurement. As another example, a measurement of 54.972 may be rounded to two decimal places as 54.97. Yet, if rounded to two significant figures, it would be 55.

Zeros that follow a decimal but precede other nonzero digits are not considered when counting significant decimal places. However, zeros following nonzero digits are considered when counting significant decimal places. For example, 0.0062 has two significant decimal places, while 24.240 has three significant decimal places.

Calculations involving measurements have specific rules for significant decimal places.

- For addition or subtraction, the number of significant decimal places used for the sum or difference is determined by the value in the expression with the fewest decimal places.

- For multiplication or division, the number of significant decimal places used for the product or quotient is determined by the value in the expression with the fewest significant decimal places.

To avoid round-off errors in calculations, it is a good idea to carry one extra significant decimal place for intermediate calculations and to only round off to the correct number of significant decimal places for the final answer.

3. PRECISION VERSUS ACCURACY

The terms "accuracy" and "precision" are often confused. *Accuracy* is nearness to the *true value*, the correct value for that which is being measured. Because the true value is generally not known in surveying, a measurement

---

[1] Note that significant decimal places are not the same as significant digits.

from a higher-order survey is typically used to evaluate the accuracy of a measurement. *Precision* is nearness to other values in the same set of measurements.

A useful analogy often used to explain the difference between precision and accuracy involves bullet holes on a target at a shooting range. A group of bullet holes that are tightly clustered but not close to the bull's eye are still precise. Yet a group of bullet holes closer to the bull's eye, but not tightly clustered, are more accurate.

4. MEASUREMENT ERRORS

A *measurement error* is the difference between the measured value of any quantity and the quantity's true value. Measurement errors have three sources: imperfections in instruments; changing conditions in the surrounding environment, such as temperature, atmospheric pressure, wind, and gravitational and magnetic fields; and personal limitations, such as an observer being unable to perfectly read a micrometer or center a level bubble. Traditionally, measurement errors are classified as blunders, systematic errors, or random errors.

Blunders are not true errors, as they result from incorrect measurements caused by carelessness or confusion on the part of the observer. Typical blunders include making mistakes while recording measurements and entering the wrong height of an instrument for a GPS measurement.

Systematic errors are errors caused by physical laws, which may be predicted and sometimes corrected for by observational procedures. *Observational procedures* include balancing backsight and foresight distances during leveling operations to compensate for the curvature of the earth and for collimation error. Systematic errors that cannot be corrected using observational procedures, such as errors caused by non-standard temperature and atmospheric pressure during electronic distance measurement, may be corrected by applying correction to the measurements based on published tabulations of the effect of meteorological conditions on the speed of light.

Random errors generally result from imperfections in the instrument or observer. They are the differences between observed values and true values that remain after blunders and systematic errors have been removed from the data. They are generally small and are as likely to have a negative mathematical relationship with the true value as they are likely to have a positive mathematical relationship with the true value. Typical random errors include not holding a level rod exactly plumb or imperfectly centering an instrument over a point. Because random errors do not follow any physical law, they must be dealt with using probability theory.

In basic terms, *probability theory* states that it is normal to see minor differences due to imperfections in equipment and measurement techniques when a measurement is made numerous times. However, those results will follow a pattern: measurement values will be more likely to be closer to the mean than far from the mean, as

measurements are generally close to the true value. Probability and descriptions of data sets are explained in more detail in Sec. 11.5.

5. DESCRIPTIVE STATISTICS

As mentioned in Sec. 11.1, redundant measurements are essential to sound surveying practices, and statistical tools describe the data sets that result from these redundant measurements. These tools include range; measures of central tendency such as mean, median, and mode; histograms; and adherence to the distribution of measurements associated with random measurements. Table 11.1 shows an example of a redundant data set, which includes a series of 26 taped distance measurements made by students in a surveying class.

Table 11.1 Sample Set of Taped Distance Measurements

no.	measurement (ft)	no.	measurement (ft)
1	18.60	14	18.70
2	18.63	15	18.70
3	18.65	16	18.71
4	18.65	17	18.71
5	18.66	18	18.72
6	18.66	19	18.73
7	18.67	20	18.73
8	18.67	21	18.74
9	18.68	22	18.74
10	18.68	23	18.75
11	18.69	24	18.76
12	18.69	25	18.77
13	18.69	26	18.80

The most basic statistical tool used in surveying to analyze a data set is the *range* or *dispersion* of the data. The range is the difference between the highest value, x_{max}, and lowest value, x_{min}, in the data set. Range allows the surveyor to see how much repeated observations vary, which allows him or her to get a feel for the precision of a single observation.

$$R = x_{max} - x_{min} \qquad 11.1$$

The *mean (average)*, $\bar{x}$, is calculated as the sum of all of the measurements, x_i, divided by the number of measurements, n. In surveying, the mean may be used to calculate the best true value if the data are not skewed.

$$\bar{x} = \frac{\sum_{i=1}^{n} x_i}{n} \qquad 11.2$$

The *median* is the measurement in the middle when the data set is arranged in increasing (or decreasing) order. In other words, half of the measurements are above this value and half are below. When the number of measurements is odd, the median is the middle measurement, and no calculation is needed. When the number of measurements is even, the median must be calculated as the average of the two middle measurements.

In surveying, the median or mode may be used to calculate the most probable true value if the data are skewed due to a systematic error or blunder.

The *mode* is the value that occurs most frequently in the data set. Although the mode is numerical, no calculation is needed. For example, the mode of the data set in Table 11.1 is 18.69 ft, as that value occurs three times, which is more often than any other value in that data set.

Example 11.1

Calculate the range, mean, and median of the data set given in Table 11.1.

Solution

Using Eq. 11.1, the range is

$$R = x_{max} - x_{min} = 18.80 \text{ ft} - 18.60 \text{ ft} = 0.20 \text{ ft}$$

From Eq. 11.2, the mean is

$$\bar{x} = \frac{\sum_{i=1}^{n} x_i}{n} = \frac{x_1 + x_2 + \cdots + x_n}{n}$$
$$= \frac{18.60 \text{ ft} + 18.63 \text{ ft} + \cdots + 18.80 \text{ ft}}{26}$$
$$= 18.70 \text{ ft}$$

Since there is an even number of observations, the median is the average of the middle two observations when the data are arranged in increasing magnitude, or

$$\frac{18.69 \text{ ft} + 18.70 \text{ ft}}{2} = 18.695 \text{ ft}$$

6. STANDARD DEVIATION

Another important statistical tool used in surveying is the *standard deviation*, σ. It is a commonly used measure of data quality that is often used as a standard for the acceptance or rejection of data. The standard deviation is a numerical descriptor that measures the data variation and represents the most probable error in a single observation. One standard deviation is equal to one sigma, $\pm\sigma$. To calculate the standard deviation, the squares of each variance are required. The *variance* is the difference between a value, x_i, and the mean, $\bar{x}$.

Generally, each variance and the squares of each variance are calculated and added to a table of values before calculating the standard deviation.

$$\sigma = \sqrt{\frac{\sum_{i=1}^{n}(x_i - \bar{x})^2}{n - 1}} \quad\quad 11.3$$

Equation 11.3 is used for analyzing a data sample. However, when an entire population is analyzed, an equivalent measure, usually called the standard error, is used.[2] The standard error is calculated using the same equation, with a denominator of n rather than $n - 1$.

Example 11.2

Calculate the standard deviation of the data set given in Table 11.1.

Solution

no.	x (ft)	$x - \bar{x}$ (ft)	$(x - \bar{x})^2$ (ft^2)
1	18.60	−0.10	0.01
2	18.63	−0.07	0.00
3	18.65	−0.05	0.00
4	18.65	−0.05	0.00
5	18.66	−0.04	0.00
6	18.66	−0.04	0.00
7	18.67	−0.03	0.00
8	18.67	−0.03	0.00
9	18.68	−0.02	0.00
10	18.68	−0.02	0.00
11	18.69	−0.01	0.00
12	18.69	−0.01	0.00
13	18.69	−0.01	0.00
14	18.70	0.00	0.00
15	18.70	0.00	0.00
16	18.71	0.01	0.00
17	18.71	0.01	0.00
18	18.72	0.02	0.00
19	18.73	0.03	0.00
20	18.73	0.03	0.00
21	18.74	0.04	0.00
22	18.74	0.04	0.00
23	18.75	0.05	0.00
24	18.76	0.06	0.00
25	18.77	0.07	0.01
26	18.80	0.10	0.01
	$\bar{x} = 18.70$		$\Sigma = 0.053$

$$\sigma = \sqrt{\frac{\sum_{i=1}^{n}(x_i - \bar{x})^2}{n - 1}} = \sqrt{\frac{0.053 \text{ ft}^2}{26 - 1}} = 0.05 \text{ ft}$$

---

[2]Confusingly, some texts also refer to this as a standard deviation.

7. ROOT-MEAN-SQUARED ERROR

A statistical tool similar to the standard deviation is the *root-mean-squared error* (RMSE) which is a measure of data quality in surveying and is used to test the accuracy of measurements by comparing them with independently measured values from a higher-order survey. This measure is frequently used to evaluate topographic elevations established by photogrammetry or light detection and ranging (LiDAR).

The RMSE includes the independently measured value, x_i', in the calculation.

$$\text{RMSE} = \sqrt{\frac{\sum_{i=1}^{n}(x_i - x_i')^2}{n}} \quad \textit{11.4}$$

Comparing Eq. 11.4 with Eq. 11.3, the RMSE is almost the same as the standard deviation. The only differences are that the independently measured value is used instead of the mean value and that the denominator is n rather than $n-1$.

Example 11.3

Using the table shown, calculate the RMSE.

	mapped elevation	check elevation
1	25.52	25.54
2	5.22	5.30
3	8.64	8.50
4	15.50	15.45
5	30.22	30.30
6	17.23	17.45
7	19.45	19.32
8	8.64	8.60
9	9.21	9.25
10	29.89	29.82
11	15.55	15.64
12	13.39	13.23
13	13.55	13.67
14	19.96	20.02
15	10.99	10.84
16	17.54	17.67
17	9.87	9.82
18	30.32	30.37
19	32.17	32.10
20	30.55	30.52

Solution

	mapped elevation	check elevation	$(x_i - x_i')$	$(x_i - x_i')^2$
1	25.52	25.54	0.02	0.00
2	5.22	5.30	0.08	0.01
3	8.64	8.50	−0.14	0.02
4	15.50	15.45	−0.05	0.00
5	30.22	30.30	0.08	0.01
6	17.23	17.45	0.22	0.05
7	19.45	19.32	−0.13	0.02
8	8.64	8.60	−0.04	0.00
9	9.21	9.25	0.04	0.00
10	29.89	29.82	−0.07	0.00
11	15.55	15.64	0.09	0.01
12	13.39	13.23	−0.16	0.03
13	13.55	13.67	0.12	0.01
14	19.96	20.02	0.06	0.00
15	10.99	10.84	−0.15	0.02
16	17.54	17.67	0.13	0.02
17	9.87	9.82	−0.05	0.00
18	30.32	30.37	0.05	0.00
19	32.17	32.10	−0.07	0.00
20	30.55	30.52	−0.03	0.00
				$\Sigma = 0.21$

$$\text{RMSE} = \sqrt{\frac{\sum_{i=1}^{n}(x_i - x_i')^2}{n}} = \sqrt{\frac{0.21}{20}} = 0.10$$

8. RANDOM ERROR THEORY

Truly random errors will follow a pattern governed by probability called a *normal distribution*. The histogram for measurements with only random errors will follow a predictable bell-shaped curve known as a *normal distribution curve*. (See Fig. 11.1.) The shape of a normal distribution curve is based on a mathematical equation that relates a value's frequency of occurrence to its closeness to the mean of the data set.

In a perfectly normal distribution, there is a predictable relationship between the distribution of errors and the standard deviation, σ. The relationship consists of the one sigma error, the two sigma error, and the three sigma error.[3] The *one sigma error* represents 68.3% of the errors (or residuals), which will fall within one standard deviation of the mean. The *two sigma error* represents 95% of the errors (or residuals), which will fall within 1.96 standard deviations of the mean.[4] The *three sigma error* represents 99.7% of the errors (or residuals), which will fall within 2.968 standard deviations of the

---

[3] As mentioned in Sec. 11.6, one standard deviation is equal to one sigma, $\pm\sigma$.
[4] The term "two sigma error" references the value of 1.96 rounded to 2.

mean.[5] Measurements with residuals that are more or less than the three sigma error can be safely rejected with 99.7% certainty that they are "wild data," or outliers, and probably the result of a blunder. Even though outliers are normal, they may vary slightly with smaller data sets.

Figure 11.1 *Normal Distribution Curve*

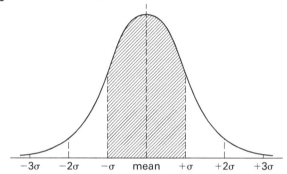

When independent observations are used to check measurements, the RMSE can be considered equivalent to the two sigma errors. So, 95% of the measurements in a data set of observations should have an accuracy less than or equal to 1.96 times the RMSE, an often-cited statistic defining the uncertainty of a data set at the 95% confidence level.

9. PRACTICE PROBLEMS

Data set for Prob. 1 through Prob. 6

angle no.	x ($''$)	angle no.	x ($''$)
1	29.3	11	24.1
2	24.0	12	26.2
3	27.9	13	30.1
4	26.8	14	29.7
5	26.1	15	24.1
6	25.9	16	26.2
7	26.1	17	27.1
8	27.8	18	24.9
9	27.2	19	25.7
10	28.0	20	25.2

1. The range of the data set is most nearly

(A) 6.0″
(B) 6.1″
(C) 12.0″
(D) 30.0″

2. The mean of the data set is most nearly

(A) 26.2″
(B) 26.6″
(C) 27.0″
(D) 27.2″

3. The median of the data set is most nearly

(A) 26.2″
(B) 26.6″
(C) 27.0″
(D) 27.2″

4. The standard deviation of the data set is most nearly

(A) 0.80″
(B) 1.80″
(C) 3.20″
(D) 3.60″

5. If the data set had a normal distribution, the percentage of measurements less than or equal to 1.96 times the standard deviation would be most nearly

(A) 50.0%
(B) 68.3%
(C) 95.0%
(D) 99.7%

6. If measurements that occur outside the three sigma error range are rejected, the angle(s) rejected from the data set will be

(A) angle 1, angle 2, angle 13, and angle 14
(B) angle 1, angle 13, and angle 14
(C) angle 13
(D) none of the angles

---

[5] The term "three sigma error" references the value of 2.968 rounded to 3.

7. According to the rules of significant decimal places, the product of 10.23 and 298.45 is expressed as

(A) 3053

(B) 3053.1

(C) 3053.14

(D) 3053.1435

8. The effect of atmospheric pressure on EDM measurements is an example of a

(A) blunder

(B) one sigma error

(C) random error

(D) systematic error

9. Interpreting a rod reading as 6.05 ft when the correct reading is 5.05 ft is an example of a

(A) blunder

(B) one sigma error

(C) random error

(D) systematic error

SOLUTIONS

1. Use Eq. 11.1 to calculate the range of the data set. In the data set, the highest value is 30.1″, and the lowest value is 24.0″.

$$R = x_{\max} - x_{\min} = 30.1'' - 24.0'' = 6.1''$$

The answer is (B).

2. Use Eq. 11.2 to calculate the mean of the data set.

$$\bar{x} = \frac{\sum_{i=1}^{n} x_i}{n} = \frac{x_1 + x_2 + \cdots + x_n}{n}$$
$$= \frac{24.0'' + 24.1'' + \cdots + 30.1''}{20}$$
$$= 26.6''$$

The answer is (B).

3. In a table, arrange the measurements from lowest value to highest value. If two equally long columns are used, one middle value will be at the bottom of the first column and the other will be at the top of the second column.

	$x('')$		$x('')$
lowest value	24.0	middle value 2	26.2
	24.1		26.8
	24.1		27.1
	24.9		27.2
	25.2		27.8
	25.7		27.9
	25.9		28.0
	26.1		29.3
	26.1		29.7
middle value 1	26.2	highest value	30.1

From this table, it can be seen that the middle values are both 26.2″. Therefore, 26.2″ is the median.

The answer is (A).

4. Create a table using the measurements from the data set with a column for the variances and a column for the squared variances. Use the mean calculated in Sol. 2 to find the variance and squared variance of each measurement.

angle	$x('')$	$x - \bar{x} =$ $x - 26.6\,('')$	$(x - \bar{x})^2 =$ $(x - 26.6)^2\,('')$
1	29.3	2.7	7.2
2	24.0	−2.6	6.9
3	27.9	1.3	1.6
4	26.8	0.2	0.0
5	26.1	−0.5	0.3
6	25.9	−0.7	0.5
7	26.1	−0.5	0.3
8	27.8	1.2	1.4
9	27.2	0.6	0.3
10	28.0	1.4	1.9
11	24.1	−2.5	6.4
12	26.2	−0.4	0.2
13	30.1	3.5	12.1
14	29.7	3.1	9.5
15	24.1	−2.5	6.4
16	26.2	−0.4	0.2
17	27.1	0.5	0.2
18	24.9	−1.7	3.0
19	25.7	−0.9	0.8
20	25.2	−1.4	2.0
			$\Sigma = 61.1$

Calculate the standard deviation using Eq. 11.3.

$$\sigma = \sqrt{\dfrac{\displaystyle\sum_{i=1}^{n}(x_i - \bar{x})^2}{n-1}} = \sqrt{\dfrac{61.1''}{20-1}} = 1.8''$$

The answer is (B).

5. With a normal distribution, 95.0% of the variances for the observations should occur between $-3.5''$ and $+3.5''$.

The answer is (C).

6. From Sol. 4, the standard deviation is $1.8''$. The three sigma error is

$$3\sigma = (3)(1.8'') = 5.4''$$

Referencing the table in Sol. 4, none of the variances are within the $-5.4''$ to $+5.4''$ range. Therefore, if the three sigma error is used to reject data, none of the angles can be rejected.

The answer is (D).

7. For multiplication or division, the resulting quantity of significant decimal places is determined by the number in the expression with the fewest significant decimal places. While a calculator may show a number with more significant decimal places, the final answer should be rounded to match the precision of the number with the fewest significant decimal places.

$$(10.23)(298.45) = 3053.1435 \quad (3053.14)$$

The correct statement of the product of 10.23 and 298.45 is 3053.14.

The answer is (C).

8. The effect of atmospheric pressure on EDM measurements is a systematic error.

The answer is (D).

9. Interpreting a rod reading as 6.05 ft when the correct reading is 5.05 ft is a blunder.

The answer is (A).

12 Accuracy Standards

1. ACCURACY VERSUS PRECISION IN STANDARDS

The distinction between accuracy and precision is an important concept in standards for surveying and mapping. As discussed in Chap. 11, *precision* is the measure of consistency between observations, while *accuracy* is the measure of nearness to the true value. In surveying, a "true" value is usually not known, as no measurement is absolutely perfect. Instead, a measurement from a higher-order survey is typically used to evaluate the accuracy of a measurement.

Many surveying standards, such as those requiring redundancy of measurements or closure of figures, are really standards for precision, since they do not provide for comparison with an accepted value. For example, while standards establishing minimum closure ratios for traverses may check the precision of the points in the traverse, the points may be inaccurate if a scale error exists in the electronic distance measurement (EDM) used.

2. NATIONAL MAP ACCURACY STANDARDS

The *United States National Map Accuracy Standards* are possibly the oldest standards for comparing the relative accuracy of published maps. These standards have been superseded by the *Federal Geographic Data Committee (FGDC) Geospatial Positioning Accuracy Standards* of 1998, but are included for historical interest.

The *U.S. National Map Accuracy Standards* were originally developed by a committee established by the American Society for Photogrammetry and Remote Sensing in 1937. That draft inspired several U.S. agencies to begin studies of standards for map accuracy. The final standards were formally issued by the U.S. Bureau of the Budget in 1941, and were revised in 1947. The standards were intended for application by all federal agencies producing maps.

The standards require that the accuracy of a map be tested by comparison of well-defined points, such as the intersection of two roads, whose location or elevation is shown on the map, with positions determined by surveys of a higher accuracy. Generally, to test a map, 20 or more well-defined points are used.

For testing of horizontal accuracy, the standards require that for maps with scales larger than 1:20,000, not more than 10% of the points tested can be in error by more than 1/30 in, as measured on the publication scale. For maps with scales of 1:20,000 or smaller, not more than 10% of the points tested can be in error by more than 1/50 in. For testing of vertical accuracy (with contours), not more than 10% of the elevations tested can be in error by more than one-half the contour interval.

As an example of the *U.S. National Map Accuracy Standards* for horizontal accuracy, for a 1:24,000 scale topographic map like those produced by the United States Geological Survey (USGS), not more than 10% percent of the points tested can be in error by more than 1/50 in. At that scale, an error of 1 in represents an actual error of

$$1 \text{ in} = \frac{24{,}000 \text{ in}}{12 \, \frac{\text{in}}{\text{ft}}} = 2000 \text{ ft}$$

Therefore, 1/50 in represents 40 ft on the maps, and that value would be the maximum horizontal error allowable under the standards for tested points.

As an example of application of that standard for vertical accuracy, for a map with a 10 ft contour interval, no more than 10% of tested points can have an error greater than 5 ft.

3. FEMA ACCURACY STANDARDS

The Federal Emergency Management Administration (FEMA) has promulgated accuracy standards for light detection and ranging (LiDAR) data used in the agency's flood plain management maps. Those standards address system calibration, flight planning, acceptable control points, and post processing. In addition, the FEMA standards address quality control and quality assurance (QC/QA). As opposed to the requirements for selection of features that can be identified on a map under the *U.S. National Map Accuracy Standards*, the test points for these standards may be selected at random since, with LiDAR mapping, a horizontal position is associated with an elevation for each measured point.

The standards regarding QC/QA require that the elevation data have a maximum *root-mean-squared error* (RMSE) of 15 cm. As used in the FEMA accuracy standards, RMSE is the square root of the average of the squared differences between the measured elevations for selected points, compared to the *check elevations*, elevations for the same points measured by an independent survey of higher accuracy. (See Chap. 11 for further discussion of the calculation of the RMSE.)

$$\text{RMSE} = \sqrt{\frac{\sum_{i=1}^{n}(x_i - x_i')^2}{n}} \qquad 12.1$$

Based on random error theory, if there is a normal distribution of errors, 95% of the errors in a sufficiently large sample of measurements should be less than 1.96 times the RMSE. This value (1.96 times the RMSE) is often called the *2 sigma error*.

The standards require that at least 20 points be sampled for each of the main categories of ground cover in the study area, allowing for one point to fail the test without compromising the 95% confidence level. Examples of categories of ground cover provided in the standards include:

- bare earth and low grass (e.g., plowed fields, lawns, golf courses)

- high grass and crops (e.g., hay fields, corn fields, wheat fields)

- brush lands and low trees (e.g., chaparrals, mesquite, mangrove swamps)

- ground fully covered by trees (e.g., hardwoods, evergreens, mixed forests)

- urban areas (e.g., high, dense manmade structures)

- sawgrass

Since the criterion for measurement accuracy is based on the assumption of a normal distribution of errors, these standards require examination of the data for possible skew, which would be suggestive of systematic error. If the errors are absolutely random, the mean of the errors is statistically likely to be zero. The FEMA standards consider a mean of the errors greater than ± 2 cm as an indicator of unacceptable skew.

Example 12.1

The table shown gives a series of elevations obtained by means of an airborne LiDAR survey. The check elevations were obtained from a real-time kinematic (RTK) GPS survey of a brush lands area. What is the RMSE for the data?

n	latitude	longitude	LiDAR, x_i (m)	GPS, x_i' (m)	$x_i - x_i'$ (m)	$(x_i - x_i')^2$ (m²)
1	9°55'10.25" N	85°20'12.45" W	17.43	17.47	−0.03	0.001
2	9°55'13.22" N	85°20'30.23" W	19.93	19.87	0.06	0.004
3	9°55'17.31" N	85°20'50.22" W	18.04	18.32	−0.28	0.078
4	9°55'21.67" N	85°21'12.67" W	21.67	21.67	0.00	0.000
5	9°55'25.50" N	85°21'29.43" W	19.08	19.05	0.03	0.001
6	9°55'28.41" N	85°21'09.76" W	18.87	19.05	−0.18	0.033
7	9°55'31.67" N	85°20'49.43" W	18.29	18.59	−0.30	0.093
8	9°55'34.93" N	85°20'41.19" W	17.40	17.47	−0.07	0.005
9	9°55'37.13" N	85°20'31.84" W	16.55	16.52	0.03	0.001
10	9°55'40.85" N	85°20'11.32" W	18.17	18.17	0.00	0.000
11	9°55'43.32" N	85°20'26.56" W	19.48	19.42	0.06	0.004
12	9°55'46.64" N	85°20'37.21" W	19.02	19.11	−0.09	0.008
13	9°55'49.25" N	85°20'50.78" W	19.96	19.84	0.12	0.014
14	9°55'52.33" N	85°21'10.62" W	20.70	20.67	0.03	0.001
15	9°55'55.27" N	85°21'28.42" W	20.79	20.82	−0.03	0.001
16	9°55'58.21" N	85°21'14.91" W	18.50	18.56	−0.06	0.004
17	9°56'00.81" N	85°20'55.52" W	20.51	20.45	0.06	0.004
18	9°56'04.54" N	85°20'40.15" W	21.79	21.92	−0.13	0.017
19	9°56'07.76" N	85°20'28.68" W	23.84	23.77	0.07	0.005
20	9°56'09.21" N	85°20'12.01" W	24.99	24.72	0.27	0.073
					$\overline{x} =$ −0.02	$\Sigma =$ 0.347

Solution

The RMSE for the data is

$$\text{RMSE} = \sqrt{\frac{\sum_{i=1}^{n}(x_i - x_i')^2}{n}} = \sqrt{\frac{0.347}{20}} = 0.13 \text{ m} \quad (13 \text{ cm})$$

The results show a maximum RMSE of less than 15 cm and a mean error of 0.02 meters (2 cm), so the survey appears to meet the FEMA accuracy standards.

4. FGDC ACCURACY STANDARDS

The Federal Geographic Data Committee (FGDC) is a U.S. government committee that promotes the coordination, use, sharing, and dissemination of geospatial data on a national basis. The committee is chaired by a person nominated by the Department of the Interior, with a nominee from the Office of Management and Budget as vice chair. Other members are representatives from various federal government agencies, including the Departments of Agriculture, Commerce, Defense, Energy, Housing and Urban Development, Interior, State, and Transportation; the Environmental Protection Agency; FEMA; the Library of Congress; the National Aeronautics and Space Administration; the National Archives and Records Administration; and the Tennessee Valley Authority. The FGDC has developed comprehensive geospatial positioning accuracy standards, known as the *FGDC Geospatial Positioning Accuracy Standards*.

Part 2: Standards for Geodetic Networks

Part 2 of the *Geospatial Positioning Accuracy Standards* applies to geodetic networks. The standards provide a common method for classifying the accuracy of geodetic control points, which are usually part of a control network for surveying and mapping. Such geodetic surveys are the basis for other surveys and must be performed to more rigorous standards than those performed for general engineering, construction, or topographic mapping purposes. These standards supersede the *Standards and Specifications for Geodetic Control Networks* promulgated by the Federal Geodetic Control Commission in 1984.

For geodetic control points, accuracy under the FGDC standards is determined for each geodetic component (i.e., horizontal control, ellipsoidal height, and orthometric height). The accuracy standards for geodetic networks are not measured by closure within a survey, but by relationships with the network datum, taking into account crustal motion, refraction, and any other systematic effects known to influence survey measurements.

Survey accuracy is checked by comparing the survey results against established controls using a minimally constrained least-squares adjustment for the 95% confidence level. This accuracy is reported as meeting a level of precision of 1, 2, or 5 mm; 1, 2, or 5 cm; or 1, 2, 5, or 10 m.

Part 3: National Standard for Spatial Data Accuracy

Part 3 of the *Geospatial Positioning Accuracy Standards* provides a methodology for estimating the positional accuracy of points on maps and in digital geospatial data, with respect to georeferenced control points of higher accuracy.

These standards are similar to the FEMA standards, in that they use the RMSE to estimate positional accuracy, comparing map values with measurements based on an independent source of higher accuracy for identical points. The standards require that a minimum of 20 check points, distributed to reflect the geographic area of interest, be tested. As with the FEMA standards, when 20 points are tested, the 95% confidence level allows one point to fail the threshold specifications.

Part 4: Architecture, Engineering, Construction, and Facilities Management

Part 4 of the *Geospatial Positioning Accuracy Standards* provides accuracy standards for engineering drawings, maps, and surveys for planning, design, construction, operation, maintenance, and management of facilities, installations, structures, transportation systems, and related projects. This includes related geographical information systems (GIS), computer-aided drafting and design (CADD), and automated mapping/facility management (AM/FM) products.

The standards require that such projects be referenced to local boundary control commonly used in the project area. Where practical and feasible, horizontal coordinates should be referenced to national coordinate systems, such as the State Plane Coordinate System based on the North American Datum of 1983 (NAD 83) and vertical coordinates based on the North American Vertical Datum of 1988 (NAVD 88). For the control surveys for such projects, two types of survey accuracies may be specified: positional accuracy and relative closure ratio accuracy.

Positional accuracy is the difference between the results of a measurement of a location of a point and the true position of that point. When positional accuracy is used to determine survey accuracy, the accuracy is checked in the manner required for geodetic control surveys, as described in Part 1 of the FGDC standards. The survey results are compared to established controls using a minimally constrained least squares adjustment for the 95% confidence level, consistent with the engineering or construction applications or specifications. Generally, horizontal and vertical control point accuracies should be twice as accurate as tolerances required for features or objects on the resulting maps.

Relative closure ratio accuracy is the linear misclosure, such as is determined by a closed traverse, divided by the total length of the traverse. Relative closure ratio accuracy is often the prevailing standard when certain closure ratios are specified in state codes and/or state minimum technical standards. No simple correlation between relative closure accuracies and 95% radial positional accuracies exists. Therefore, knowing the closure is not sufficient to determine whether a survey meets a specified feature accuracy requirement. As a result, positional accuracy standards should be used instead of closure accuracy standards where practical and allowable. Relative closure ratio accuracy could be considered a measure of precision, rather than accuracy, but is included to allow use of surveys meeting various state standards.

For the resulting maps of this type, the survey accuracy is checked in the manner required for maps under Part 2 of the FGDC standards by testing both horizontal positions of planimetric features and elevations against measurements from a source of higher accuracy and determining the RMSE.

Part 5: Standards for Nautical Charting Hydrographic Surveys

Hydrographic surveys, as covered by Part 5 of the *Geospatial Positioning Accuracy Standards*, are defined as surveys conducted to determine the configurations of the bottoms of water bodies, and to identify and locate all natural and man-made features that may affect navigation. Defining the accuracy of such surveys is complex, due to the fact that several components are involved. These include the accuracy of the horizontal positioning of each sounding; the accuracy of the determination of the vertical datum used in the survey; the accuracy of any corrections that are applied to the soundings, such as tide, draft, settlement, and squat corrections; and the accuracy of the soundings themselves. As a result, several considerably different approaches are used for defining the accuracy of hydrographic surveys.

Typically, both the horizontal position of each sounding and the vertical measurement at that position are indirect measurements determined by mathematical relationships to direct measurements. For example, horizontal positions may be derived from direct angle and distance measurements establishing control points on the shore, together with direct GPS measurements on those control points and on the sounding positions. Similarly, the vertical measurement (depth) may be derived from direct tide gauge measurements establishing the vertical datum for the survey and for tide corrections, direct fathometer measurements, and direct measurement of other corrections for the soundings.

As a result of the nature of measurements associated with hydrographic surveys, the FGDC standards require that all contributions to the total error be propagated (combined) using a statistical method. One method used for this purpose for less complex relationships is that of using the square root of the sum of the squares of all of the contributing errors. For example, if three direct components of an indirect measurement have errors of 0.01, 0.02, and 0.03 ft, the total error combined by this method would be

$$(0.01 \text{ ft})^2 + (0.02 \text{ ft})^2 + (0.03 \text{ ft})^2 = \sqrt{0.0014 \text{ ft}} = 0.04 \text{ ft}$$

The propagation of errors for hydrographic surveys is especially complex with the increased use of multi-beam fathometers. With such equipment, roll, pitch, heading, refraction, and other errors associated with both the horizontal and vertical positions of each sounding becomes critical. Formal procedures for calculation of the total propagated error (TPE) for multi-beam hydrographic surveys have been developed, but these procedures are beyond the scope of this chapter. For horizontal errors, those procedures take into account refraction errors; roll, pitch, and heading errors; positioning system errors; sensor coordinate offset errors; transducer head misalignment errors; and latency errors between sensors. For vertical errors, the calculation of TPE takes into account sounder measurement error; sound speed correction errors; dynamic draft measurement errors; heave measurement errors; tide or water-level errors; roll, pitch, and heading errors; and transducer head misalignment errors.

Classifications of Hydrographic Surveys

To accommodate different accuracy requirements in a systematic manner, the FGDC standards define four orders of survey, as shown in Table 12.1.

Table 12.1 Minimum Standards for Hydrographic Surveys

classification	special order	first order	second order	third order
typical application	harbors, berths, & critical depth areas	channels & coastal areas ≤ 100 m depth	< 200 m depth	offshore & > 200 m depth
horizontal accuracy	2 m	5 m + 5% of depth	20 m + 5% of depth	150 m + 5% of depth
depth accuracy	$a = 0.25$ m $b = 0.0075$	$a = 0.5$ m $b = 0.013$	$a = 1.0$ m $b = 0.023$	$a = 1.0$ m $b = 0.023$
maximum line spacing	100% coverage required	greater of 25 m or 3 × average depth, 100% coverage in some places	greater of 200 m or 4 × average depth	4 × average depth

Note: Accuracies refer to accuracy at the 95% confidence level in reference to the datum reference frame.

Datums for Hydrographic Surveys

The *Geospatial Positioning Accuracy Standards* require that the North American Datum of 1983 (NAD 83) be used as the horizontal datum for hydrographic surveys, and that depths be referenced to the applicable chart datum, rather than geodetic vertical datums such as the North American Vertical Datum of 1988 (NAVD 88) or the National Geodetic Vertical Datum of 1929 (NGVD 29). The Mean Lower Low Water (MLLW) datum is typically used for Atlantic, Pacific, and Gulf coast charts, and the International Great Lakes Datum (1985) is typically used for the Great Lakes. Other water level-based datums are used for lakes and rivers.

Horizontal Accuracy

The *Geospatial Positioning Accuracy Standards* require primary shore control points to be located by ground survey methods to a relative accuracy of 1 part in 100,000. When such points are established by satellite positioning, the error should not exceed 10 cm at the 95% confidence level. Secondary shore control points (those not used for extending the control) should have errors not exceeding 1 part in 10,000 for ground survey techniques or 50 cm using satellite positioning.

Under these standards, *horizontal spatial accuracy* is defined as the two-dimensional circular error of horizontal coordinates at the 95% confidence level. Errors in latitude and longitude can be combined to create a one-dimensional radial error, such as a circle of equal probability.

Vertical Accuracy

In determining the depth accuracy of the reduced depths, the sources of individual errors should be quantified and statistically combined to obtain a TPE at the 95% confidence level. As an additional check on data quality, redundant depths from crossline intersections should be measured.

For wrecks and obstructions with less than 40 m clearance, the least depth over them should be determined by high definition sonar examination or physical examination (i.e., diving). Mechanical sweeping may be used when guaranteeing a minimum safe clearance depth.

Due to the rapidly increasing size of commercial vessels, there is a demand for more accurate and reliable knowledge of the water depths, in order to safely exploit the maximum cargo capabilities. Therefore, depth accuracy standards in critical areas are more stringent than those established in the past.

Tidal Observation Accuracy

Unless 3-D positioning is used, which would eliminate the need for tide corrections, tidal height observations should be made throughout the course of a survey, and for a sufficient period to provide tidal reductions for soundings and data for tidal analysis. The standards require sufficient observations be made to ensure that the total measurement error at the tide gauge, including timing errors, does not exceed ± 5 cm at 95% for special order surveys or ± 10 cm for other classes of surveys.

Bottom Sampling

The seabed should be sampled to determine the nature of (rocky, sandy, etc.). The nature of the seabed may also be inferred from other sensors, up to the depth required by local anchoring or trawling conditions. Normally, sampling is not required for depths greater than 200 m. Bottom sample spacing should normally be 10 times that of the main scheme line spacing. In anchorages, density of sampling should be increased.

Soundings

Sounding systems are required to be able to detect cubic features greater than 1 m in any dimension for special order surveys or first- and second-order surveys in water with depths less than 40 m. In addition to the requirements for line spacing in Table 12.1, full bottom coverage is required in selected areas where the bottom characteristics and the risk of obstructions are potentially hazardous to vessels. Crosslines (and other redundant lines) intersecting the principal sounding lines are required to confirm the accuracy of positioning, sounding, and tidal reductions. The interval between crosslines should normally be no more than 15 times that between the principal sounding lines.

5. ALTA/ACSM ACCURACY STANDARDS FOR LAND TITLE SURVEYS

Although primarily directed at the legal aspects of surveying, another set of accuracy standards are given in the standards for land title surveys developed by the American Land Title Association (ALTA) and the American Congress on Surveying and Mapping (ACSM). These standards include a requirement for land title surveys to have a relative positional accuracy of 0.07 feet (or 20 mm) + 50 ppm. Those standards define relative positional accuracy as the uncertainty due to random errors in measurements in the location of any point on the same survey at the 95 percent confidence level.

Under the ALTA and ACSM standards, relative positional accuracy may be tested by comparing the relative location of surveyed points, as measured by an independent survey of higher accuracy, and calculating the uncertainty as detailed in the FEMA standards; or by use of a minimally constrained, correctly weighted least-squares adjustment of the survey.

6. PRACTICE PROBLEMS

1. Under the Federal Geographic Data Committee accuracy standards, which vertical datum should be used for hydrographic surveys in the Gulf of Mexico?

(A) NGVD 29

(B) NAVD 88

(C) MLLW

(D) any of the above

2. The depth accuracy required for a first-order hydrographic survey in waters with a 20 m average depth is most nearly

(A) 0.01 m

(B) 0.1 m

(C) 0.6 m

(D) 1.0 m

3. What is the maximum line spacing allowable under the Federal Geographic Data Committee accuracy standards for a first-order hydrographic survey in waters with a 20 m average depth?

(A) 25 m

(B) 50 m

(C) 60 m

(D) 100 m

4. Which of the following criteria is the accuracy standard used to determine if boundary surveys meet the American Land Title Association and American Congress on Surveying and Mapping standards?

(A) standard error

(B) root-mean-squared error

(C) standard deviation

(D) relative positional accuracy

5. The table shown provides elevations for a series of bare earth points in a pine forest. The elevations are based on a LiDAR survey and check elevations for the same points measured by a real-time kinematic (RTK) global positioning system.

n	elevation from LiDAR (ft)	elevation from GPS (ft)	n	elevation from LiDAR (ft)	elevation from GPS (ft)
1	118.5	119.1	11	113.3	113.4
2	119.4	119.3	12	117.7	117.6
3	121.0	121.0	13	111.2	111.2
4	116.5	115.9	14	111.7	111.6
5	116.1	116.2	15	110.6	110.9
6	117.9	117.7	16	107.0	106.9
7	120.9	121.0	17	103.1	103.1
8	115.3	115.4	18	115.1	115.4
9	115.0	114.8	19	117.7	117.5
10	113.2	113.3	20	118.9	118.8

The root-mean-squared error for this test of the accuracy is most nearly

(A) 0.08 ft

(B) 0.11 ft

(C) 0.16 ft

(D) 0.24 ft

6. The 2 sigma error for a survey with a root-mean-squared error of 0.08 ft is most nearly

(A) 0.08 ft

(B) 0.11 ft

(C) 0.16 ft

(D) 0.24 ft

7. Which accuracy standards control for the creation of a digital elevation model (DEM)?

(A) National Map Accuracy Standards

(B) FGDC standards

(C) FGDC AM/FM standards

(D) ALTA/ACSM standards

SOLUTIONS

1. Under the Federal Geographic Data Committee accuracy standards, the local Mean Lower Low Water (MLLW) datum should be used as a vertical datum for hydrographic surveys in the Gulf of Mexico, as well as in the Atlantic and Pacific Oceans.

The answer is (C).

2. Under the Federal Geographic Data Committee accuracy standards, for a first-order hydrographic survey, the tabulated factors are

$$a = 0.5 \text{ m}$$
$$b = 0.013$$

The required depth accuracy for an average depth of 20 m is

$$\sqrt{a^2 + (bd)^2} = \sqrt{(0.5 \text{ m})^2 + \big((0.013)(20 \text{ m})\big)^2}$$
$$= 0.56 \text{ m} \quad (0.6 \text{ m})$$

The answer is (C).

3. The maximum line spacing allowable under the Federal Geographic Data Committee accuracy standards for the first-order hydrographic survey in waters with a 20 m average depth is the greater of 25 m or 3 times the average depth. With an average depth of 20 m, the maximum line spacing would be $(3)(20 \text{ m}) = 60 \text{ m}$.

The answer is (C).

4. The American Land Title Association and American Congress on Surveying and Mapping standards use the relative positional accuracy as derived from a minimally constrained, correctly weighted least-squares adjustment of the survey to determine if boundary surveys meet the standards.

The answer is (D).

5. Find the difference between the elevation and the check elevation and the square of the difference between those elevations for each point surveyed, then sum those values to find $\bar{x}$ and $\sum$.

n	LiDAR, x_i (ft)	GPS, x_i' (ft)	$x_i - x_i'$ (ft)	$(x_i - x_i')^2$ (ft)
1	118.5	119.1	−0.6	0.4
2	119.4	119.3	0.1	0.0
3	121.0	121.0	0.0	0.0
4	116.5	115.9	0.6	0.4
5	116.1	116.2	−0.1	0.0
6	117.9	117.7	0.2	0.0
7	120.9	121.0	−0.1	0.0
8	115.3	115.4	−0.1	0.0
9	115.0	114.8	0.2	0.0
10	113.2	113.3	−0.1	0.0
11	113.3	113.4	−0.1	0.0
12	117.7	117.6	0.1	0.0
13	111.2	111.2	0.0	0.0
14	111.7	111.6	0.1	0.0
15	110.6	110.9	−0.3	0.1
16	107.0	106.9	0.1	0.0
17	103.1	103.3	−0.2	0.0
18	115.1	115.4	−0.3	0.1
19	117.7	117.5	0.2	0.0
20	118.9	118.8	0.1	0.0
			$\bar{x} = 0.0$	$\sum = 1.0$

The root-mean-squared error is

$$\text{RMSE} = \sqrt{\frac{\displaystyle\sum_{i=1}^{n}(x_i - x_i')^2}{n}} = \sqrt{\frac{1.0}{20}} = 0.22 \text{ ft} \quad (0.24 \text{ ft})$$

The answer is (D).

6. The 2 sigma error for a survey is 1.96 times the root-mean-squared error (RMSE), so the 2 sigma error for a survey with an RMSE of 0.08 ft is

$$(1.96)(0.08 \text{ ft}) = 0.1568 \text{ ft} \quad (0.16 \text{ ft})$$

95% of the errors should be less than 0.16 ft.

The answer is (C).

7. The Federal Geographic Data Committee (FGDC) standards control for the creation of a digital elevation model.

The answer is (B).

Topic II: Field Data Acquisition

13 Taping

Nomenclature

a	cross-sectional area	in^2	cm^2
C	correction	ft	m
E	elastic modulus	lbf/in^2	kg/cm^2
H	horizontal distance	ft	m
L	length	ft	m
P	applied tension	lbf	kg
P_0	standardized tension	lbf	kg
S	slope distance	ft	m
T	temperature		
V	vertical distance	ft	m
w	weight per unit length	lbf/ft	kg/m
W	weight	lbf	kg

Subscripts

0	standardized
s	between supports
S	sag
t	total

1. LINEAR MEASUREMENT

The distance between two points can be determined by pacing, taping, electronic distance measurement (EDM), tacheometry (with stadia rods or other tacheometers), using an odometer, or scaling on a map.[1] Of these methods, only taping is discussed in this chapter.

The process of *taping* consists of aligning the tape, pulling the tape tight, using the plumb bob on unlevel ground, marking tape lengths, and reading the tape.

2. GUNTER'S CHAIN

A Gunter's chain is 66 ft long and consists of 100 links, each 7.92 in long. One chain = 1/80 mi, and 10 square chains = $(10)(66\ ft)^2 = 43{,}560\ ft^2$.

The Gunter's chain was once used extensively in surveying the public lands of the United States, but that is no longer the case. However, the term *chaining* is still used to mean taping, and surveyors skilled at taping are still called *chainpersons*. Knowledge of the Gunter's chain and its uses is important when retracing old surveys in which the Gunter's chain was used.

3. STEEL TAPES

Steel tapes are available in widths from 3/8 in to 5/16 in, with thicknesses varying from 0.016 in to 0.025 in. They are available in lengths of 50 ft, 100 ft, 200 ft, 300 ft, and 500 ft. A 100 ft tape is most common for surveying. Steel tapes are also made in 30 m, 50 m, and 100 m lengths.

4. INVAR TAPES

Invar tapes are made of a steel alloy containing 35% nickel. The expansion and contraction of an Invar tape due to changes in temperature is only about 3% of the change of a steel tape. These tapes are used when measurements of extreme accuracy are required, such as measuring baselines or calibrating steel tapes. Invar tapes are, however, too fragile for normal use.

---

[1] In surveying, the distance between two points is the horizontal distance, regardless of the slope.

5. TYPES OF STEEL TAPES

Some 100 ft tapes measure 100 ft from the outer edges of the end loops. Most tapes, however, are in excess of 100 ft from end loop to end loop and have graduations for every foot from 0 to 100 ft. An *add tape* has an extra graduated foot beyond the zero mark; the extra foot is usually graduated in tenths of a foot, but it is sometimes graduated in tenths and hundredths of a foot. A *cut tape*, or a *subtract tape*, does not have the extra graduated foot, but the last foot at each end is graduated in tenths of a foot or in tenths and hundredths of a foot.

Tapes are not guaranteed by their manufacturers to be of exact length. The National Bureau of Standards will, for a fee, compare any tape with a standard tape or distance, and it will certify the exact length of the tape under the conditions under which it was tested. Figure 13.1 shows a certificate from the National Bureau of Standards.

Figure 13.1 *National Bureau of Standards Certificate*

UNITED STATES DEPARTMENT OF COMMERCE
WASHINGTON

National Bureau of Standards

Certificate
100-Foot
Steel Tape

Maker: Keuffel & Esser Co.   Submitted by   NBS No. 10565

This tape has been compared with the standards of the United States, and the intervals indicated have the following lengths at 68° Fahrenheit (20° centigrade) under the conditions given below:

Supported on a horizontal flat surface

Tension (pounds)	Interval (feet)	Length (feet)
10	0 to 100	100.002
5 1/2	0 to 100	100.000

Supported at the 0, 50, and 100-foot points

Tension (pounds)	Interval (feet)	Length (feet)
30	0 to 100	100.000

Supported at the 0 and 100-foot points

Tension (pounds)	Interval (feet)	Length (feet)
38 1/2	0 to 100	100.000

See Note 3(a) on the reverse side of this certificate.

the Director
National Bureau of Standards

Test No. 2.4/1426
Date: June 13, 1955

Lewis V. Judson

Lewis V. Judson
Chief Length Section
Optics and Metrology Division

Tapes are standardized under two conditions: supported throughout, and supported only at the ends. In the United States, steel tapes are standardized for use at 68°F. The standard pull for a 100 ft tape with the tape supported throughout its length is usually 10 lbf, but can vary with the cross-sectional area of the tape. When supported only at the ends, the pull is usually 30 lbf, but this can be varied on request.

6. CHAINING PINS

Chaining pins, also known as *marking pins*, are used to mark tape lengths. They are usually 12 in to 18 in long and made of $\frac{3}{16}$ in width steel; they are sharpened on one end and have a ring approximately $2\frac{1}{2}$ in in diameter at the other end. A set of chaining pins contains 11 pins.

7. TAPING PROCEDURE

When measuring with a tape supported throughout its length, such as on level ground, the following procedure should be followed. At the back end of the tape, the tape is held firmly with the end mark exactly on the pin. At the forward end of the tape, the tape is aligned to the line of sight and pulled with the proper tension, and a pin is placed exactly at the end mark. At both ends of the tape, personnel handling the tape should work from the sides of the tape to keep the line of sight clear.

When taping on a slope and going downhill, the same procedure as with level ground is followed at the back end of the tape. At the forward end, the tape is held as horizontal as possible, with a plumb bob string looped over the tape and the plumb bob itself slightly above ground. After making sure that the tape is on the line of sight, the tape is pulled to the proper tension, the plumb bob is dropped, and the point is marked with a pin. When taping on a slope and going uphill, the opposite procedure is followed, with the rear of the tape held up and a plumb bob held over the pin or point on the ground and the forward end of the tape held at ground level.

When measuring a segment with a length of less than a hundred feet, an even foot mark on the tape is held on or over the beginning point. With an add tape, the tenths and hundredths reading at the forward end of the tape is added to the even foot reading at the beginning point. With a cut tape, the tenths and hundredths reading at the forward end of the tape is subtracted from the even foot reading at the beginning point.

Figure 13.2 shows the distance between point A and point B measured using both an add tape and a cut tape. The distance using the add tape is 26 ft + 0.18 ft = 26.18 ft; the distance using the cut tape is 27 ft − 0.82 ft = 26.18.

Figure 13.2 *Add and Cut Tapes*

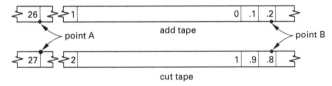

8. BREAKING TAPE

Where the slope is so great that a 100 ft length of the tape cannot be held horizontally without the lower end of the tape being held above the shoulders, a procedure known as *breaking tape* can be used.

The tape is pulled forward a full length on the line, as usual. The person handling the forward end of the tape then walks back to a point where the tape can be held horizontal below the shoulder level. A foot mark at that point, ideally ending in 0 or 5, is selected and held over a point on the line. The tape is then held horizontally and with the proper tension applied, that distance is marked at the front end of the tape, and a pin is placed.

9. STATIONING WITH CHAINING PINS

A *station* is 100 ft (one standard tape) in length. In route surveying, stationing is carried along continuously from a starting point designated as sta 0+00. When using the taping process to set stationing, chaining pins are used to keep track of stationing.

The process begins with one pin in the ground at the beginning of the line being measured, and with the person who is working the forward end of the tape holding the remaining ten pins in the eleven-pin set. As each 100 ft segment is measured, a pin is placed at the end mark of the tape. Then, the pin at the rear end of the tape is pulled by the person handling that end of the tape as he or she moves on to measuring the next station.

After the tenth station is measured, the person handling the rear end of the tape should have ten pins in hand, confirming that 100 ft have been measured. The ten pins are passed to the person handling the forward end of the tape, and the process is then repeated as necessary. When a segment of less than 1000 ft (ten stations) is being stationed, the number of pins in hand represents the number of stations that have been measured.

10. TAPING AT AN OCCUPIED STATION

When taping at a station that is occupied by an instrument, personnel who are handling the tape must be extremely careful not to hit the leg of the instrument. If a plumb bob is needed at the point, a plumb bob string hanging from the instrument can be used. In some cases, it may be necessary to use a point on top of the instrument and on the instrument's vertical axis as a measuring point.

11. CARE OF THE TAPE

The tape will not be broken by pulling on it unless there is a kink (loop) in it. The tape is easily broken if it is pulled when there is a kink in it. Chainpersons should always be alert to kinking.

If the tape has been used in wet grass or mud, it should be cleaned and oiled lightly by pulling it through an oily rag.

12. CORRECTION FOR SLOPE MEASUREMENTS

On ground where the slope is steep but uniform, it is sometimes easier to determine the slope and make corrections for the slope measurement rather than to break tape every few feet. To determine horizontal distance, the correction is subtracted from the slope distance.

In Fig. 13.3, H is the horizontal distance from A to B, S is the slope distance from A to B, V is the difference in elevation from A to B, and C is the correction. Thus,

$$\begin{aligned} V^2 &= S^2 - H^2 \\ &= (S-H)(S+H) \end{aligned} \qquad 13.1$$

Where the slope difference is small, $S + H$ is approximately $2S$. Therefore,

$$V^2 = 2S(S-H) \qquad 13.2$$

Figure 13.3 *Measurements on Slope*

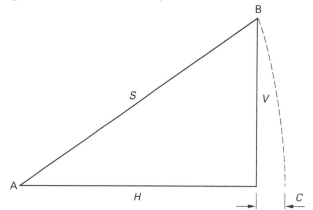

Because $S - H$ equals the correction C,

$$C \approx \frac{V^2}{2S} \qquad 13.3$$

Where $S = 100$ ft (one tape length),

$$C \approx \frac{V^2}{200 \text{ ft}} \qquad 13.4$$

The approximate value will be within 0.007 ft of the actual value when the difference in elevation per 100 ft of slope distance is not more than 15 ft. This amount is typically insignificant in taping and can usually be ignored. For steeper slopes, Eq. 13.1 should be used.

Example 13.1

The difference in elevation between two points is 4.0 ft, and the slope distance is 100 ft. What is the horizontal distance between the points?

Solution

The slope distance is 100 ft, so use Eq. 13.4. The correction is

$$C \approx \frac{V^2}{200 \text{ ft}} = \frac{(4 \text{ ft})^2}{200 \text{ ft}}$$
$$= 0.08 \text{ ft}$$

The horizontal distance between the points is

$$H = 100 \text{ ft} - 0.08 \text{ ft}$$
$$= 99.92 \text{ ft}$$

13. TENSION

When a tape is pulled with more or less than the standard amount of tension (10 lbf when supported, 30 lbf when unsupported), the effective length may be more or less than that reported from the standardization process. The correction for variation in tension for a steel tape may be calculated using Eq. 13.5.

$$C = \frac{(P - P_0)L}{aE} \qquad \text{13.5}$$

The modulus of elasticity for a steel tape is about 30×10^6 lbf/in^2 (2.1×10^6 kg/cm^2). The cross-sectional area is about 0.003 in^2 for light steel (1 lbf) 100 ft tapes and about 0.009 in^2 for heavy steel (3 lbf) 100 ft tapes. Cross-sectional areas for light and heavy 30 m tapes are 0.019 cm^2 and 0.058 cm^2, respectively.

By using spring-balance handles, chainpersons can get the feel of a 10 lbf pull so that, in ordinary taping with the tape supported, the error caused by variation in tension is negligible.

Example 13.2

A light 100 ft steel tape has been standardized at a 10 lbf pull, but is stretched with a 40 lbf pull. What is the corrected length of the tape?

Solution

Use Eq. 13.5. The cross-sectional area of a light 100 ft steel tape is 0.003 in^2, and the modulus of elasticity is 30×10^6 lbf/in^2.

$$
\begin{aligned}
C &= \frac{(P - P_0)L}{aE} \\
&= \frac{(40 \text{ lbf} - 10 \text{ lbf})(100 \text{ ft})}{(0.003 \text{ in}^2)\left(30 \times 10^6 \, \dfrac{\text{lbf}}{\text{in}^2}\right)} \\
&= 0.033 \text{ ft}
\end{aligned}
$$

14. CORRECTION FOR SAG

When the tape is supported only at the ends, it sags and takes the form of a catenary. The correction for sag may be determined using Eq. 13.6.

$$C_S = \frac{w^2 L^3}{24 P^2} = \frac{W^2 L}{24 P^2} \qquad \text{13.6}$$

As an alternative to Eq. 13.6, the error due to sag can be offset by increased tension. For a medium-weight tape standardized for a pull of 10 lbf, a pull of 30 lbf will offset the difference in length caused by sag. If the tape is supported at 25 ft intervals, the pull only need be 14 lbf.

15. EFFECT OF TEMPERATURE ON TAPING

Steel tapes are standardized for 68°F in the United States. If there is a change in temperature of 15°F, a steel tape will undergo a change in length of about 0.01 ft, introducing an error of about 0.5 ft per mile.

The coefficient of thermal expansion for steel is approximately 0.00000645 ft/°F. For a 100 ft tape where T is the temperature (in °F) at time of measurement, the correction in length C due to change in temperature is

$$C = \left(0.00000645 \, \frac{\text{ft}}{\text{°F}}\right)(T_{\text{°F}} - 68°)L \qquad \text{13.7}$$

Example 13.3

A steel tape is used to measure the distance between two points at a temperature of 30°F. The distance is measured to be 675.48 ft. What is the corrected measurement after taking into account change in tape length due to temperature?

Solution

The correction due to change in temperature is

$$C = \left(0.00000645 \ \frac{\text{ft}}{°\text{F}}\right)(T_{°\text{F}} - 68°)L$$
$$= \left(0.00000645 \ \frac{\text{ft}}{°\text{F}}\right)(30° - 68°)(675.48 \ \text{ft})$$
$$= -0.17 \ \text{ft}$$

The corrected length is

$$L = 675.48 \ \text{ft} - 0.17 \ \text{ft} = 675.31 \ \text{ft}$$

16. CORRECTION FOR INCORRECT LENGTH OF TAPE

A standardized tape can be used to check other tapes. If a 100 ft tape is known to be of incorrect length, the correction factor for measurements that have been made with the incorrect tape is

$$C = \left(L_{\text{tape measurement}} - 100 \ \text{ft}\right)\left(\frac{L_{\text{line measurement}}}{100 \ \text{ft}}\right) \quad \textit{13.8}$$

For Eq. 13.8, a line measured with a tape that is longer than 100 ft is actually longer than the measurement shown by the tape. A line measured with a tape that is shorter than 100 ft is actually shorter than the measurement shown by the tape. A rule to remember is "for a tape too long, add; for a tape too short, subtract." (This rule can also be applied to temperature correction.)

Example 13.4

A distance between two points is measured as 662.35 ft with a 100 ft tape. The 100 ft tape is later found to be 100.02 ft long. What is the actual distance between the two points?

Solution

From Eq. 13.8, the correction factor is

$$C = \left(L_{\text{tape measurement}} - 100 \ \text{ft}\right)\left(\frac{L_{\text{line measurement}}}{100 \ \text{ft}}\right)$$
$$= (100.02 \ \text{ft} - 100 \ \text{ft})\left(\frac{662.35 \ \text{ft}}{100 \ \text{ft}}\right)$$
$$= 0.13 \ \text{ft}$$

The corrected measurement is 662.35 ft + 0.13 ft = 662.48 ft.

17. CORRECTION FOR IMPROPER ALIGNMENT

Improper alignment is probably the least important error in taping. Many instrument operators and persons handling the tape spend time aligning that is not justified by the effect of improper alignment. The linear error when one end of the tape is off-line can be found in the same way slope correction is found. For example, for a 100 ft tape with one end off-line by 1.0 ft, the correction is

$$C = \frac{V^2}{200 \ \text{ft}} = \frac{1.0 \ \text{ft}^2}{200 \ \text{ft}} = 0.005 \ \text{ft}$$

When the error in alignment is 0.5 ft, the linear error is 0.001 ft per tape length, or about 0.05 ft per mile.

18. COMBINED CORRECTIONS

Corrections for incorrect length of tape, temperature, and slope can be combined algebraically, as illustrated in Ex. 13.5.

Example 13.5

A line is measured with a 100 ft tape and found to be 1238.22 ft long. The tape is later measured when the temperature is 18°F and found to be 100.03 ft long. What is the corrected length of the line?

Solution

From Eq. 13.7, the correction factor for temperature is

$$C = \left(0.00000645 \ \frac{\text{ft}}{°\text{F}}\right)(T_{°\text{F}} - 68°)L$$
$$= \left(0.00000645 \ \frac{\text{ft}}{°\text{F}}\right)(18° - 68°)(1238.22 \ \text{ft})$$
$$= -0.40 \ \text{ft}$$

From Eq. 13.8, the correction factor for the length of the tape is

$$C = \left(L_{\text{tape measurement}} - 100 \ \text{ft}\right)\left(\frac{L_{\text{line measurement}}}{100 \ \text{ft}}\right)$$
$$= (100.03 \ \text{ft} - 100 \ \text{ft})\left(\frac{1238.22 \ \text{ft}}{100 \ \text{ft}}\right)$$
$$= 0.37 \ \text{ft}$$

The corrected measurement is

$$1238.22 \ \text{ft} - 0.40 \ \text{ft} + 0.37 \ \text{ft} = 1238.19 \ \text{ft}$$

Field Data Acquisition

19. PRACTICE PROBLEMS

1. How much does an Invar tape expand and contract due to change in temperature, compared to a standard steel tape under the same conditions?

(A)   An Invar tape expands and contracts much more than a steel tape.

(B)   An Invar tape expands and contracts slightly less than a steel tape.

(C)   An Invar tape expands and contracts much less than a steel tape.

(D)   The two tapes expand and contract about the same amount.

2. How many pins are in a set of chaining pins?

(A)   10

(B)   11

(C)   12

(D)   20

3. What features distinguish an add tape?

(A)   The last foot of each end is graduated in tenths or hundredths of a foot.

(B)   There is an extra graduated foot beyond the zero mark.

(C)   There is an extra graduated foot at the 100 foot end of the tape.

(D)   There is an extra graduated foot at both ends of the tape.

4. How long is a Gunter's chain?

(A)   30 ft

(B)   53 ft

(C)   66 ft

(D)   130 ft

5. The area of a square one Gunter's chain on a side is most nearly

(A)   250 ft^2

(B)   4400 ft^2

(C)   10,000 ft^2

(D)   44,000 ft^2

6. Most nearly, what is the area of a rectangle 6 Gunter's chains long by 5 Gunter's chains wide?

(A)   3 ac

(B)   7 ac

(C)   20 ac

(D)   30 ac

7. A distance of 100 ft is measured along a line where the difference in elevation from sta 0+00 to sta 10+00 is 11 ft. The measurement is taken at a temperature of 68°F, with a tape calibrated to be 100.0 ft in length. Most nearly, what is the corrected measurement?

(A)   98 ft

(B)   99 ft

(C)   100 ft

(D)   101 ft

8. A distance of 787.35 ft is measured on level ground with a steel tape 100 ft in length. The tape was calibrated for 68°F, and the temperature at the time of the measurement is 98°F. Most nearly, what is the corrected measurement?

(A)   787.0 ft

(B)   787.2 ft

(C)   787.5 ft

(D)   787.6 ft

9. A line is measured on level ground with a 100 ft tape and is found to be 582.32 ft long. Later, the tape is calibrated and found to be 100.03 ft in length. The temperature at the time of measurement was 68°F. Most nearly, what is the corrected length of the line?

(A)   582.1 ft

(B)   582.3 ft

(C)   582.5 ft

(D)   582.7 ft

10. Most nearly, what is the length of a standardized 100 ft steel tape at 28°F?

(A)   99.97 ft

(B)   99.99 ft

(C)   100.01 ft

(D)   100.03 ft

11. A tape that is 100.02 ft long at 68°F is used to measure a line along sloping ground when the temperature is 28°F. The difference in elevation from the beginning to the end of the line was 10 ft. The line is recorded as 196.44 ft. Most nearly, what is the corrected length of the line?

(A) 196.2 ft

(B) 196.3 ft

(C) 196.5 ft

(D) 196.7 ft

12. A 100 ft tape that was tested to be 99.97 ft long is used to measure a line when the temperature is 98°F. The length of the line is recorded as 713.19 ft. Most nearly, what is the corrected measurement?

(A) 712.9 ft

(B) 713.1 ft

(C) 713.3 ft

(D) 713.5 ft

SOLUTIONS

1. An Invar tape will expand and contract much less than a steel tape when exposed to the same change in temperature.

The answer is (C).

2. There are 11 pins in a set of chaining pins.

The answer is (B).

3. An add tape has an extra graduated foot beyond the zero mark.

The answer is (B).

4. A Gunter's chain is 66 ft long.

The answer is (C).

5. A square one Gunter's chain on a side has an area of

$$(66 \text{ ft})^2 = 4356 \text{ ft}^2 \quad (4400 \text{ ft}^2)$$

The answer is (B).

6. A Gunter's chain is 66 ft long. The length of the rectangle is

$$\frac{\big((6)(66 \text{ ft})\big)\big((5)(66 \text{ ft})\big)}{43{,}560 \dfrac{\text{ft}^2}{\text{ac}}} = 3 \text{ ac}$$

The answer is (A).

7. From Eq. 13.3, the correction factor for the slope is

$$C \approx \frac{V^2}{2S} = \frac{(11 \text{ ft})^2}{(2)(1000 \text{ ft})} = 0.06$$

The corrected length for 100 ft

$$L = 100 \text{ ft} - (\tfrac{100}{1000} \times 0.06)$$
$$= 99.994 \text{ ft} \quad (100 \text{ ft})$$

The answer is (C).

8. From Eq. 13.7, the correction factor for the temperature is

$$C = \left(0.00000645 \frac{\text{ft}}{\text{°F}}\right)(T_{\text{°F}} - 68°)L$$
$$= \left(0.00000645 \frac{\text{ft}}{\text{°F}}\right)(98° - 68°)(787.35 \text{ ft})$$
$$= 0.152 \text{ ft}$$

The corrected length is

$$L = 787.35 \text{ ft} + 0.152 \text{ ft} = 787.502 \text{ ft} \quad (787.5 \text{ ft})$$

The answer is (C).

9. From Eq. 13.8, the correction factor for the incorrect tape length is

$$C = \left(L_{\text{tape measurement}} - 100 \text{ ft} \right) \left(\frac{L_{\text{line measurement}}}{100 \text{ ft}} \right)$$

$$= (100.03 \text{ ft} - 100 \text{ ft}) \left(\frac{582.32 \text{ ft}}{100 \text{ ft}} \right)$$

$$= 0.17 \text{ ft}$$

The corrected length is

$$L = 582.32 \text{ ft} + 0.17 \text{ ft} = 582.49 \text{ ft} \quad (582.5 \text{ ft})$$

The answer is (C).

10. From Eq. 13.7, the change in the length of the tape due to a temperature of 28°F is

$$C = \left(0.00000645 \ \frac{\text{ft}}{\text{°F}} \right) (T_{\text{°F}} - 68°) L$$

$$= \left(0.00000645 \ \frac{\text{ft}}{\text{°F}} \right) (28° - 68°)(100 \text{ ft})$$

$$= -0.03 \text{ ft}$$

The corrected length of the tape is

$$L = 100 \text{ ft} - 0.03 \text{ ft} = 99.97 \text{ ft}$$

The answer is (A).

11. From Eq. 13.7, the correction factor for the temperature is

$$C = \left(0.00000645 \ \frac{\text{ft}}{\text{°F}} \right) (T_{\text{°F}} - 68°) L$$

$$= \left(0.00000645 \ \frac{\text{ft}}{\text{°F}} \right) (28° - 68°)(196.44 \text{ ft})$$

$$= -0.05 \text{ ft}$$

From Eq. 13.3, the correction factor for the slope is

$$C \approx \frac{V^2}{2S} = \frac{(10 \text{ ft})^2}{(2)(196.44 \text{ ft})} = 0.25 \text{ ft}$$

From Eq. 13.8, the correction for length of the tape is

$$C = \left(L_{\text{tape measurement}} - 100 \text{ ft} \right) \left(\frac{\text{line measurement}}{100 \text{ ft}} \right)$$

$$= (100.02 \text{ ft} - 100 \text{ ft}) \left(\frac{196.44 \text{ ft}}{100 \text{ ft}} \right)$$

$$= 0.04 \text{ ft}$$

The corrected measurement is

$$L = 196.44 \text{ ft} - 0.05 \text{ ft} - 0.25 \text{ ft} + 0.04 = 196.18 \text{ ft} \quad (196.2 \text{ ft})$$

The answer is (A).

12. From Eq. 13.7, the correction factor for the temperature is

$$C = \left(0.00000645 \ \frac{\text{ft}}{\text{°F}} \right) (T_{\text{°F}} - 68°) L$$

$$= \left(0.00000645 \ \frac{\text{ft}}{\text{°F}} \right) (98° - 68°)(713.19 \text{ ft})$$

$$= 0.14 \text{ ft}$$

From Eq. 13.8, the correction factor for the incorrect tape length is

$$C = \left(L_{\text{tape measurement}} - 100 \text{ ft} \right) \left(\frac{L_{\text{line measurement}}}{100 \text{ ft}} \right)$$

$$= (99.97 \text{ ft} - 100 \text{ ft}) \left(\frac{713.19 \text{ ft}}{100 \text{ ft}} \right)$$

$$= -0.21 \text{ ft}$$

The corrected length is

$$L = 713.19 \text{ ft} + 0.14 \text{ ft} - 0.21 \text{ ft} = 713.12 \text{ ft} \quad (713.1 \text{ ft})$$

The answer is (B).

14 Leveling

Nomenclature

BM	bench mark	ft	m
BS	backsight	ft	m
c	correction	ft/ft	m/m
d	distance	ft	m
e	error	ft	m
FS	foresight	ft	m
h	elevation	ft	m
H	orthometric elevation	ft	m
HI	height of instrument	ft	m
N	geoidal separation	ft	m
R	reading	ft	m
s	stadia hair reading	ft	m
TP	turning point	ft	m

Subscript

a	average
b	bottom
cl	error of closure
co	collimation error
L	long
R	sum of readings
S	short
t	top

Symbols

θ	angle	deg

1. SURVEYOR'S LEVELS

Although relatively simple in concept, the modern surveyor's level is a complex instrument that allows remarkable precision in differential leveling. Like other surveying equipment, levels have evolved considerably in recent years from the relatively simple spirit levels to modern self-leveling digital levels. In the following sections, various types of surveyor's levels are described.

Spirit Levels

Historically, a surveyor used a *spirit level* to measure differences in elevation. (See Fig. 14.1.) A spirit level is an instrument equipped with an elongated, slightly curved glass tube usually filled with alcohol. By centering a bubble within the liquid-filled tube, a level horizontal line can be established. Although still available today, spirit levels are time-consuming to set up, as the instrument has to be carefully adjusted to ensure that it is level when pointed in any direction.

Figure 14.1 *Spirit Level*

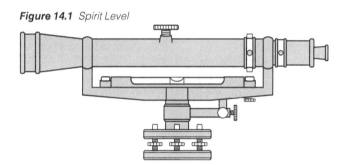

Self-Leveling Levels

Today, surveyors typically use a *self-leveling level*, a light, quick, and accurate instrument that is easy to handle and set up. A self-leveling level is equipped with a simple *bull's-eye level*, a circular level vial with concentric circles and a suspended bubble used to approximately level the instrument in two dimensions, and a *compensator*, a device consisting of a system of prisms and mirrors that is suspended as a pendulum within the level. (See Fig. 14.2.) When approximately level, typically within ± 10 minutes of arc of the true vertical, the compensator swings freely and establishes a horizontal line of sight.

Figure 14.2 *Compensator in a Self-Leveling Level*

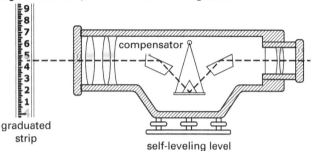

graduated strip

self-leveling level

Digital Levels

The most advanced level in use by surveyors is the *digital level.* (See Fig. 14.3.) Like a self-leveling level, a digital level is equipped with a compensator and is self-leveling. A digital level utilizes solid-state and electronic image processing. The level rod for a digital level has barcode graduations. When readings of the rod are made, the barcode graduations are scanned by the digital level and automatically recorded, a process that eliminates observation errors and recording errors.

Figure 14.3 *Digital Level*

 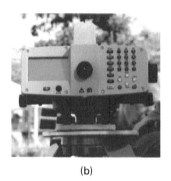

(a) (b)

(a) Members of Las Flores geodetic leveling team from the Department of Civil Engineering and Surveying, University of Puerto Rico, demonstrate the use of a digital level.
(b) The Leica DNA03 Digital Level has a precision of 0.3 mm/km.

2. LEVELING RODS

As described previously, a level rod is merely a graduated rod used to measure the difference in elevation between a point and the line of sight.

Philadelphia Rods

The most common type of level rod in current use is the *Philadelphia rod.* It consists of two sliding parts. The base part typically has graduations extending from zero to seven feet. When the sliding upper part is extended, the rod has continuous graduations from 0 ft to 13 ft. The rod's ability to extend or retract makes it easy to transport and suitable for a variety of terrains and projects. Most Philadelphia rods are graduated in feet, tenths, and hundreds, with graduations running continuously from zero to 13 feet. (See Fig. 14.4 for examples of graduations on a

Philadelphia rod.) Each full foot is marked with a red number (shown in gray in Fig. 14.4) and each tenth of a foot is marked with black numbers. The hundredth graduations alternate from black to white. For ease of reading, the black graduations are pointed at each tenth of a foot (0.10 ft, 0.20 ft, 0.30 ft, etc.) and at each half of a tenth of a foot (0.05 ft, 0.15 ft, 0.25 ft, etc.). When greater precision is needed, the thousandth of a foot graduation is interpolated.

Figure 14.4 *Graduations on a Philadelphia Rod*

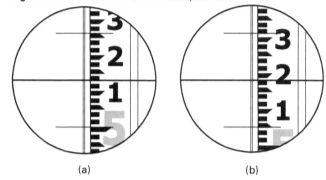

(a) (b)

Example 14.1

What is the correct rod reading for illustrations (a) and (b) of Fig. 14.4?

Solution

1. The center horizontal crosshair indicates where the reading should be made. The grey (normally red) number indicates the number of feet. The reading is over 5 ft, because the center crosshair in Fig. 14.4(a) is above the gray five. The black numbers indicate tenths of a foot, so the reading is between 5.1 ft and 5.2 ft, because the center crosshair is between the one and the two. The graduations indicate hundredths of a foot, so the correct reading is 5.13 ft, because the center crosshair is three graduations above the one.

2. The center horizontal crosshair indicates where the reading should be made. Figure 14.4(b) shows that the center crosshair is over 5 ft and two hundredths of a foot below the two tenths mark. The correct reading is 5.18 ft.

Invar Rods

An *Invar rod* consists of a continuous graduated strip made of Invar, a nickel-iron alloy with a low coefficient of thermal expansion, supported in a protective rod. It is more precise than a Philadelphia rod for four reasons.

1. Because Invar has a low coefficient of expansion, the contraction and expansion of the graduated strip due to changing temperature is minimized.

2. The graduated strip is supported at a constant tension, which standardizes measurements.

3. An Invar rod is one continuous piece, which eliminates errors due to the join of a Philadelphia rod when extended.

4. Typically, Invar rods are equipped with a level bubble and prop poles (see Fig. 14.5) to ensure that the rods are perfectly plumb when read.

Figure 14.5 *Invar Rod with Prop Pole*

3. STADIA

Distances between the level and the level rods may be measured by use of the *stadia hairs* (horizontal crosshairs) in the telescope. (See Fig. 14.6.) One stadia hair is above and one below the principal horizontal crosshair. Distance measurement by this method is based on the principle that the interval between the two stadia hairs is proportional to the distance between the instrument and the rod.

Most levels have a stadia constant of 100. The distance to the rod, d, may be determined by multiplying 100 by the interval between the top stadia hair reading, s_t, and the bottom stadia hair reading, s_b.

$$d = 100(s_t - s_b) \qquad 14.1$$

Example 14.2

What is the distance from the instrument to the rod in Fig. 14.6?

Solution

Use Eq. 14.1 and interpolate to thousandths place with the stadia hair readings shown.

$$d = 100(s_t - s_b) = (100)(5.248 \text{ ft} - 5.013 \text{ ft}) = 23.5 \text{ ft}$$

Figure 14.6 *Stadia*

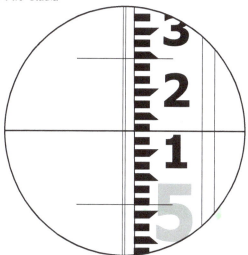

Note that stadia may also be used for topographic surveys with theodolites or transits. That application is covered in Chap. 16.

4. VERTICAL DATUMS

Surveys are usually referenced to a common framework, or *datum*, so that survey data can be compared with other geospatial data. Typically, horizontal coordinates for points in a survey are referenced to a datum based on an *ellipsoid* of revolution that approximates the figure of the earth, whereas elevations are referenced to a geoidal datum.[1] A *geoid* is a horizontal surface based on the earth's gravitational force that approximates the typical elevation of the stilled sea level around the world. This concept is ideally suited as a datum for elevations, because it represents an elevation with equal gravitational forces. The geoid is a somewhat irregular surface when compared with the predictable, smooth, and mathematically perfect shape of the ellipsoid (see Fig. 14.7), but elevations referenced to the geoid are preferable for most applications. The most important reason for this preference is that water could theoretically run "uphill" in terrain measured relative to the ellipsoid in areas where the geoidal separation is changing. This would be the case if the elevation above the ellipsoid increases between two points, but the elevation above the geoid decreases. In such cases, the respective measurements are not invalid, but elevations expressed relative to the ellipsoid do not consider the effect of gravity as do elevations expressed relative to the geoid. Additionally, the ellipsoid can measure 50 m (164 ft) or more above sea level in some parts of the United States, a perspective that can cause some confusion.

The primary geoidal datum used in the United States, the *North American Vertical Datum of 1988* (NAV 88), was established using a minimal constraint adjustment,

---

[1] Traditionally, elevations were referenced to an average stage of the tide, such as mean sea level (msl). However, the use of a tidal datum created challenges when dealing with large areas because of the geographic variability of tidal datums.

Figure 14.7 *Orthometric versus Ellipsoidal Elevations*

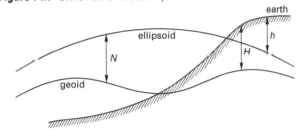

H = orthometric height (above geoid)
h = height above ellipsoid
N = geoidal separation (usually negative in U.S.)

holding the elevation of mean sea level as fixed at just one point, Father Point in Quebec, Canada.[2] A similar geoidal datum has since been established on the island of Puerto Rico, known as the *Puerto Rico Vertical Datum of 2002* (PRVD02), by holding the elevation of mean sea level at San Juan as fixed. In addition, similar geoidal datums have been established in American Samoa (ASVD02), Guam (GUVD04), the Northern Marianas (NMVD03), and the Virgin Islands (VIVD09). Each of these datums has either a single datum origin point based on tidal observations or one such point per island in the coverage area.

Elevations referenced to the geoidal datum are considered *orthometric elevations*, defined as the distance between the geoid and the point as measured along a plumb line passing through the point. The orthometric elevation above the geoid, H, is calculated by subtracting the geoidal separation, N, (which is usually negative in the United States) at that location from the elevation above the ellipsoid, h, using a model of the geoid. In the United States, most elevations used in topographic mapping, engineering studies, geographic information systems, and geodetic or construction surveys are orthometric elevations referenced to NAVD 88.

$$H = h - N \qquad 14.2$$

It is important to understand the difference between ellipsoidal and orthometric elevations when dealing with elevations derived from GPS observations. GPS orbital information is referenced to the ellipsoid, so raw elevations determined by GPS observations are relative to the ellipsoid. For applications such as topographic mapping, elevations determined by GPS observations are adjusted to orthometric elevations using Eq. 14.2. Within the United States, models of the geoid have been developed using a combination of geodetic leveling, GPS observations, and gravity observations. Some less-developed areas of the world do not have precise geoidal models.

In addition to the vertical datums already discussed, local tidal datums, such as *mean high water* (MHW), *mean low water* (MLW), and *mean lower low water* (MLLW), are used for some applications. Typical applications include water boundary surveys and hydrographic surveys.

5. DIFFERENTIAL LEVELING

As the term implies, *differential leveling* is a process of measuring the difference in elevation between points. Typically, this process is used to determine the elevation of a point (or series of points) based on measurements between the point(s) and a *bench mark*, a permanent object with a known elevation. A surveyor's level and level rod are used in this process. A level rod is merely a piece of wood, fiberglass, or metal marked off in meters or feet and fractional parts of those units. The level is a telescope equipped with crosshairs and attached to some type of device allowing the telescope to remain at the same height while turning in a horizontal plane. Thus, the level establishes a line of sight in a horizontal plane from which measurements can be made with the rod. As the observer focuses on the level rod, the graduation of the rod coincident with the line of sight can be read, thus reflecting the distance between the line of sight and the point on which the rod is resting.

Figure 14.8 illustrates the use of a level and rod to measure the difference between two points. In that illustration, the bench mark (BM) has an elevation of 436.27 ft above the datum being used. A *turning point* (TP 1) is a temporary object, such as the top of a stake, for which an elevation is to be measured. For the measurement, the level is set up in a location allowing visibility of both objects and the following two-step process is followed:

step 1: With the rod on BM 1, a reading is made and recorded. This reading is known as a *backsight* (BS) and is added to the elevation of the bench mark to determine the *height of instrument* (HI), which is the elevation of the line of sight.

$$\text{HI} = \text{elev}_{\text{BM}} + \text{BS} \qquad 14.3$$

step 2: The rod is then placed on TP 1 and a reading made and recorded. This reading is known as the *foresight* (FS) and is subtracted from the height of instrument to determine the elevation of TP 1.

$$\text{elev}_{\text{TP}} = \text{HI} - \text{FS} \qquad 14.4$$

Using the process described, the elevations of a continuing line of points (or two points separated by a greater distance or greater difference in elevation than can be covered by one setup of the level) may be determined by moving the level between setups as illustrated in

---

[2] The geoidal datum used prior to 1991 in the United States was the *National Geodetic Vertical Datum of 1929* (NGVD 29), formerly called the *Sea Level Datum of 1929*. The NGVD 29 datum considered the elevations of mean sea level at 26 locations along the coastlines of the United States and Canada as fixed and used a least squares adjustment of the level lines connecting those points. With increased understanding of the geographic variability of sea level, the current geoidal datum was established using a different approach.

Figure 14.8 *Differential Leveling*

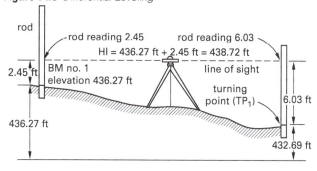

Fig. 14.9. With that process, a "leap frog" process is used between the rod and level. After a foresight is taken, the level rod remains at its location while the level is moved ahead to the next setup and then a backsight is taken to the point that served as the previous foresight. The level then remains at the new setup while the rod is moved to the next point for the foresight. Table 14.1 shows field notes for the differential leveling setups shown in Fig. 14.9.

Table 14.1 *Differential Leveling Field Notes (all distances in ft)*

point	BS(+)	HI	FS(−)	elev
BM 3				441.72
	8.56	450.28	1.10	
TP 1				449.18
	10.34	459.52	1.37	
TP 2				458.15
	7.75	465.90	1.83	
TP 3				464.07
	8.89	472.96	3.46	
BM 4				469.50

6. PROFILE LEVELING

Profile leveling is the process of determining the elevations of points at measured intervals along a fixed line. This process is commonly used for planning highways, canals, pipelines, and other linear projects where a vertical section, or *profile*, of the earth is needed to determine the location of the centerline of the project. Profile leveling is similar to differential leveling except that multiple foresight readings are typically taken from one backsight reading.

At each setup of the level, foresight readings are taken along the centerline of the profile for each full station and for any *break*, a sudden, significant change in slope of the topography. The elevation of each of these points can then be determined by subtracting the foresight reading from the height of instrument at the setup.

Table 14.2 shows typical field notes for the profile leveling process. The resulting profile from the notes is shown in Fig. 14.10. Sightings on the bench mark and turning point are read to a higher order of precision than sightings on the ground.

Table 14.2 *Profile Leveling Field Notes (all distances in ft)*

sta	BS	HI	FS	elev
BM 4				478.26
	4.87	483.13		
32+00			11.5	471.6
33+00			9.4	473.7
33+75			10.1	473.0
34+00			8.2	474.9
35+00			3	480.1
35+15			1.9	481.2
35+70			2.3	480.8
36+00			5.2	477.9
36+50			6.8	476.3
37+00			5.9	477.2
38+00			13.3	469.8
TP			10.72	472.41
	4.54	476.95		
38+60			13.2	463.8
39+00			12.0	465.0
40+00			3.9	473.1
41+00			1.2	475.8
42+00			0.8	476.2
42+70			0.7	476.3
42+80			1.5	475.5
43+00			0.4	476.6
BM 5			0.17	476.78

7. THREE-WIRE LEVELING

Leveling precision can be increased using the process of *three-wire leveling*. In this process, the average of three readings—the top and bottom stadia hair readings and the center horizontal crosshair reading—is used for each reading of the rod. By comparing the average of the three readings with the center horizontal crosshair reading, blunders may be detected immediately, and those readings may be repeated to obtain accurate readings. This process is also ideal for determining if the distances from the level to the foresight and backsight are *balanced* for elimination of collimation errors or correction of observations after field work is performed to eliminate such errors (see Sec. 14.10 for information about collimation errors). Notes are typically kept as shown in Table 14.3. The FS elevation is calculated as

$$\text{FS elevation} = \text{elevation} + \text{BS}_a - \text{FS}_a \qquad 14.5$$

Field Data Acquisition

Figure 14.9 *Continuous Differential Leveling*

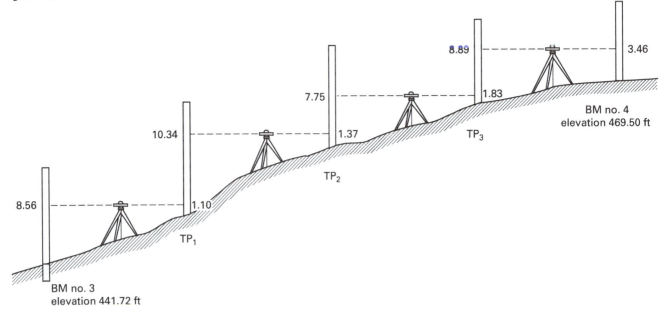

Figure 14.10 *Profile Plotted from Leveling Notes in Table 14.2*

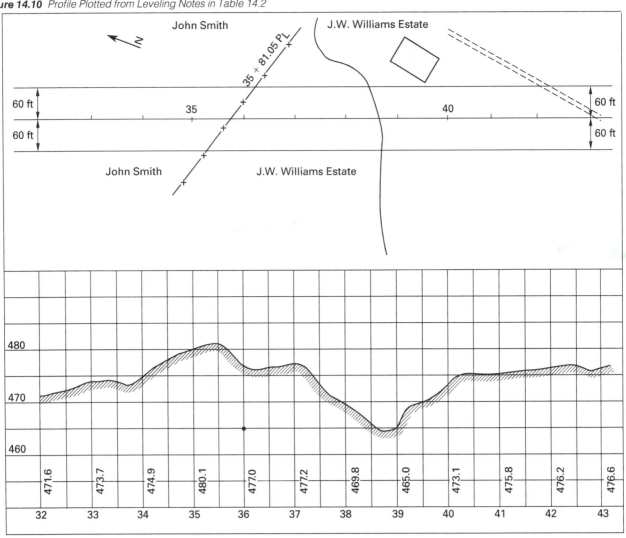

Table 14.3 *Three-Wire Leveling Run Notes (all distances in ft)*

	BS	FS	elev	FS dist	BS dist
			10.000		
top	3.000	1.500			
mid	2.500	0.995			
bot	2.000	0.490			
average	+2.500	−0.995	11.505	100	101
top	5.255	2.962			
mid	4.555	2.254			
bot	3.853	1.546			
average	+4.554	−2.254	13.805	140	142

8. INDIRECT LEVELING

In addition to the differential leveling process described in this chapter, elevations may also be determined by indirect methods.

One such method is *barometric leveling*. That process makes use of the fact that differences in elevation are proportional to differences in atmospheric pressure. Therefore, the method involves reading the atmospheric pressure with a barometer at various points on the earth's surface. Elevations determined by this method are typically several feet in error, so it is not used where precise elevations are needed. Rather, it is limited to reconnaissance surveys in areas, such as mountainous terrain, where differences in elevation are large.

Another indirect leveling process is *trigonometric leveling.* This process is typically performed using a total station to measure the vertical angle and distance to a distant point, and then calculating the elevation of the point using basic trigonometry This process can produce fairly precise elevations, and is often used for topographic surveys since it can determine both the horizontal position and elevation by the same process. See Chap. 16 for examples of these calculations.

Another indirect leveling process widely used today is *GPS leveling.* This method uses distance measurements to a constellation of satellites with known positions to calculate horizontal and vertical positions. This process has the capability of producing very precise (sub-centimeter precision) coordinates for positions based on the ellipsoid. For areas, such as most of the United States, with a precise model of the geoid's relationship to the ellipsoid, it can also produce the precise, centimeter-level geoidal elevations.

9. ADJUSTMENT OF LEVELING LINES

To ensure the accuracy of sightings, good leveling practice calls for either redundant leveling lines or closed leveling lines (a leveling line is closed when it starts and ends at the same bench mark) so that observational blunders can be detected and measurements can be adjusted.

Double-Run Levels

The preferred practice for leveling is to *double run* each segment in the leveling line; that is, obtaining two independent sets of measurements by making one forward run and one back run. This method provides redundancy as well as closure checks.

With the double-run process, there are typically some differences in the forward and back runs, which allow determination of a closure error. In addition, since each segment in the line is double run, this allows distribution of corrections to the appropriate segment by averaging the forward and back runs. The process for determining error of closure and abstracting double-run levels is illustrated in Table 14.4.

Table 14.4 *Abstract of Double-Run Levels (all distances in ft)*

BM	direction	difference in elev	avg	elev
1				10.000
	F	1.092		
	B	−1.089	1.091	
2				11.091
	F	5.336		
	B	−5.340	5.338	
3				16.429
	F	−2.786		
	B	2.779	−2.783	
4				13.646
	F	−5.443		
	B	5.447	−5.445	
5				8.201
total forward run	−1.801			
total back run	1.797			
error of closure	0.004			

Closed Circuit Levels

Closure errors may also be determined for leveling lines that are not double run, but close on either the beginning bench mark or on a bench mark with known elevation. This is known as a *single-run closed leveling circuit.* The error of closure may be determined either by comparing the difference between the beginning elevation and ending elevation when closing on the beginning bench mark or by comparing the difference between the known elevation and measured elevation when closing on a bench mark with known elevation.

Since the location of any error is not obvious with this method, as it is for double-run levels, an alternate method must be used to distribute the error. Because random errors in leveling are usually in proportion to the number of instrument setups, the error of closure should be distributed to the segments of the level line in proportion to the number of setups in each segment. Table 14.5 shows the process for adjusting and abstracting a typical closed leveling circuit that closes on the beginning bench mark. The correction that is applied to each segment is calculated as

$$c_{\text{cl}} = \left(\frac{\text{number of setups in segment}}{\text{total number of setups}} \right) \times (\text{error of closure}) \quad \textit{14.6}$$

10. COLLIMATION ERROR

The most common cause of errors in leveling, other than reading blunders, is the imperfect adjustment of the level. Although almost all observed lines of sight will have a slight difference from a truly horizontal line, called the *collimation error* (see Fig. 14.11), this error can be eliminated by ensuring that the backsight and foresight distances are equal in length. By frequently performing collimation checks on the level and adjusting as needed, this error can be minimized even in unbalanced shots.

Figure 14.11 *Collimation Error*

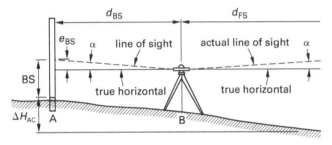

Collimation checks may be made by setting two points about 200 feet apart. The level is first set up about 20 feet from the first point on a line between the two points. Rod readings for all three horizontal crosshairs are read and recorded with the rod on both the near and far points. The level is then moved along the same line to a location about 20 feet from the second point and rod readings for all three horizontal crosshairs read and recorded. The correction for the collimation error, c,[3] may then be calculated as the difference between the sum of the short readings, R_S, minus the sum of the long readings, R_L, divided by the difference between the long distances, d_L, and the difference between the short distances, d_S.

$$c = \frac{\left(R_{S,1} + R_{S,2} \right) - \left(R_{L,1} + R_{L,2} \right)}{\left(d_{L,1} + d_{L,2} \right) - \left(d_{S,1} + d_{S,2} \right)} \quad \textit{14.7}$$

If c and the distances to the rod for each setup are known, post-correction of the leveling data can eliminate the collimation error. If c is large (generally considered to be over 0.01 ft/ft), the level should be adjusted. The level can be adjusted while it is still set up at the second position by adjusting the alignment screw until the center horizontal crosshair for the short reading aligns with the corrected value for R_S.

$$R_{S,2,\text{corrected}} = R_{S,2} + cd_{S,2} \quad \textit{14.8}$$

11. CURVATURE ERROR

The curvature of the earth creates an effect where a level reading on a distant rod appears larger than it would were the level reading taken on a plane (see Fig. 14.12). Equation 14.9, Eq. 14.10, and Eq. 14.11 can be used to find the *curvature of the earth effect*, C, in meters or feet. K is the distance from the level to the rod in kilometers, M is the distance from the level to the rod in miles, and F is the distance from the level to the rod in thousands of feet.

$$C = 0.079K^2 \quad \textit{14.9}$$

$$C = 0.667M^2 \quad \textit{14.10}$$

$$C = 0.024F^2 \quad \textit{14.11}$$

Due to the large distances involved, the curvature of the earth error is not a significant error for short sightings. The error can, however, become a factor for long sightings. This error can be eliminated by having balanced backsights and foresights.

Figure 14.12 *Curvature Error*

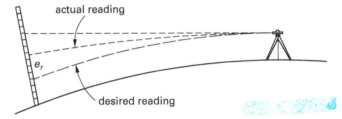

12. REFRACTION ERROR

Refraction causes light rays to bend due to differences in air temperature, and a leveling error occurs when this happens within the line of sight. The *refraction error* is proportional to the distance along the line of sight between the level and the rod. Because the refraction error is generally in the opposite direction of the curvature error, it cancels out some of the curvature error. Temperature gradients are usually greatest close to the

---

[3] The correction for the collimation error is also known as the *"C" factor.*

Table 14.5 Abstract of a Closed Leveling Circuit (all distances in ft)

segment	difference in elev	elev	setups	correction	adjusted elev
BM 1		10.000			10.000
BM 1–BM 2	1.092		3	−0.006	
BM 2		11.092			11.086
BM 2–BM 3	5.336		2	−0.004	
BM 3		16.428			16.418
BM 3–BM 4	−2.786		3	−0.006	
BM 4		13.642			13.626
BM 4–BM 1	−3.618		4	−0.008	
BM 1		10.024			10.000
			total		
error of closure = 10.024 − 10.000 = 0.024			12		

earth's surface, and the refraction error can be minimized by avoiding sightings where the line of sight passes close to the ground. Generally, lines of sight passing less than a few feet (0.5 m) should be avoided. The refraction error can also be minimized by having short, balanced sightings and by making the backsight and foresight readings in quick succession.

Research by the National Geodetic Survey (NGS) has resulted in a refraction modeling process that can be used to determine the refraction error. For precise geodetic leveling, tripods for levels are equipped with temperature sensors mounted at two heights, usually 0.3 m and 1.3 m above the ground. Temperatures are then recorded for each sighting for post-correction purposes.

When leveling between two widely separated points, such as across a river, errors associated with both refraction and curvature of the earth can be significant. For such situations, *reciprocal leveling* is sometimes used. That process involves the use of simultaneous observations in both directions to minimize the effect of both phenomena.

13. ROD PLUMB ERROR

One of the most common and preventable errors is the *level rod plumb error*, an error caused by the level rod not being plumb at the time of sighting. This error can be prevented if the person holding the rod is especially alert to the plumb of the rod. Some rods are equipped with rod bubbles, a bull's-eye level built into the rod that shows when the rod is plumb. The true reading of an inclined rod is the product of the rod reading and the cosine of the angle between the rod and the true vertical, θ. (See Fig. 14.13.)

$$\text{true reading} = (\text{rod reading})\cos\theta \qquad 14.12$$

Figure 14.13 Level Rod Plumb Error

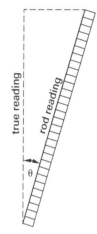

14. PARALLAX ERROR

Parallax is the apparent change in the position of the crosshairs as viewed through the telescope. Because the *reticle*, the ring that holds the crosshairs in the telescope, is stationary, the distance between it and the eyepiece must be adjusted to suit the eye of each individual observer. The eyepiece is turned slowly until the crosshair is as black as possible. After the eyepiece is adjusted, the object viewed should be brought into sharp focus by adjusting the focusing knob for the objective lens. If the crosshairs seem to move across the object when the viewer moves his or her eye slightly, parallax exists. Parallax is eliminated by carefully adjusting the eyepiece and the objective lens. If parallax is not eliminated, it can affect the accuracy of the rod readings.

Field Data
Acquisition

15. OTHER LEVELING ERRORS

Human error, such as imperfect observations and recording blunders (see Chap. 11), accounts for many other common errors in leveling. Both can be eliminated through the use of digital levels. Another common source of human error is the choice of imperfect-turning points. Each turning point should be symmetrical on top and exceptionally stable so that the same elevation is reflected for the foresight and backsight on the point.

Other errors include index errors, expansion errors, and length errors of the level rod. *Index errors* occur when the bottom of the rod does not precisely correspond with zero on the rod scale. *Expansion errors* are due to the expansion and contraction of the metallic strip on the rod due to temperature changes. *Length errors* occur when the graduations of the rod do not precisely match national length standards. For precise leveling, rods are usually calibrated in a testing laboratory against national standard lengths maintained by the U.S. Bureau of Standards.

16. PRACTICE PROBLEMS

1. The term "height of instrument," as used in leveling, means

(A) the distance from the ground to the axis of the telescope

(B) the elevation of the line of sight above the datum being used

(C) the height of the line of sight above a bench mark

(D) all of the above

2. Balancing distances to backsights and foresights eliminates

   I.   collimation error

  II.   curvature of the earth errors

 III.   rod plumb errors

(A) I only

(B) II only

(C) I and III only

(D) I and II only

3. At a leveling setup, the rod's backsight point is a monument with an elevation of 10.000 ft. The backsight reading is 2.456 ft, and the foresight reading is 4.232 ft. The elevation of the foresight point is most nearly

(A) 7.5 ft

(B) 8.2 ft

(C) 12 ft

(D) 17 ft

4. The curvature of the earth error associated with a leveling shot of 1000 ft is most nearly

(A) 0.024 ft

(B) 0.032 ft

(C) 0.035 ft

(D) 0.055 ft

5. Assuming a typical stadia constant and rod graduations in feet.

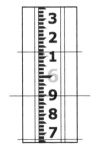

Most nearly, what is the distance between the level and the rod?

(A) 23 ft

(B) 46 ft

(C) 59 ft

(D) 67 ft

6. What is an advantage of an Invar leveling rod?

(A) low coefficient of expansion

(B) more visible graduations

(C) top portion of rod can be detached

(D) visibility at longer distances

7. What is the rod reading for the typical backsight shown in this illustration?

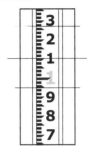

(A)   0.93 ft

(B)   1.0 ft

(C)   1.1 ft

(D)   1.3 ft

8. For a typical modern level, the device that defines the line of sight is the

(A)   bull's-eye bubble

(B)   compensator

(C)   plumb bob

(D)   spirit level bubble

9. The measurements for a closed leveling run are as shown.

setup	BS (ft)	FS (ft)
BM 1–BM 2	2.543	5.222
BM 2–BM 3	3.278	4.362
BM 3–BM 2	4.135	3.053
BM 2–BM 1	5.047	2.359

The error of closure is most nearly

(A)   0.005 ft

(B)   0.006 ft

(C)   0.007 ft

(D)   0.008 ft

10. The measurements for a closed leveling run are as shown. The published elevation for BM 1 is 10.000 ft, and the published elevation for BM 5 is 4.291 ft.

setup	BS (ft)	FS (ft)
BM 1–BM 2	3.124	5.667
BM 2–BM 3	3.487	4.223
BM 3–BM 4	5.213	5.673
BM 4–BM 5	2.413	4.392

Most nearly, what is the error of closure?

(A)   0.009 ft

(B)   0.011 ft

(C)   0.012 ft

(D)   0.015 ft

11. A level rod reading is 6.000 ft, and the level rod is 5° off vertical. The corrected reading is most nearly

(A)   5.977 ft

(B)   5.986 ft

(C)   6.002 ft

(D)   6.023 ft

12. An observer sights on a bench mark with an elevation of 5.25 ft, and reads a backsight of 2.52 ft. The foresight rod reading that is necessary to find a point on the ground with an elevation of 2.00 ft is most nearly

(A)   2.73 ft

(B)   5.25 ft

(C)   5.77 ft

(D)   7.77 ft

13. Three-wire field notes are provided in the table shown.

		BS (ft)	FS (ft)	elev (ft)
BM 1				10
	top	3.477	4.566	
	mid	2.965	4.044	
	bot	2.456	3.521	
BM 2				

Most nearly, what is the elevation of BM 2?

(A)   7.078 ft

(B)   8.905 ft

(C)   8.922 ft

(D)   9.812 ft

14. Field notes for a collimation error test are shown.

setup	short point (ft)	long point (ft)
	4.621	4.236
1	4.516	3.320
	4.412	2.403
	4.400	3.929
2	4.302	3.010
	4.199	2.091

The collimation correction for determining the collimation error is most nearly

(A) 0.002 ft/ft

(B) 0.008 ft/ft

(C) 0.011 ft/ft

(D) 0.015 ft/ft

15. To minimize leveling error due to refraction, surveyors should

I. balance backsights and foresights

II. keep the line of sight above the ground

III. take backsights and foresights in quick succession

(A) I only

(B) II only

(C) I and III only

(D) I, II, and III

SOLUTIONS

1. In leveling, the term "height of instrument" refers to the elevation of the line of sight above the datum being used.

The answer is (B).

2. By balancing distances to backsights and foresights, both curvature of the earth errors and collimation errors are eliminated.

The answer is (D).

3. To find the elevation of the foresight point, first find the height of instrument by adding the backsight reading to the elevation of BM 1. From Eq. 14.3,

$$HI = elev_{BM} + BS$$
$$= 10 \text{ ft} + 2.456 \text{ ft}$$
$$= 12.456 \text{ ft}$$

From Eq. 14.4, the elevation of the foresight point is

$$elev_{TP} = HI - FS$$
$$= 12.456 \text{ ft} - 4.232 \text{ ft}$$
$$= 8.224 \text{ ft} \quad (8.2 \text{ ft})$$

setup	BS (ft)	HI (ft)	FS (ft)	elev (ft)
BM 1				10.000
	2.456	12.456	4.232	
				HI − FS = 12.456 − 4.232 = 8.224

The answer is (B).

4. Because curvature is typically calculated based on measurements in thousands of feet, use Eq. 14.11. The effect of the curvature of the earth is

$$C = 0.024F^2$$
$$= (0.024)(1 \text{ ft})^2$$
$$= 0.024 \text{ ft}$$

The answer is (A).

5. Use Eq. 14.1. The top stadia hair is 6.13 ft, and the bottom stadia hair is 5.67 ft, so the stadia distance is

$$d = 100(s_t - s_b) = (100)(6.13 \text{ ft} - 5.67 \text{ ft}) = 46 \text{ ft}$$

The answer is (B).

6. An advantage of an Invar leveling rod is that it has a low coefficient of expansion, which reduces errors caused by temperature differences.

The answer is (A).

7. The center horizontal crosshair is above the large number one (which is normally red), which indicates 1.000 ft, so the reading is over 1.000 ft. The center horizontal crosshair is on the pointed graduation next to the smaller number one, which indicates 1/10 ft (0.100 ft). The rod reading is 1.000 ft + 0.100 ft = 1.100 ft.

The answer is (C).

8. Most modern levels use a compensator to define the line of sight.

The answer is (B).

9. The closed leveling run starts and ends at BM 1, so the error of closure is the first measured elevation of BM 1 minus the second measured elevation of BM 1. Because only the differences of the measurements are needed to calculate the error of closure, any elevation may be used for BM 1. (For the table shown, BM 1 = 10.000 ft.)

Calculate the elevation and height of instrument at each bench mark according to the data from the forward run. From Eq. 14.3, the height of instrument at BM 1 is

$$HI_{BM1} = elev_{BM1} + BS_{BM1}$$
$$= 10 \text{ ft} + 2.543 \text{ ft}$$
$$= 12.543 \text{ ft}$$

From Eq. 14.4, the elevation at BM 2 is

$$elev_{BM2} = HI_{BM1} - FS_{BM1}$$
$$= 12.543 \text{ ft} - 5.222 \text{ ft}$$
$$= 7.321 \text{ ft}$$

The height of instrument at BM 2 is

$$HI_{BM2} = elev_{BM2} + BS_{BM2}$$
$$= 7.321 \text{ ft} + 3.278 \text{ ft}$$
$$= 10.599 \text{ ft}$$

The elevation at BM 3 is

$$elev_{BM3} = HI_{BM2} - FS_{BM2}$$
$$= 10.599 \text{ ft} - 4.362 \text{ ft}$$
$$= 6.237 \text{ ft}$$

Calculate the height of instrument and elevation at each bench mark according to the data from the backward run. The height of instrument at BM 3 is

$$HI_{BM3} = elev_{BM3} + BS_{BM3}$$
$$= 6.237 \text{ ft} + 4.135 \text{ ft}$$
$$= 10.372 \text{ ft}$$

The elevation at BM 2 is

$$elev_{BM2} = HI_{BM3} - FS_{BM3}$$
$$= 10.372 \text{ ft} - 3.053 \text{ ft}$$
$$= 7.319 \text{ ft}$$

The height of instrument at BM 2 is

$$HI_{BM2} = elev_{BM2} + BS_{BM2}$$
$$= 7.319 \text{ ft} + 5.047 \text{ ft}$$
$$= 12.366 \text{ ft}$$

The elevation at BM 1 is

$$elev_{BM1} = HI_{BM2} - FS_{BM2}$$
$$= 12.366 \text{ ft} - 2.359 \text{ ft}$$
$$= 10.007 \text{ ft}$$

The height of instrument (HI) for each setup is calculated by adding the backsight reading (BS) to the previous elevation. Then, the next elevation is calculated by subtracting the foresight reading (FS) from the height of instrument.

setup	BS (ft)	HI (ft)	FS (ft)	elev (ft)
BM 1				10.000
	2.543	12.543	5.222	
BM 2				7.321
	3.278	10.599	4.362	
BM 3				6.237
	4.135	10.372	3.053	
BM 2				7.319
	5.047	12.366	2.359	
BM 1				10.007

The error of closure is

$$closure error = elev_{BM1,1} - elev_{BM1,2}$$
$$= 10.007 \text{ ft} - 10.000 \text{ ft}$$
$$= 0.007 \text{ ft}$$

$$closure error = h_{BM1,1} - h_{BM1,2}$$
$$= 10.007 \text{ ft} - 10.000 \text{ ft}$$
$$= 0.007 \text{ ft}$$

The answer is (C).

10. The leveling run starts at BM 1 and ends at BM 5, both of which have a known elevation, so the error of closure can be found by subtracting the measured elevation of BM 5 from the known elevation of BM 5.

Calculate the elevation and height of instrument at each bench mark. From Eq. 14.3, the height of instrument at BM 1 is

$$\text{HI}_{\text{BM1}} = \text{elev}_{\text{BM1}} + \text{BS}_{\text{BM1}}$$
$$= 10 \text{ ft} + 3.124 \text{ ft}$$
$$= 13.124 \text{ ft}$$

From Eq. 14.4, the elevation at BM 2 is

$$\text{elev}_{\text{BM2}} = \text{HI}_{\text{BM1}} - \text{FS}_{\text{BM1}}$$
$$= 13.124 \text{ ft} - 5.667 \text{ ft}$$
$$= 7.457 \text{ ft}$$

The height of instrument at BM 2 is

$$\text{HI}_{\text{BM2}} = \text{elev}_{\text{BM2}} + \text{BS}_{\text{BM2}}$$
$$= 7.457 \text{ ft} + 3.487 \text{ ft}$$
$$= 10.944 \text{ ft}$$

The elevation at BM 3 is

$$\text{elev}_{\text{BM3}} = \text{HI}_{\text{BM2}} - \text{FS}_{\text{BM2}}$$
$$= 10.944 \text{ ft} - 4.223 \text{ ft}$$
$$= 6.721 \text{ ft}$$

The height of instrument at BM 3 is

$$\text{HI}_{\text{BM3}} = \text{elev}_{\text{BM3}} + \text{BS}_{\text{BM3}}$$
$$= 6.721 \text{ ft} + 5.213 \text{ ft}$$
$$= 11.934 \text{ ft}$$

The elevation at BM 4 is

$$\text{elev}_{\text{BM4}} = \text{HI}_{\text{BM3}} - \text{FS}_{\text{BM3}}$$
$$= 11.934 \text{ ft} - 5.673 \text{ ft}$$
$$= 6.261 \text{ ft}$$

The height of instrument at BM 4 is

$$\text{HI}_{\text{BM4}} = \text{elev}_{\text{BM4}} + \text{BS}_{\text{BM4}}$$
$$= 6.261 \text{ ft} + 2.413 \text{ ft}$$
$$= 8.674 \text{ ft}$$

The elevation at BM 5 is

$$\text{elev}_{\text{BM5}} = \text{HI}_{\text{BM4}} - \text{FS}_{\text{BM4}}$$
$$= 8.674 \text{ ft} - 4.392 \text{ ft}$$
$$= 4.282 \text{ ft}$$

setup	BS (ft)	HI (ft)	FS (ft)	elev (ft)
BM 1				10.000
	3.124	13.124	5.667	
BM 2				7.457
	3.487	10.944	4.223	
BM 3				6.721
	5.213	11.934	5.673	
BM 4				6.261
	2.413	8.674	4.392	
BM 5				4.282

The error of closure is

$$\text{closure error} = h_{\text{BM5,published}} - h_{\text{BM5,measured}}$$
$$= 4.291 \text{ ft} - 4.282 \text{ ft}$$
$$= 0.009 \text{ ft}$$

The answer is (A).

11. From Eq. 14.12, the true reading is

$$\text{true reading} = (\text{rod reading})\cos\theta$$
$$= (6.000 \text{ ft})\cos 5°$$
$$= 5.977 \text{ ft}$$

The answer is (A).

12. Find the height of instrument using Eq. 14.3.

$$\text{HI} = h_{\text{BM}} + \text{BS}$$
$$= 5.25 \text{ ft} + 2.52 \text{ ft}$$
$$= 7.77 \text{ ft}$$

To find a contour of 2.00 ft, the required rod reading is

$$\text{FS} = \text{HI} - 2.00 \text{ ft}$$
$$= 7.77 \text{ ft} - 2.00 \text{ ft}$$
$$= 5.77 \text{ ft}$$

The answer is (C).

13. Averaging the readings, the backsight at BM 1 is

$$\text{BS}_{\text{BM1}} = \frac{3.477 \text{ ft} + 2.965 \text{ ft} + 2.456 \text{ ft}}{3} = 2.966 \text{ ft}$$

Averaging the readings, the foresight at BM 1 is

$$\text{FS}_{\text{BM1}} = \frac{4.566 \text{ ft} + 4.044 \text{ ft} + 3.521 \text{ ft}}{3} = 4.044 \text{ ft}$$

From Eq. 14.3, the height of instrument at BM 1 is

$$\text{HI}_{\text{BM1}} = \text{elev}_{\text{BM1}} + \text{BS}_{\text{BM1}}$$
$$= 10 \text{ ft} + 2.966 \text{ ft}$$
$$= 12.966 \text{ ft}$$

The elevation at BM 2 is

$$\text{elev}_{BM2} = \text{HI}_{BM1} - \text{FS}_{BM1}$$
$$= 12.966 \text{ ft} - 4.044 \text{ ft}$$
$$= 8.922 \text{ ft}$$

Table for Solution 13

	BS	avg BS	HI	FS	avg FS	elev
BM 1						10.000
top	3.477			4.566		
mid	2.965	2.966	12.966	4.044	4.044	
bot	2.456			3.521		
BM 2						HI – FS =
						12.966 – 4.044
						= 8.922

The answer is (C).

14. Calculate the average reading for each sighting. The average readings at the short point are

$$R_{S,1} = \frac{4.621 \text{ ft} + 4.516 \text{ ft} + 4.412 \text{ ft}}{3} = 4.516 \text{ ft}$$

$$R_{S,2} = \frac{4.400 \text{ ft} + 4.302 \text{ ft} + 4.199 \text{ ft}}{3} = 4.3 \text{ ft}$$

The average readings at the long point are

$$R_{L,1} = \frac{4.236 \text{ ft} + 3.320 \text{ ft} + 2.403 \text{ ft}}{3} = 3.320 \text{ ft}$$

$$R_{L,2} = \frac{3.929 \text{ ft} + 3.010 \text{ ft} + 2.091 \text{ ft}}{3} = 3.010 \text{ ft}$$

From Eq. 14.1, the distance of the level from each point is

$$d_{L,1} = 100(s_t - s_b)$$
$$= (100)(4.236 \text{ ft} - 2.403 \text{ ft})$$
$$= 183.3 \text{ ft}$$

$$d_{L,2} = 100(s_t - s_b)$$
$$= (100)(3.929 \text{ ft} - 2.091 \text{ ft})$$
$$= 183.8 \text{ ft}$$

$$d_{S,1} = 100(s_t - s_b)$$
$$= (100)(4.621 \text{ ft} - 4.412 \text{ ft})$$
$$= 20.9 \text{ ft}$$

$$d_{S,2} = 100(s_t - s_b)$$
$$= (100)(4.400 \text{ ft} - 4.199 \text{ ft})$$
$$= 20.1 \text{ ft}$$

From Eq. 14.7, the correction for collimation error is

$$c = \frac{(R_{S,1} + R_{S,2}) - (R_{L,1} + R_{L,2})}{(d_{L,1} + d_{L,2}) - (d_{S,1} + d_{S,2})}$$
$$= \frac{(4.516 \text{ ft} + 4.3 \text{ ft}) - (3.320 \text{ ft} + 3.010 \text{ ft})}{(183.3 \text{ ft} + 183.8 \text{ ft}) - (20.9 \text{ ft} + 20.1 \text{ ft})}$$
$$= 0.008 \text{ ft/ft}$$

The answer is (B).

15. To minimize refraction errors in leveling, observers should balance backsights and foresights, should take backsights and foresights in quick succession, and should not allow the line of sight to be close to the ground.

The answer is (D).

15 Electronic Positional Measurements

Nomenclature

d	distance	ft	m
D	distance between two points	ft	m
f	frequency	Hz	Hz
h	height	ft	m
HI	height of instrument	ft	m
n	quantity		
N	geoid separation	ft	m
r	refractive index		
v	velocity	ft/sec	m/s
v_0	velocity of light in a vacuum, 186,282.05 (299 792.5)	mi/sec	km/s

Symbols

θ	angle	deg	deg
λ	wavelength	ft	m

Subscripts

H	horizontal
I	instrument
R	rod
S	slope
V	vertical
Z	zenith

1. ELECTRONIC POSITIONAL MEASUREMENTS

There have been major changes in surveying technology over the last hundred years that have made significant changes in surveying practice. Most of these changes have been due to the development of electronics.

Probably the first example of this was the fathometer. That device, invented in 1919 by Reginald Fessenden, measures the distance from a vessel to the ocean floor by observing the time it takes a sound wave to be reflected and returned. The fathometer made the surveying of deeper waters possible for the first time. (See Chap. 18 for a discussion on its use in hydrographic surveys.)

Following that development, the development of the electronic distance measurement (EDM) instruments resulted in a rapid change in surveying practice. EDM instruments, together with the development of electronic theodolites, led to total stations, which are now in widespread use.

Digital levels have also made significant changes in surveying practice. One of the most recent significant changes was the development of the Global Positioning System (GPS), which has become one of the indispensable tools of surveying as well as impacting many other aspects of modern society.

The latest development in electronic positioning is the laser scanner. That instrument has allowed the creation of virtual models of a landscape and has transferred many of the functions of field surveying to the in-office manipulation of collected data. The laser scanner is rapidly automating a large part of field surveying measurements.

The net result of all of these developments is that electronic positional measurement, as described in the following sections, has changed, and is continuing to greatly change, the practice of surveying.

2. ELECTRONIC DISTANCE MEASUREMENTS

For over two millennia, surveyors have used taping for the measurement of distance. The advent of *electronic distance measurement* (EDM) in the 20th century not only represented a significant change in the method of distance measurement, it also marked the beginning of a movement toward the use of electronics in all aspects of surveying, which has revolutionized the profession.

The earliest electronic instruments for distance measurement on land were developed in Sweden in the early 1950s. Typical of these was the geodimeter. This device transmits a modulated light beam to a mirror at the end of the line being measured, and then makes phase comparisons between the transmitted and reflected pulses. The geodimeter measures the elapsed time and calculates the distance based on the velocity of light. Under optimal conditions, the first generation of electro-optical instruments had maximum ranges of 2–3 mi (3–5 km) during the day and 15–20 mi (25–30 km) at night. The units were quite heavy and bulky, as well as expensive.

Shortly thereafter, during the late 1950s, EDM instruments that used microwave signals were developed in South Africa. Typical of these was the tellurometer. This device consists of two identical instruments, one set up at each end of the line being measured. The sending unit transmits a series of microwaves that are received by the other unit and sent back to allow measurement of the elapsed time. This process determines the distance based on the velocity of the microwaves.

These early electromagnetic instruments could measure distances up to 50 mi (80 km) under favorable conditions. They were somewhat lighter than the early electro-optical instruments. They did not require intervisibility, so they could be used in weather conditions that would hinder electro-optical equipment. A tellurometer is shown in Fig. 15.1.

Figure 15.1 *Early Use of a Tellurometer by U.S. Coast and Geodetic Survey Personnel*

Source: U.S. National Geodetic Survey

Due to their cost and size, early EDM instruments were used primarily for geodetic surveys performed by government agencies. But beginning in the mid-1960s and continuing to the present, solid-state electronic components and mass production allowed the development of smaller, lighter instruments that used much less power and were considerably less expensive. This allowed the adoption of EDM technology by the land surveying profession as the standard mode for distance measuring.

Principles of EDM Operation

Most modern EDM instruments are electro-optical and use a laser or infrared light source. The units are portable, have low power requirements, and are easily operated.

These devices generally have shorter ranges than the first-generation units, typically 2 mi (5 km), and are thus geared to the practicing land surveyor market. A majority of EDM devices manufactured today are incorporated into total stations to provide angulation and distance measurement in one instrument.

Some of the latest developments in EDM technology are "reflectorless" units with typical ranges of 600–1000 ft that measure the distance to objects without using retro-prisms. The laser or infrared light used by most electro-optical EDM instruments is modulated into wavelengths of a certain frequency. Measurement of a line is accomplished by setting up the transmitter at one end and a reflector (usually a prism) at the other end. The reflected light is converted to an electrical signal by the EDM to allow phase comparison of the transmitted and reflected signals.

The integral number of wavelengths in the double path is determined by transmitting pulses on multiple frequencies and comparing the results. The phase difference is then used to determine the length of the line, which would be equivalent to one-half of the sum of the number of wavelengths in the double path distance plus the partial wavelength represented by the phase difference. This calculation is represented by Eq. 15.1. D is the length of the line, n is the integral number of wavelengths in the double path of the light, λ is the wavelength of the modulated beam, and d is the distance representing the phase difference. The measurement process is illustrated by Fig. 15.2.

$$D = \tfrac{1}{2}(n\lambda + d) \qquad \textit{15.1}$$

For the measurement process, the wavelength, λ, is determined from the modulation frequency of the EDM using Eq. 15.2. f is the modulation frequency in hertz (cycles per second), and v is the velocity of the light.

$$\lambda = \frac{v}{f} \qquad \textit{15.2}$$

The reflectors usually consist of retrodirective prisms that cause the incident light to be reflected parallel to itself. This is achieved by having the reflective faces of the prism at right angles, as illustrated in Fig. 15.3. As may be seen, the path of the light beam within the prism is the sum of distances a, b, and c, which is equal to twice the depth of the prism. Due to the refractive properties of the glass in the prism, the equivalent travel in air would be $1.57(a + b + c)$. Therefore, the ideal prism would be mounted so that the front face of the prism is set at a distance of 1.57 times twice the depth of the prism forward of the survey point. Yet that

Figure 15.2 *Measurement Process for Electro-optical EDM*

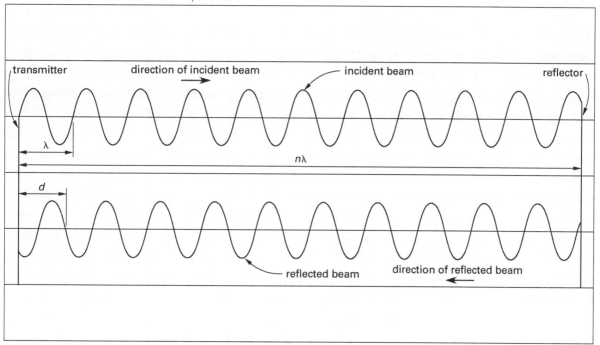

distance is so far behind the prism that mounting it at that point would create an unbalanced prism. Therefore, most prisms manufactured today are created with a small *prism offset* that can be compensated for by an adjustment in the EDM instrument. This standard prism offset is typically either 30 mm or 0 mm.

Figure 15.3 *Retrodirective Prism*

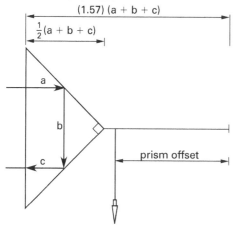

Example 15.1

An EDM instrument with a wavelength of 60 ft measures a distance resulting in a double path length of 50.25 cycles. What is the length of the line measured?

Solution

Using Eq. 15.1,

$$D = \tfrac{1}{2}n\lambda + d$$
$$= \left(\frac{1}{2}\right)((50)(60 \text{ ft}) + (0.25)(60 \text{ ft}))$$
$$= 1507.5 \text{ ft}$$

Sources of EDM Error

The accuracy of modern electro-optical EDM devices is usually stated as plus or minus the sum of a constant plus a certain number of parts per million (ppm) for the standard error. A typical high-end EDM device might have an accuracy of $\pm(2 \text{ mm} + 2 \text{ ppm})$. This is equivalent to $\pm(0.007 \text{ ft} + 2 \text{ ppm})$. Achieving the stated standard accuracy depends on proper observation procedures, calibration, and care in avoiding errors. A surveyor needs a thorough knowledge of the major causes of error when using an EDM instrument in order to avoid mistakes.

Atmospheric conditions that affect the speed of light are a major source of error in EDM. The speed of light for a given set of atmospheric conditions is defined by Eq. 15.3. The parameter v_0 is the velocity of light in a vacuum (186,282.05 mi/sec or 299 792.5 km/s). r is the refractive index for the defined atmospheric conditions.

$$v = \frac{v_0}{r} \qquad\qquad 15.3$$

The refractive index r varies with atmospheric pressure, air temperature, and, to a lesser degree, vapor pressure. It is important that meteorological parameters be measured and a correction be applied. Fortunately, most EDM instruments will calculate the correction value automatically when the meteorological readings are entered. An important precaution is to avoid measurements along lines where the light beam passes close to the ground. In such areas there can be considerable variation in temperature.

Uncertainty in the offset correction for the prism is another frequently encountered error. As discussed earlier in this chapter, most prisms have an offset that requires a correction to measurements. Fortunately, prisms typically use a standard offset of either 30 mm or 0 mm and have the offset marked on the prism assembly. However, even with the correct offset, errors can appear under certain conditions. Where distances are measured on a slope, light rays do not strike the face of a vertical prism in a direction perpendicular to its front face. This changes the position of the effective center of the prism, causing significant errors with steep slope distances. In such conditions, use a reflector that can be adjusted to an angle perpendicular to the slope. If such a reflector is not available, a correction to compensate for the incident angle may be applied to the measurements.

Frequency drift is another source of error. Periodic calibration of the EDM instrument against a known distance ensures accurate and consistent results. Another frequent source of error in EDM is *tribrach error*. If the optical plummet on the tribrach supporting either the EDM instrument or the prism is misadjusted, this could result in the instrument or the prism being off the point. In this situation, the line being measured is not the line between occupied points.

Example 15.2

An EDM has a standard error of $\pm(0.007 \text{ ft} + 2 \text{ ppm})$. What is the expected error in measuring a 3000 ft line?

Solution

$$\text{standard error} = \pm(0.007 \text{ ft} + 2 \text{ ppm}) \times (\text{measured distance})$$
$$= \pm\left(0.007 \text{ ft} + \frac{2}{10^6}\right)(3000 \text{ ft})$$
$$= \pm 0.013 \text{ ft}$$

EDM Calibration

Periodic calibration of EDM instruments against a known distance is essential for obtaining consistent and accurate measurements. A calibration is best performed on a precise baseline established specifically for this purpose. Most states have a number of such baselines, typically established in cooperative programs with the National Geodetic Survey, and data for such sites are available on that agency's website. A typical baseline configuration includes four marks set at 0 m, 150 m,

800 m, and 1500 m. It is recommended that each mark be occupied and that redundant measurements be made to each of the other marks.

Slope Distance Reduction

Unlike taping, where the normal procedure is to make horizontal distance measurements, electronic measurements are typically made using slope distances. Figure 15.4 illustrates the need to reduce slope distance to horizontal distance.

Many total stations automatically apply slope distance reduction, so care should be taken to apply a correction but not apply it twice. Equation 15.4 and Eq. 15.5 are used to reduce slope distances to horizontal distances using zenith or vertical angles. D_H is the horizontal distance, D_S is the slope distance, θ_Z is the zenith angle, and θ_V is the vertical angle.

$$D_H = D_S \sin \theta_Z \qquad 15.4$$

$$D_H = D_S \cos \theta_V \qquad 15.5$$

The reduction of slope distance to horizontal distance may also be accomplished by finding elevations at both ends of the measured line, instead of vertical angles, and using Eq. 15.6. The elevation at the rod is represented by elev_R, the elevation at the instrument by elev_I, the height of instrument at the rod by HI_R, and the height of instrument at the instrument by HI_I.

$$D_H = \sqrt{D_S^2 - \left((\text{elev}_R + \text{HI}_R) - (\text{elev}_I + \text{HI}_I)\right)^2} \qquad 15.6$$

Example 15.3

For a measured slope distance of 662.5 ft and a zenith angle of 92°20′, what is the corresponding horizontal distance?

Solution

Using Eq. 15.4,

$$D_H = D_S \sin \theta_Z$$
$$= (662.5 \text{ ft})(\sin 92°20′)$$
$$= (662.5 \text{ ft})(0.99917)$$
$$= 661.9 \text{ ft}$$

Example 15.4

An EDM instrument is set up with a height of instrument of 4.00 ft over a point with an elevation of 205.25 ft. The reflector is set at a height of 6.00 ft over a point with an elevation of 159.15 ft. The measured slope distance between the points is 1260.50 ft. What is the horizontal distance?

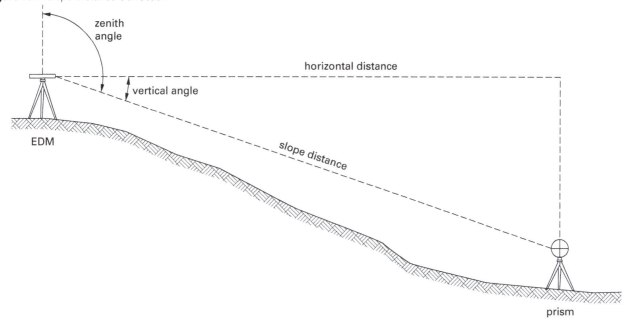

Figure 15.4 *Slope Distance Correction*

Solution

Using Eq. 15.6,

$$D_H = \sqrt{D_S^2 - \big((\text{elev}_R + \text{HI}_R) - (\text{elev}_I + \text{HI}_I)\big)^2}$$

$$= \sqrt{\begin{array}{c}(1260.50 \text{ ft})^2 - \big((159.15 \text{ ft} + 6.00 \text{ ft}) \\ - (205.25 \text{ ft} + 4.00 \text{ ft})\big)^2\end{array}}$$

$$= 1259.73 \text{ ft}$$

3. TOTAL STATIONS

Traditionally, different equipment has been used for measuring direction (or angles) and for measuring distance. As an example, surveyors in the early U.S. Public Land Survey used magnetic compasses for measuring direction and Gunter's chains for measuring distance. With time, transits and theodolites replaced the compass. Then, EDMs replaced the chain. Subsequently, as the use of EDMs expanded, theodolite manufacturers began including attachments for mounting EDMs over the telescopes of theodolites. Then, as digital theodolites were developed, the EDM and digital theodolite were combined into one instrument—the *total station*.

The total station is a digital theodolite integrated with an EDM to allow the measurement of angles and distances with one instrument. The optics of the angular-measuring portion of the instrument are essentially the same as the traditional theodolite, with the primary difference being that the engraved angular plates have been replaced with glass plates designed to be read with light-emitting diodes and computer circuitry. The distance-measuring portion is the same as the EDM described in the previous section of this chapter. The

difference is that the horizontal angle, the vertical angle, and the slope distance can all be measured, displayed, and typically recorded with the push of a button—thus eliminating human error in the measurements as well as greatly reducing the time for observations and recordation of the measurements. Further, most systems also have the capability to correct the slope distances to horizontal distances and annotate the measurements to essentially create digital field notes.

A recent innovation in this area is the *robotic total station*. That feature adds the capability of remote operation of the device. This allows one operator to walk around with a reflector while the total station follows the operator's location. The operator may then remotely make measurements to various points over which the reflector is held, as well as enter descriptive information regarding the points. Essentially, this allows a survey to be conducted with a one-person crew.

4. DIGITAL LEVELS

Electronics has also made a significant impact in differential leveling. As discussed in Chap. 13, the *digital level* is the most advanced level in use today. Like a self-leveling level, a digital level is equipped with a compensator and is self-leveling. A digital level utilizes solid-state and electronic image processing. The level rod for a digital level has barcode graduations. As with total stations, the advantages of the technology are the elimination of human reading and recording errors and considerable saving of observational time. As a result, digital levels are rapidly replacing conventional levels, especially for high-precision leveling.

5. GLOBAL POSITIONING SYSTEM

Principles of GPS Operation

Surveys for relating various landmarks over large areas to a common datum have traditionally been accomplished through the process of triangulation. In the United States, a high-accuracy triangulation network with stations located on mountain peaks and tall, steel observing towers served as the geodetic foundation for surveying and mapping systems for many years. That network was created by the U.S. Coast and Geodetic Survey, founded in 1807 under the administration of president Thomas Jefferson. Originally charged with mapping the coasts, that agency realized early on that comprehensive mapping was impossible without a "skeleton" or control network. Similar networks have existed for a long time within other developed countries and continents, allowing accurate surveying to a common datum within each network.

That type of network requires line of sight between stations, so the networks typically have been limited in coverage to continents or islands, with the relationship between networks only roughly known prior. Absolute positioning using conventional astronomic observations was adequate for coarse navigation, but it did not allow precise positioning. By the middle of the 20th century, the world had a number of fairly precise local networks with only coarsely known interrelationships.

With the advent of the space age, a new tool for surveyors emerged: *satellite geodesy*. After initial programs conducted by the U.S. Coast and Geodetic Survey involving simultaneous observation of reflective satellites to tie continental networks, satellite geodesy made a leap forward in the early 1980s with the *Global Positioning System* (GPS). The system was developed by the Department of Defense to allow continuous determination of geographic position anywhere in the world.

The system currently consists of a number of operating satellites together with tracking stations located around the world. The satellites are spaced to allow visibility of several satellites from any point on the earth at all times, to allow continuous use and to continually transmit radio signals that may be used for positioning determination. Other nations have also established similar systems, which have created dense satellite coverage for positioning, resulting in what is known as the *Global Navigation Satellite System* (GNSS).

In addition to the intended navigational uses of GPS, the system has revolutionized the field of surveying. It has allowed for the first time the measurement of centimeter- and even sub-centimeter-level geographic positions without line-of-sight measurements to established control points. It provides a practical means of determining the geographical positions of almost all boundaries. As a result, surveying's role has rapidly changed from determining positions relative to local landmarks to determining geodetic positions relative to a global network.

The basic concept of GPS is to simultaneously determine the distances, or ranges, between a point on earth and a number of satellites with known positions. Thus, the system uses the principle of *resection*. The distances are determined by measuring the time between signal generation at the satellites and reception at the receiver. The knowledge of this time and the velocity of the signal allows the calculation of the range.

Ranges from three satellites will provide a single solution for a 3-D point in space, but since the time at the satellite and receiver may not be synchronized, a range from a fourth satellite is necessary to determine a corrected position. Four satellites must therefore be visible to the receiver to determine a precise position.

GPS Receivers

Survey-grade GPS receivers are available in a wide range of capabilities, so the intended application should be carefully considered in the selection of a receiver. The technology is changing rapidly, so it is difficult to state best features. Yet, a key performance feature for GPS receivers intended for geodetic use is dual-frequency reception. As the radio waves generated by the GPS satellites travel through the earth's atmosphere, they are deflected and slowed by *atmospheric effects*, resulting in receiver errors as large as several meters. Since the deflection is frequency dependent, the reception of both GPS frequencies allows the receiver to calculate and remove such errors in the solution for the position. Dual-frequency receivers require shorter occupation times and provide greater accuracy than single-frequency receivers. However, some single-frequency receivers have the capability to use the atmospheric information contained in the navigation message to remove some of the errors due to atmospheric deflection.

Another desirable feature is the receiver's ability to reduce or eliminate interfering signals. Even though the frequencies used by GPS are dedicated to that use, electrical interference still occurs. Better receivers can collect good observational data despite such interference. There are a number of other features to consider, depending on the application of the receiver. These include the ability to track a large number of satellites simultaneously, the ability to receive differential corrections, the ability to set elevation masks to exclude observations from satellites low on the horizon where atmospheric effects are greatest, and the ability to output data in a format suitable for software programs or navigation.

GPS Surveying Processes

Autonomous GPS is the use of a standalone receiver without differential corrections or post-processing. Such an application is capable of accuracies of a few meters.

Field Data Acquisition

While this application may provide sufficient accuracies for navigation, it is usually considered inadequate for survey applications.

Differential GPS requires the use of a second set of GPS observations, taken simultaneously with the first at a known position that is used as a base station. Individual corrections are determined from signals from all satellites tracked at the base station and applied to observations at the subordinate point or points. Differential corrections may be applied by post-processing, in which the base and unknown station points are processed simultaneously by one of various software programs designed for this purpose.

One approach to differential GPS uses data from *continuously operating reference stations* (CORS), operated to provide accurate GPS observation data. The stations are operated by both public and private entities. *Real-time differential GPS* is a variation of differential GPS that eliminates all or most post-processing. The process involves equipping the base station with a transmitter to broadcast the corrections and equipping the subordinate point with a radio receiver to accept such corrections. Many GPS receivers are equipped with modems and receivers for such an application.

[handwritten: a permanently recording GPS station]

Another differential GPS survey process involves the *Online Positioning User Service* (OPUS) of the National Geodetic Survey. Under that program, standalone GPS observations with durations of at least two hours may be submitted to the National Geodetic Survey via email. The observations are automatically processed by that agency against that agency's network of base stations, with the corrected observations immediately returned. With longer observational periods of four to five hours, the process reportedly provides centimeter-range accuracy.

Elevation Determination with GPS

The concepts of geoid and ellipsoid heights are essential to understanding the application of GPS to elevations. GPS positions are determined in reference to the ellipsoid, which is a mathematically smooth surface created by rotating an ellipse about its smaller axis. In contrast, vertical data on surveys and maps are typically expressed in terms of *orthometric heights*.

Orthometric heights refer to the geoid, which is a surface of consistent gravitational values closely coincident with the elevation of mean sea level. The ellipsoid is a predictable, mathematically perfect shape, while the geoid is an irregular surface that depends on local observations for its definition and has considerable local variation. It is necessary to know the *geoid separation* (the distance between the ellipsoid and the geoid) to convert a GPS elevation to an orthometric elevation at any given point. Those separations vary globally by well over 328 ft (100 m). Several models of the geoid are available to allow such conversion. Ellipsoid, geoid, and orthometric heights are illustrated in Fig. 15.5.

[handwritten: measures the geometry of the satellite constellation]

With a knowledge of the geoid separation, an orthometric height may be calculated from a GPS-derived ellipsoid height using Eq. 15.7, where N is the geoid separation, and h is height.

$$h_{\text{ortho}} = h_{\text{ellipsoid}} - N \qquad 15.7$$

Example 15.5

With an ellipsoid height for a point derived by GPS of 407.5 ft and a geoid separation of 66.6 ft, what is the orthometric elevation for the point?

Solution

Using Eq. 15.7,

$$h_{\text{ortho}} = h_{\text{ellipsoid}} - N$$
$$= 407.5 \text{ ft} - 66.6 \text{ ft}$$
$$= 340.9 \text{ ft}$$

Sources of GPS Error

Users of GPS for surveying applications should have an understanding of possible errors in the system. While GPS can be an extremely useful tool, the surveyor needs to take precautions against error and make necessary corrections to ensure the results meet the required accuracy standards.

Poor *satellite geometry* is a frequent source of errors. The greatest accuracy results when the GPS satellites are widely separated in the sky to provide well-defined geometry, also known as *strength of figure*. If the satellites are tightly grouped, ambiguity in the intersections of the resulting distances may contribute to a weak figure. Most survey-grade receivers provide a measurement of the strength of figure called *position dilution of precision* (PDOP). Ideally, observations should be made when the PDOP is low and the number of available satellites is high. Most receivers allow an operator to set a maximum acceptable value for PDOP and will not accept observations if the PDOP exceeds that value. As a general guideline, observations should not be made with a PDOP greater than four. Prediction programs are available that allow an operator to select observation times when optimum conditions exist.

Multipath error is caused by deflection of the satellite signal by nearby objects. This can be especially problematic in urban areas with tall structures and in mountainous areas. The result of such deflections is that the measured distance between the satellite and receiver is longer than the actual distance. This type of error may be avoided by selecting locations that are clear of tall objects and terrain. An antenna equipped with a ground plane may reduce this effect, as will extended observation times that allow sampling of satellite signals from a wide range of directions. (See Fig. 15.6.)

[handwritten: Dilution of Precision (DOP) in GPS-]

Figure 15.5 Ellipsoid, Geoid, and Orthometric Heights

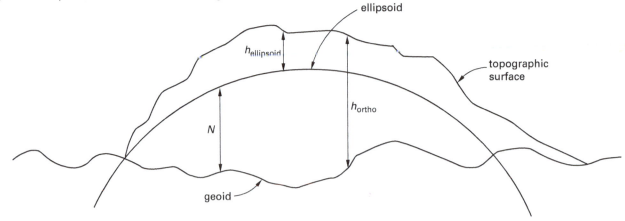

Figure 15.6 Multipath Error

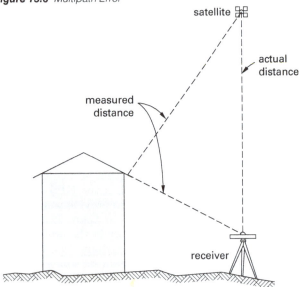

As previously mentioned in this chapter, *atmospheric effects* are another source of errors in GPS observations. These errors may be corrected using differential GPS to make sure the paths of the signals to the unknown points are consistent with those to the base station. The use of dual-frequency receivers is another means of reducing atmospheric errors.

All these errors may be reduced through the use of conservative standards for GPS observations. There is no substitute for redundant observations under a variety of conditions, appropriate limits for baseline length, proximity of base stations, and observational locations with clear horizons.

6. LASER SCANNING

Laser scanning, also known as *high-definition surveying* and *three-dimensional imaging*, is the process of using a laser to measure existing conditions in the built or natural environment. It enables high-definition mapping of topography, transportation byways, structures, mechanical systems, overhead power lines, and other surveyed objects requiring great detail.

The product resulting from laser scanning is a dense *point cloud* representing the surface of a surveyed area. The resulting point cloud is a three-dimensional model with each point defined by precise x-, y-, and z-coordinates. Although the points composing the model may have been scanned from only a few locations on the ground, the model can be viewed from any vantage point. Figure 15.7, an example of a point cloud defining a bridge, shows the level of detail of a virtual model created by a laser scanning point cloud. Although the image resembles a photograph, it is a digital model with every point associated with precise 3-D coordinates. Further, unlike a photograph, the image may be viewed from any perspective, very much like a physical model of the bridge.

Figure 15.7 Typical Point Cloud

Image courtesy of Nobles Consulting Group.

Lasers can measure thousands of points per second with a high level of detail and precision, significantly reducing the time required to produce 3-D digital models of surveyed objects or topography. Thus, it is ideal for topo surveys for site design and for as-built surveys. Further, laser scanning requires minimal personnel for data acquisition, allowing roadways, cliff sides, bridges, buildings, and similar hazardous areas to be mapped without endangering personnel. Thus it has quickly become an essential process in surveying.

Principles of Laser Scanning

A laser scanner is essentially a programmable, automated total station equipped with a reflectorless electronic distance meter (EDM). (See Fig. 15.8.) It uses an active sensing system with its own energy source. Because it does not rely on reflected or naturally emitted radiation, it can be operated in daylight. Unlike photography, it can measure distances to places in shadows and even in underground locations, such as mines and tunnels.

Figure 15.8 *Typical Laser Scanner*

As with a conventional total station, a laser scanner measures horizontal and vertical angles to specific points as well as measuring distances to those points. A major difference between a traditional total station and a laser scanner is that the laser scanner is automated. With a traditional total station, the surveyor aims the instrument at an object and then takes the angular and distance measurements to that object, repeating the process for each point needing location. With a laser scanner, angles and distances are measured on a tight grid at thousands of points per second without the need for an operator to point the device.

Laser scanners rotate around the horizon and measure angles and distances to points on a customizable grid. As a laser scanner rotates, a scanning mirror deflects the line of sight up and down, taking measurements at a rate of hundreds of thousands of points per second. Most laser scanners will scan a complete 360° around the horizon and 90° or more vertically. Distance, horizontal angles, and vertical angles are recorded for each measurement, similar to a total station. These measurements, together with knowledge of the position and directional orientation of the instrument, allow calculation of x-, y-, and z-coordinates for each point measured. Although the designs vary by manufacturer, the orientation of the measurement axes in some laser scanners is based on a compensator, similar to the compensators used on levels and conventional total stations.

Laser Scanning Data Acquisition

For most situations, a single scan will not produce a complete model of the subject. Multiple scans from several different vantage points are usually required to measure all sides of the subject. Setup locations should be carefully selected to minimize shadows, to ensure a clear view and optimum range of desired features and targeted points, and to ensure an adequate angle of incidence. An adequate angle of incidence is an important factor in obtaining quality data. Since laser scanners are typically set up at eye height, the angle of incidence a few hundred feet away is rather small.

Also, in planning setup locations, consideration must be given to how the scans from the different vantage points will be merged into a single model. There are three main approaches to the merging process: cloud-to-cloud point matching, three-point resection from additional independent control points, and traversing.[1] *Cloud-to-cloud point matching* involves using at least three points common to both scans. Such points do not need to have coordinates established by an independent process; they can be targets set on random points. The coordinates of those points from the first scan allow the point cloud for the second scan to be translated to the datum of the first. *Traversing* requires laser scanners to be equipped with a compensator, and then the laser scanner can be used for traditional traverse, with control being carried from one laser scanning setup to the next. Typically, positional and directional orientation, as well as provision for point cloud matching, is provided by special targets that can be recognized by the laser scanning system (such as the ones shown in Fig. 15.9) set on control points.

Figure 15.9 *Typical Laser Scanning Targets*

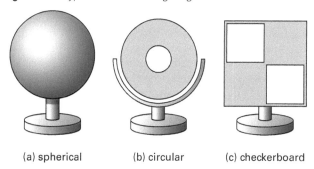

(a) spherical   (b) circular   (c) checkerboard

Laser Scanning Data Processing

Laser scanning collects a large amount of data in a relatively short time. However, while the data collection time is significantly reduced, the vast amount of data

---

[1] For subsequent scans, control points with coordinates from an independent source may not be necessary, depending upon the approach used for extending control.

that is collected shifts the burden of work from the field to the office, although modern software can facilitate the move from a raw point cloud to a fully parametric model.

The major function of the data processing is merging the separate scans into a single model with a common reference system. This process is called *registration of the scans*. Once the scans are registered, the point clouds produced by three-dimensional laser scanners can be used directly for measurement and visualization.

Software is available that allows point clouds to be viewed from any vantage point. The surveyor can virtually walk through the surveyed site and extract coordinates for desired features. Additionally, measurements can be taken for the creation of CADD drawings. For example, points on features, such as the curb, gutter, and pavement edge of roadways, may be selected visually to create CADD drawings of those features.

Laser scanner data should not be used for precise visual selection of the edges of features. Laser scanning samples points on a preset grid, so the exact location of the edge may be missed. A modeling process is typically used to fit the three-dimensional data to surfaces for the CADD drawings. For example, with a scan of a building, points covering each side of the building can be fit to planes whose intersection would accurately depict the limits of the building for CADD drawings. Similarly, the center of a round utility pole can be determined by fitting the points on the surface of the pole to a cylinder.

A specific group of points belonging to a defined surface is selected. Then, cloud fitting algorithms are used to create the best-fit geometric shape in CADD format. Most scan processing software includes modeling tools to fit points to a wide variety of surfaces, such as planes, spheres, cones, and cylinders.

Quality Control

Evaluation of point cloud data relies on identifiable points for which three-dimensional coordinates from a point cloud can be compared to independently measured coordinates. This requires the use of either formal targets, for which point cloud coordinates and measured coordinates can be determined, or other identifiable points, for which point cloud coordinates can be determined by modeling and for which independently measured selected points should be readily identifiable points on surfaces that can be modeled using the point cloud data.

Examples of points meeting these criteria include the center of a circular object, the center of a spherical object, the intersection of three planes (e.g., the intersection of a corner of a building with a concrete slab), and the intersection of the center line of a cylinder with a plane, such as the intersection of a cylindrical pole with a grade. Most specifications recommend that a minimum of 20 points be checked to measure the quality of a point cloud. These can be a combination of formal

targets and other identifiable points. The test points should be carefully selected to obtain a sample representative of the study area. Ideally, the points would be randomly spaced over the study area and over different planes.

Mobile Laser Scanning

Laser scanners are sometimes used from moving platforms (see Fig. 15.10), especially for linear projects such as roadway and coastal mapping projects. Such applications add complexity to processing in that the sensor is at a different location for almost every point that is measured, as opposed to being in one location during the entire scan. To compensate, this process requires the addition of a high-accuracy GPS device to determine the position at the time of each pulse, as well as an inertial measurement unit (IMU) for measuring acceleration and rotational changes in the three axes as the platform vehicle moves. The equipment is considerably more expensive, but linear projects can be mapped much more quickly.

Figure 15.10 *Typical Mobile Scanning Setup*

7. PRACTICE PROBLEMS

1. If an EDM instrument has a standard error of $\pm(5 \text{ mm} + 5 \text{ ppm})$, what would be the expected error in measuring a 2000 ft line?

(A) ± 0.03 ft

(B) ± 0.04 ft

(C) ± 0.05 ft

(D) ± 0.06 ft

2. For a measured slope distance of 3364.5 ft and a zenith angle of $84°40'$, what is the corresponding horizontal distance?

(A) 3349.9 ft

(B) 3355.7 ft

(C) 3361.2 ft

(D) 3402.3 ft

3. An EDM instrument is set up with a height of instrument of 4.25 ft over a point with an elevation of 100.50 ft. The reflector is set at a height of 5.00 ft over a point with an elevation of 150.25 ft. If the measured slope distance between the points is 1500.15 ft, what is the horizontal distance?

(A) 1499.12 ft

(B) 1499.22 ft

(C) 1499.30 ft

(D) 1500.00 ft

4. If the vertical angle from point 1 to point 2 is $2°30'$ and the slope distance as measured by an EDM instrument is 5604.32 ft, what is the horizontal distance?

(A) 5317.21 ft

(B) 5598.99 ft

(C) 5602.84 ft

(D) 5687.14 ft

5. Elevations determined by GPS are referenced to which of the following?

(A) mean sea level

(B) geoid

(C) NAVD88

(D) ellipsoid

6. Using GPS, a point's ellipsoidal elevation is determined to be 592.8 ft. From the GEOID03 model, the geoid separation for that point is -139.4 ft. What is the orthometric height above the geoid at that point?

(A) 453 ft

(B) 515 ft

(C) 627 ft

(D) 732 ft

7. Which of the following conditions would be expected with an increasing value for PDOP?

(A) more multipath errors

(B) a decrease in positional accuracy

(C) a greater number of visible satellites

(D) an increase in positional accuracy

8. Compared with single-frequency receivers, dual-frequency receivers generally are associated with

(A) lower PDOP values

(B) the ability to track a greater number of satellites

(C) fewer errors due to atmospheric effects

(D) fewer ephemeris errors

9. In an area where the geoid is considerably above the earth's surface, which of the following is probably true?

(A) The area has a significant deflection of the vertical.

(B) The area is mountainous or on a high plateau.

(C) The area is below sea level.

(D) The ellipsoid is above the geoid.

SOLUTIONS

1. The standard error is

$$\text{standard error} = \pm(5 \text{ mm} + 5 \text{ ppm}) \times (\text{measured distance})$$
$$= \pm(5 \text{ mm} + 5 \text{ ppm}) \times (2000 \text{ ft})$$
$$= \pm\left(5 \text{ mm} + \frac{5}{1{,}000{,}000}\right) \times (2000 \text{ ft})$$
$$= \pm 0.03 \text{ ft}$$

The answer is (A).

2. From Eq. 15.4,

$$D_H = D_S \sin\theta_Z$$
$$= (3364.5 \text{ ft})(\sin 84°40')$$
$$= 3349.9 \text{ ft}$$

The answer is (A).

3. From Eq. 15.6,

$$D_H = \sqrt{D_S^2 - \left((\text{elev}_R + \text{HI}_R) - (\text{elev}_I + \text{HI}_I)\right)^2}$$
$$= \sqrt{\begin{aligned}(1500.15 \text{ ft})^2 - ((150.25 \text{ ft} + 5.00 \text{ ft})\\ -(100.50 \text{ ft} + 4.25 \text{ ft}))^2\end{aligned}}$$
$$= 1499.30 \text{ ft}$$

The answer is (C).

4. From Eq. 15.5,

$$D_H = D_S \cos\theta_V$$
$$= (5604.32 \text{ ft})(\cos 2°30')$$
$$= 5598.99 \text{ ft}$$

The answer is (B).

5. Elevations determined by GPS are referenced to the ellipsoid.

The answer is (D).

6. Using Eq. 15.7,

$$h_{\text{ortho}} = h_{\text{ellipsoid}} - N$$
$$= 592.8 \text{ ft} - (-139.4 \text{ ft})$$
$$= 732.2 \text{ ft} \quad (732 \text{ ft})$$

The answer is (D).

7. With increasing PDOP, a decrease in positional accuracy would be expected.

The answer is (B).

8. Dual-frequency receivers generally have fewer errors due to atmospheric effects than single-frequency receivers.

The answer is (C).

9. The geoid is the equipotential surface that best fits mean sea level. Therefore, where the earth's surface is considerably lower than the geoid, the area is probably below sea level.

The answer is (C).

16 Topographic Surveying and Mapping

1. TOPOGRAPHIC MAPS

The planning of most construction begins with the *topographic map*, sometimes referred to as the *contour map*. Topographic maps provide a plan view of a portion of the earth's surface showing natural and constructed features, such as rivers, lakes, roads, buildings, and canals. The shape, or *relief*, of the area is shown by contour lines, hachures, or shading. A study of a topographic map should precede the planning of highways, canals, subdivisions, shopping centers, airports, golf courses, and other improvements.

2. TOPOGRAPHIC SURVEYS

Topographic surveys are made to determine the relative positions of points and objects so that the mapmaker can accurately represent their positions on the map. Of great importance in topographic surveys is horizontal and vertical control. *Control* is the means of transferring the relative positions of points and objects on the surface of the earth to the surface of the map. Relative position in the horizontal plane is maintained by *horizontal control*. Horizontal control consists of a series of points accurately fixed in position by distance and direction in the horizontal plane, related to a common datum. Although an assumed datum may be used, horizontal positions for most topographic surveys today are commonly related to the North American Datum of 1983. Relative position in the vertical plane can be maintained by *vertical control*, which may be related to an assumed vertical datum or, more commonly, to the North American Vertical Datum of 1988.

3. POSITIONING METHODS FOR TOPOGRAPHIC SURVEYS

Although measurements using the Global Positioning System (GPS), as well as laser scanning, are frequently used for establishing both horizontal and vertical positions for a topographic survey, conventional methods are still commonly used and are often the most efficient approach. This is especially the case with transportation corridors.

For area maps, the closed traverse is often used to establish a horizontal control network. After the traverse is closed to the required specifications, objects that are to be included on the map are then *tied* to the traverse. These *horizontal ties* are sometimes called the *detailing*. Measurements required to tie one point to the *control* traverse may consist of two horizontal distances, an angle and a horizontal distance, or two angles. There are several methods used to locate a point in the closed traverse method. Only the four most common will be discussed.

The *right-angle offset method* of ties is the most common method used in route surveying for preparing strip maps. The ties are made after the centerline (or traverse line) has been established. Usually, stakes are driven at each station on the centerline, a 100 ft steel tape is stretched between successive stations with the 100 ft mark on the tape forward, and points on either side of the tape are tied to the traverse before the tape is moved forward to the next two stakes. To tie in the corners of the house shown in Fig. 16.1, surveyors move along the tape to a point on the line where they estimate a perpendicular line from the traverse line would strike a corner of the building. They observe the plus at this point by glancing at the tape on the ground, measure the distance from the traverse line to the corner with another tape, and record both measurements. With the 100 ft mark of the steel tape forward, pluses are read directly. This procedure is repeated for the next corner. All sides of the house, including the side between the tied corners, are then measured with the tape. A sketch of the house showing the dimensions of all sides of the house is made.

Unless the scale of the map is very large, measurements for ties are recorded to the nearest foot. It is usually impossible to scale a distance on the map for a tenth of a foot. A right-angle mirror prism is convenient in establishing right angles. As a less accurate method, the surveyor may stand on the transit line facing the point to

Field Data Acquisition

Figure 16.1 Right-Angle Offsets

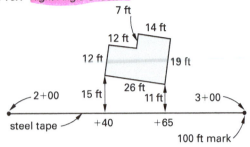

Figure 16.2 Angle/Distance Ties and Two-Distances Ties

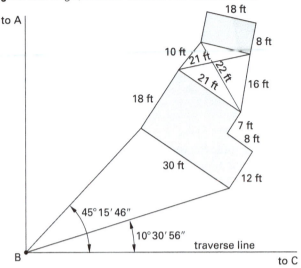

be tied, with arms outstretched on each side pointing along the traverse line, then bring both arms to the front of the body. If they do not point to the object to be tied, the surveyor should move along the traverse line until they do.

The *angle and distance method* of ties, also called the *azimuth-stadia method*, is the most common of those used in preparation of area maps. The azimuth-stadia method allows direction and distance measurements to be made almost simultaneously. If the object is a house or building, two corners must be tied to the traverse, and all sides must be measured and recorded. The transit does not have to be confined to the traverse stations. Intermediate stations can be set from the traverse stations and ties made to the intermediate station. This method, using stadia for horizontal distance, is usually the most efficient.

The *two-distances method* of ties can be used in conjunction with the angle and distance method. Where barns or outbuildings lie behind a house and are obscured from view, the house can be tied to the traverse by the angle and distance method. The outbuilding can be tied to the house by the two-distances method. Two horizontal distances are required to locate one corner of the outbuilding. Two corners must be tied to the house, as shown in Fig. 16.2. All sides of the outbuilding are measured and recorded.

The *two-angle method* of ties is used in special cases where the object to be tied is inaccessible because it lies across a river, lake, or busy highway. The object is tied to the traverse by turning an angle to the object from two different points on the traverse.

When making horizontal ties, certain practices can be followed to reduce the error in locating points on the map. These are as follows:

• When the two-distances method is used, the two distances should be as nearly at right angles as possible.

• When the two-angle method is used, the two lines of sight to the object should be as nearly at right angles as possible.

Vertical ties can be made simultaneously with horizontal ties by stadia or by leveling. Leveling is used for strip maps with *cross-sectioning* as the usual method. For area maps, the *grid system* is common. This consists of laying out the area into a grid with 50 ft or 100 ft intervals and determining elevations at the grid intersections.

No one method of making ties excludes the possible use of others. The azimuth-stadia method is very efficient, does not require a large party, and is accurate enough for most work.[1] When employing this method to obtain horizontal distances, the horizontal circle of the transit or theodolite indicates direction, and the vertical circle and a level rod determine elevation. Horizontal and vertical ties to the traverse can therefore be made simultaneously.

Combinations of methods may be used. Where ditches or streams run through the map area, a combination of azimuth-stadia and cross-sectioning may be economical. The size of the area, slope of the terrain, and amount and size of vegetation also influence the selection of a method.

It is sometimes impossible for the rod handler to place the rod exactly on the point to be located, such as a corner of a house with a wide roof overhang. In this case, the rod is held near the house in such a position that the rod is the same distance from the transit as the house corner, the intercept is read at this point, and the house corner is used to determine horizontal direction.

For long shots where the intercept does not fall entirely on the rod or portion of the rod observed, the intercept between the upper crosshair and the middle crosshair can be observed and doubled for the intercept. On long shots, a stadia rod with bold markings and different colors is more suitable than a level rod.

---

[1] In surveying, the term *stadia* is used to denote a system for measuring horizontal distances based on the optics of the transit telescope, theodolite, or level. This system eliminates the need for horizontal taping, and while not as accurate, it is satisfactory for making the horizontal measurements for topographic maps.

4. NOTE-KEEPING FOR TOPOGRAPHIC SURVEYS

Examples of field notes for a right-angle offset survey are shown in Fig. 16.3. Most measurements are shown on the right. Transit stations and full stations are shown on the left. It is not necessary that the right half be drawn to scale, but it is often very convenient to do so. Each line space on the left represents 20 ft, and the smallest line space on the right represents 10 ft. This scale makes for rapid plotting, but different topographic details require different scales. It is accepted practice to vary the scale from page to page if the amount of necessary detailing varies.

Figure 16.3 *Typical Field Notes for Right-Angle Offset Survey*

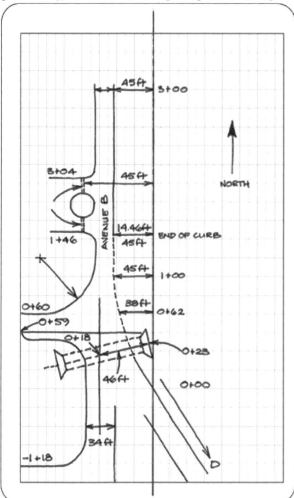

5. MAPPING CONTOURS FOR TOPOGRAPHIC SURVEYS

A *contour* is an imaginary line on the surface of the earth that connects points of equal elevation. Contours are located on the map by assuming that there is a uniform slope between any two points that have been recorded in the notes. Therefore, to ensure that the slope between any two points is uniform, shots must be taken in the field at certain key points. *Key points* are any points that will show breaks in the slope of the ground, just as in cross-sectioning. The most important of these are

- summits or peaks
- stream beds or valleys
- saddles (between two summits)
- depressions
- ridge lines
- ditch bottoms and tops of cuts
- tops of embankments and toes of slope

Several methods of surveying contours are used in topographic mapping. They are the grid method, controlling points method, cross-section method, and tracing contours method. All methods depend on the assumption that there is a uniform slope between any two ground points located in the field.

The *grid method* is very effective in locating contours in a relatively small area of fairly uniform slope. The area is divided into squares or rectangles of 25 ft to 100 ft, depending on the scale of the map and the contour interval desired. Stakes are set at each intersection and at any points of slope change, such as at ridge lines or valleys.

The location of the point where the contour line crosses each side of each square is determined by interpolation (either by estimation or mathematical proportion). In Fig. 16.4(a), contours are to be plotted on a 2 ft interval. Starting at A-1 in Fig. 16.4(b), it can be seen that the 440 ft contour will cross between A-1 and B-1. The vertical distance between A-1 and B-1 is 4 ft, so the 440 ft contour will cross halfway between A-1 and B-1. The 442 ft contour will cross at B-1, so a mark is placed at each of these points.

The 444 ft contour will cross between B-1 and C-1. The vertical distance between B-1 and C-1 is $444.3 \text{ ft} - 442.0 \text{ ft} = 2.3 \text{ ft}$. The vertical distance between B-1 and the 444 ft contour is 2.0 ft. Therefore, the horizontal distance will be 2.0/2.3, or about 0.9, of the way between B-1 and C-1. A mark is made at this point. The 440 ft, 442 ft, and 444 ft contours will cross between A-2 and B-2. The crossing points are found in a similar manner and marked.

After the crossing points are located by interpolation, the crossing points for each contour are connected as shown in Fig. 16.4(c). After all crossing points are connected, the contour lines are smoothed. Small irregularities are taken out so that the contour lines are more like the contours on the ground.

Figure 16.4 *Grid Method of Contour Location*

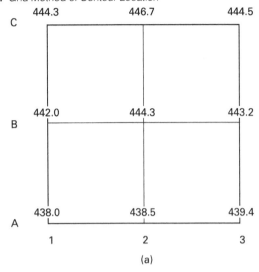

(a)

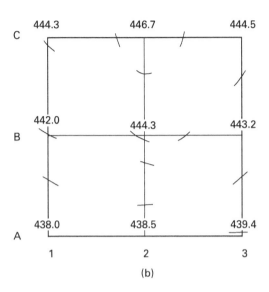

(b)

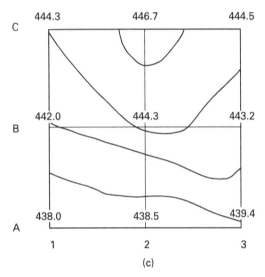

(c)

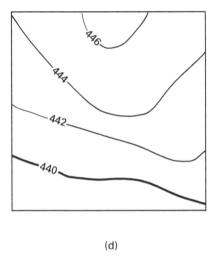

(d)

In following a particular contour line using the grid method, an inspection must be made of each grid line between each intersection to see if the contour line can cross. If it cannot, another line must be inspected. For example, in Fig. 16.4 (c), the 444 ft contour line can cross between B-1 and C-1, between B-1 and B-2, between A-2 and B-2, or between B-2 and B-3, but not between B-2 and C-2. Each contour line must close or reach the border of the map at two points.

After contour lines are smoothed, index contour lines must be made heavier than the other lines, and the elevation of the index contour must be written in a break in the line.

The *controlling points method* is suitable for maps of large area and small scale. The selection of ground points is very important. The accuracy of the contours depends on the knowledge and experience of the survey party. Shots should be taken at stream junctions, at intermediate points in stream beds between junctions, and along ridge lines. Field notes should indicate these points so that ridges and streams can be plotted before interpolations are made.

Interpolations are made in much the same way as they are made using the grid method. Figure 16.5 shows the progressive steps in plotting contour lines by the controlling points method.

The *cross-section method* is satisfactory for the preparation of strip maps. It can be accomplished by using a level and tape or azimuth-stadia. Cross sections are taken at right angles to a centerline or baseline. Elevations of each cross-section shot are written on the strip map, and interpolation of contour lines is performed as in the grid and controlling points methods.

The *tracing contours method* is used when the exact location of a particular contour line is needed. It is effectively performed by use of the plane table, but can be done by the azimuth-stadia survey method.

Figure 16.5 *Controlling Points for Contour Location*

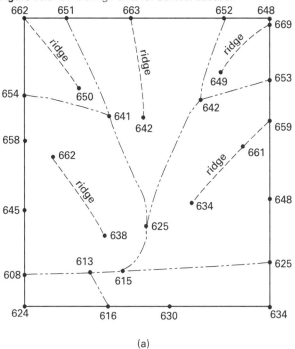

(a)

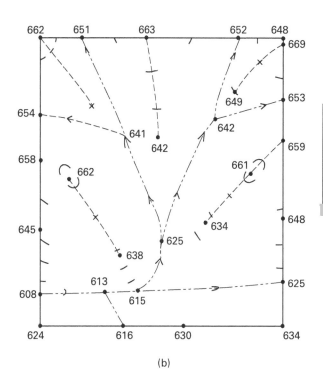

(b)

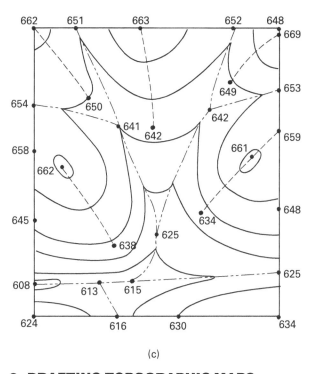

(c)

(d)

Field Data
Acquisition

6. DRAFTING TOPOGRAPHIC MAPS

The first step in preparing a topographic map is the selection of a scale. This selection is influenced by the size of the sheet to be used, the purpose of the map, and the required accuracy. Once scale is established, the horizontal control is generally plotted, but before control is plotted, the sheet is laid out with perpendicular grid lines. The distance between grid lines can be 50 ft, 100 ft, 500 ft, 1000 ft, or any other multiple that suits the scale of the map.

As an example, Fig. 16.6 shows a grid system laid out to a scale of 1 in = 500 ft with grid lines 1000 ft apart. The point P has the coordinates $x = 1660$, $y = 4705$. In plotting the point, the 50 scale is laid on the paper horizontally so that the 10 mark on the scale lines up with the vertical grid line marked 1000. A pencil dot is then made at 1660 on the scale. With a straight edge, a temporary vertical line is drawn through the pencil dot. The scale is then laid vertically along this vertical line with the 40 mark on the scale lined up with the horizontal line on the paper marked 4000. A pencil dot is carefully made on the vertical line at 4705 on the scale (as close as can be read).

Figure 16.6 *Coordinates of a Traverse Point*

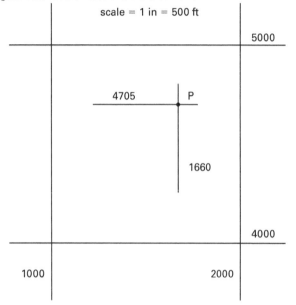

In plotting the control, the traverse can be plotted by the coordinate method, the tangent method, or the protractor method.

The *coordinate method* is the most accurate for plotting a traverse. Any error in plotting one point does not affect the location of the other points. Each point is plotted independently of the others. Coordinates of each traverse station are computed prior to plotting. Each point is plotted on the grid system using the coordinates. After all points of the traverse are plotted, they are connected with lines. Distances between points are scaled and checked against distances that were recorded in the field. Detail points can be plotted with a protractor, using grid lines to orient the protractor.

The *tangent method* is very convenient and accurate where deflection angles have been turned for a route survey. At any traverse station, the back line is produced past the points. A convenient distance (such as 10 in) is laid off from the traverse point along this prolongation. A perpendicular line is determined at this point, and the tangent distance for the deflection angle is marked on this line. A line from the traverse point defines the next leg of the traverse. Any error made in this plotting will be carried on to the next plotting.

The *protractor method* is the fastest but least accurate method of plotting a traverse. Any error in plotting an angle or a distance will be carried on throughout the traverse. The protractor is commonly used for detailing, and for this it is sufficiently accurate.

Selection of the contour interval is an important decision in creating a topographic map. The *contour interval* of a map is the vertical distance between contour lines. In flat country, the interval may be 1 ft, and in mountainous country it may be 100 ft, depending on the scale of the map and the character of the terrain. If the contour interval is too small, the map can be a maze of lines that are not legible. If the contour interval is too large, the true relief may not be visible. The intended use of the map is a basic consideration in the selection of the contour interval.

Figure 16.7 shows that the vertical distance between contour lines is constant, but the horizontal distance varies with the steepness of the ground.

Figure 16.7 *Contour Line Intervals*

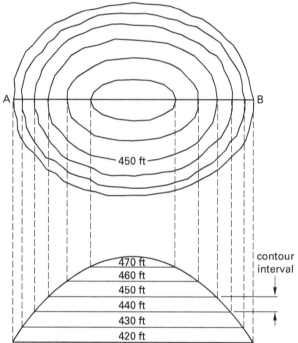

To facilitate reading of a topographic map, it is common practice to darken every fifth contour line. The elevation of that contour line is written in a break in the line, as shown in Fig. 16.8. There are five spaces between any two heavier lines, called *index contours*, so that the contour interval can be computed by dividing the difference in elevation between two index contours by five. In Fig. 16.8, there are five spaces between the 700 ft contour and the 750 ft contour. The contour interval is

$$\frac{750 \text{ ft} - 700 \text{ ft}}{5} = 10 \text{ ft}$$

Figure 16.8 *Typical Contour Map*

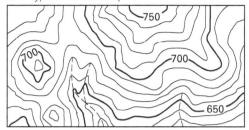

7. USING TOPOGRAPHIC MAPS

Certain fundamental characteristics of contours should be kept in mind when using a topographic map. These are as follows and are illustrated in Fig. 16.9.

- A contour line cannot end abruptly on a map, since each contour line must close upon itself either within or outside the borders of the map.

- Contour lines cannot cross or meet, except in unusual cases of waterfalls or cliff overhangs. (If it were possible for two contour lines to cross, the intersection would represent two elevations for the same point.)

- A series of closed contour lines represents either a hill or depression. The elevations of the index contour lines will indicate which series represents a hill and which represents a depression. For a hill, the elevations increase as the contour lines become shorter. Depressions are often indicated by short hachures on the downslope side of the contour line.

- Contour lines crossing a stream form a "V" that points upstream.

- Contour lines crossing a ridge form a "U" that points down the ridge.

- Contour lines tend to parallel streams. Rivers usually have a flatter gradient than do intermittent streams. Therefore, contour lines along rivers will be more nearly parallel than contours along intermittent streams, and they will run parallel for a longer distance.

- Contour lines form an "M" just above stream junctions.

- Contour lines are uniformly spaced on uniformly sloping ground.

- Irregularly spaced contour lines represent rough, rugged ground.

- The horizontal distance between contour lines indicates the slope of the ground. Closely spaced contour lines represent steeper ground than widely spaced contour lines.

- Contours are perpendicular to the direction of maximum slope. The direction of rainfall run-off in a map area can be determined from this characteristic.

8. USGS TOPOGRAPHIC MAPS

Topographic maps of various scales are available from the United States Geological Survey (USGS). USGS quadrangle series maps cover areas bounded by parallels of latitude and meridians of longitude. The standard edition maps are produced at 1:24,000 scale in 7.5 ft × 7.5 ft format. That scale (1:24,000) is very convenient because 24,000 in = 2000 ft. Therefore, 1 in on the map represents 2000 ft on the ground. The contour intervals may be 5 ft or 10 ft, depending on the terrain.

Field Data Acquisition

Figure 16.9 *Typical Contour Map Features*

9. PRACTICE PROBLEMS

1. Complete the topography field notes for the area shown using the right-angle offset method for ties to the road, stream, and buildings. Consider the enclosed area to be the right half of the page in the field notes. The vertical line is the baseline of the survey. The scale is $\frac{1}{2}$ in = 100 ft.

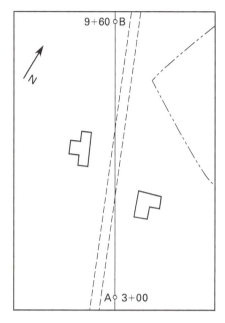

2. Complete the topography field notes for the area shown using the angle and distance method and the two-distances method. Use the two-distances method to tie the small building to the adjacent building; use the angle and distance method for other ties. Consider the transit to be set up at station A on the baseline, with the foresight on station B on the baseline. The scale is $\frac{1}{2}$ in = 100 ft.

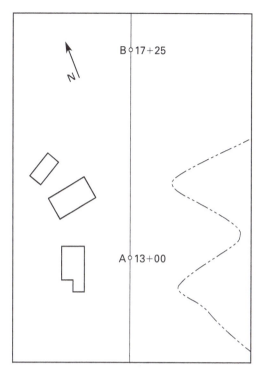

3. Tie the streets to the baseline by the right-angle offset method. The scale is $\frac{1}{2}$ in = 100 ft.

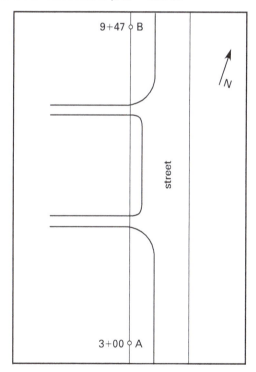

4. Plot 1 ft contours in the illustration shown. Indicate the index contours.

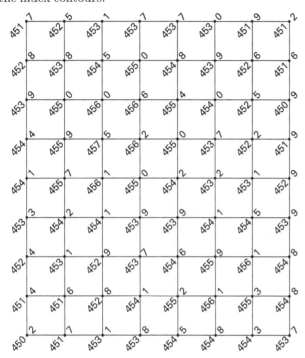

5. Plot 5 ft contours in the illustration shown.

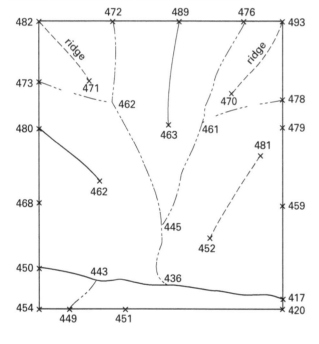

SOLUTIONS

1. Use an engineer's scale to determine distances.

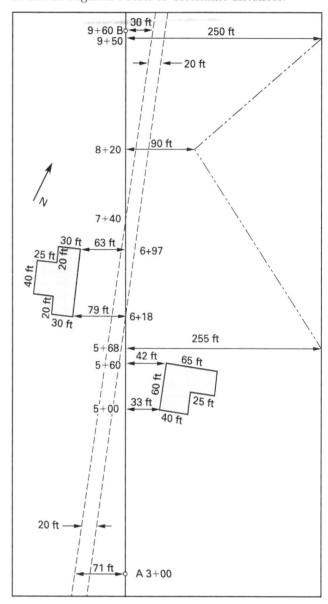

(not to scale)

Right angle offset method

2. Measure angles with a protractor; measure distances with an engineer's scale. *angle & distance Method*

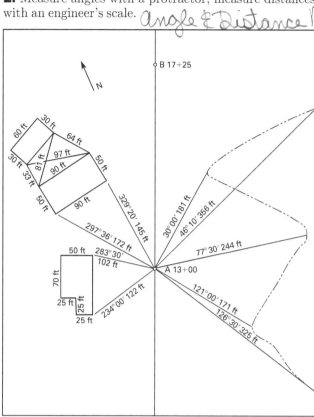

(not to scale)

3. Use an engineer's scale for distances.

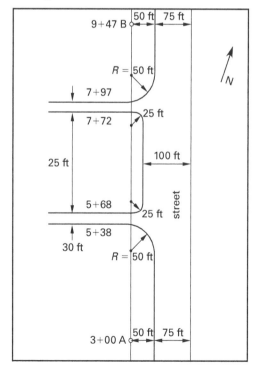

4.

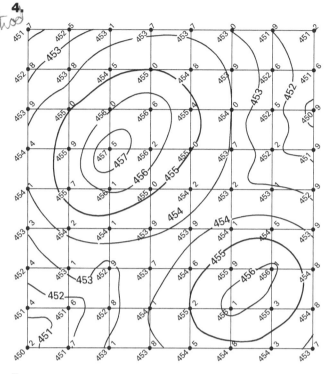

5.

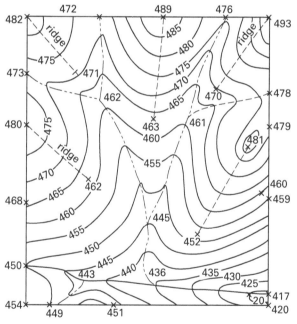

17 Aerial Mapping

Nomenclature*

d	photo displacement	in	mm
D	ground displacement or length of ground shadow	ft	m
f	focal length	in	mm
h	object height	ft	m
H	mean flying height AGL (above ground level)	ft	m
R	radial distance from isocenter	ft	m
S	photo scale	ft/ft	m/m
t	tilt angle	deg	deg
y	separation of image point and isocenter	in	mm

*Generally, uppercase letters are used to represent locations and distance on the earth, while lowercase letters are used to represent points and distances on the photograph or within the camera.

Subscripts

t	tilt
p	principal

1. INTRODUCTION TO AERIAL MAPPING

Aerial mapping is the process of creating maps from measurements made with photography or other types of remote sensing from an airborne platform. While it is usually considered a recent surveying and mapping innovation, aerial mapping has been used for almost two centuries. Today, it is widely used for topographic surveys since it is generally more economical than ground surveys except for relatively small projects.

The first aerial mapping photographs were taken from hot-air balloons in the late nineteenth century. Kites were also used as platforms for aerial photography during this period. For example, after San Francisco's destruction in the 1906 earthquake, the city was mapped with a camera suspended from seven kites.

Fortunately today, more controllable aerial platforms are available, including both aircraft and satellites. In recent years, *unmanned aerial vehicles* (UAVs) have been increasingly used for all modes of airborne remote sensing, although regulation of their use is still evolving. The versatility and economy of UAVs make them an ideal aerial platform for modern miniaturized equipment.

Data acquisition has also evolved, advancing from relatively crude cameras to photographic systems with advanced optics as well as digital cameras, airborne LiDAR (see Sec. 17.8), and other remote sensing systems. Although some aerial mapping still uses traditional photographic film technology for data acquisition, the use of digital cameras and airborne LiDAR sensors is becoming more and more prevalent. The processing of aerial mapping data today, whether acquired on film or as digital data, is generally performed using digital methods. Typically, when film is used, it is scanned to allow digital processing. This trend to digital processing avoids the costly optical-mechanical components previously associated with aerial mapping and allows maximum use of modern computers. In addition, having both the original data and resulting map data in digital format allows more efficient use of GIS and other computer applications.

2. PHOTOGRAPHIC OPTICS

Photographic optics are based on the *refraction of light*, an optical effect used since the Middle Ages to project images. When light rays pass from one medium into another, they change speed and bend, or refract. Light rays will bend toward the normal if the speed of light is less in the second medium (such as when passing from air into a camera's glass lens). This refraction produces a reduced-scale copy of the image as it passes through the lens, which can be projected onto a plane at a focal length dependant on the lens' characteristics.

3. PLANNING FOR DATA ACQUISITION

To obtain good data, it is important to consider exactly when and where aerial photography will take place. When mapping bare-earth elevations in temperate zones with deciduous trees, aerial photography should be scheduled during the winter months when there is better visibility because trees have fewer leaves, and there are smoother flying conditions and clearer air because cooler, dryer air prevails. In tropical zones, aerial photography is typically scheduled during the dry season, generally January through March in the northern hemisphere, so that photographs are not obstructed by rain.

In addition to the time of year, time of day must also be considered for aerial photography. A relatively high sun angle minimizes large shadows, though small shadows help delineate detail and generally increase the photograph's quality. A 45° sun angle is generally considered most desirable. Solar angular altitude diagrams and several computer programs are available to assist in selecting the most appropriate schedule for mapping a given location. Time of day is not a concern with active remote sensors such as LiDAR since sunlight and shadows are not factors with those systems.

It is also necessary to plan the route of the mapping aircraft. Aerial photography is generally taken along a series of preplanned flight lines where successive photographs are exposed such that a predetermined amount of overlap occurs. Generally, 60% forward overlap and 20–40% (30% nominal average) side lap is used.

4. SCALE

The *scale*, *S*, of a photograph is typically expressed as the ratio of the dimension of the image of an object on the photograph to the dimension of that object on the ground. As an example, the scale of 1:24,000 indicates that an object measuring one inch in length on a photograph would measure 24,000 inches or 2000 feet on the ground. If the focal length, *f*, of the camera and the flying height, *H*, is known, the scale of a photograph may be calculated as the ratio of those parameters.

$$S = \frac{f}{H} \qquad \text{17.1}$$

When the desired scale of the photography and the focal length of the camera are known, the required *flying height* can be calculated by rearranging Eq. 17.1.

$$H = \frac{f}{S} \qquad \text{17.2}$$

Example 17.1

What is the required flying height necessary to produce a 1:24,000 negative scale if a land surveyor is using a camera with a 6 in focal length?

Solution

Using Eq. 17.2,

$$H = \frac{f}{S} = \frac{(6\ \text{in})\left(\dfrac{1\ \text{ft}}{12\ \text{in}}\right)}{\dfrac{1}{24,000}} = 12,000\ \text{ft}$$

5. TILT AND RELIEF DISPLACEMENT

When photogrammetric measurements are made from an aerial photograph, not all images will have the correct spatial relationship to each other. Tilt in the photograph, as well as variation in terrain elevation, will cause image displacement as well as scale variation. Figure 17.1 shows *tilt displacement*, d_t, of a point A, which causes points downslope of the isocenter to be displaced radially away from the isocenter, while points upslope would be displaced toward the isocenter. The tilt displacement for a specific point is calculated as

$$d_t = \frac{y^2}{\left(\dfrac{f}{\sin\ t}\right) - y} \qquad \text{17.3}$$

t is the tilt angle, *f* is camera lens focal length, and *y* is the distance between the image point and the isocenter. The *principal line* is the line of intersection between the plane in which the tilt angle is measured and the plane of the tilted photograph.

Figure 17.1 *Image Displacement Due to Tilt in Aerial Photograph*

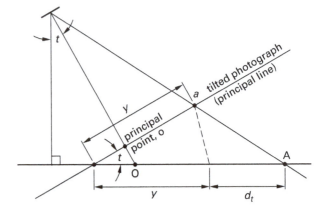

In Fig. 17.2, the principal point, O, is the only point in a truly vertical photograph where there is no *relief displacement*. Relief displacement is especially visible when tall buildings or towers are visible in aerial photographs. When viewing those images, the radial displacement from the photo center of the tops may be readily seen. Relief displacement, *d*, is calculated as

$$d = \frac{Df}{H} \qquad \text{17.4}$$

Figure 17.2 *Image Displacement Due to Relief*

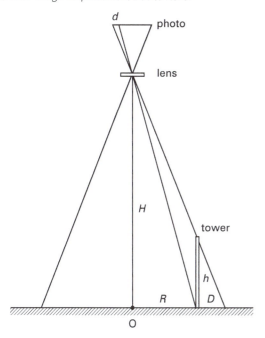

Similarly, Eq. 17.4 can be used to calculate the actual height, D, of a vertical object from its measured (on the photograph) distance (displacement) from its top and bottom, d. f is the focal length of the camera, H is the elevation of the camera, and D is the length of the ground image's shadow. D can be calculated by

$$D = \frac{hR}{H-h} \qquad \text{17.5}$$

H is the camera elevation above mean terrain, h is the object's ground elevation above mean terrain, and R is the radial distance on the ground from the principal point to the object.

Example 17.2

A tower's height is 300 ft above mean terrain. It has a 1000 ft radial distance from the center of a vertical aerial photograph. The photograph was taken 2000 ft above mean terrain using a camera with a 6 in focal length. What is the relief displacement for the top of the tower?

Solution

Use Eq. 17.5 to calculate the length of the tower's shadow on the ground.

$$D = \frac{hR}{H-h}$$
$$= \frac{(300 \text{ ft})(1000 \text{ ft})}{2000 \text{ ft} - 300 \text{ ft}}$$
$$= 176.47 \text{ ft}$$

Use Eq. 17.4 to find the relief displacement.

$$d = \frac{Df}{H}$$
$$= \frac{(176.47 \text{ ft})(6 \text{ in})\left(\dfrac{1 \text{ ft}}{12 \text{ in}}\right)}{2000 \text{ ft}}$$
$$= 0.441 \text{ ft}$$

6. MAP COMPILATION

Photogrammetric mapping is traditionally performed using a *stereoplotter,* which is an optical-mechanical instrument that reconstructs the depicted terrain's spatial geometry in overlapping aerial photographs. It allows the measurement of a terrain's planimetric and topographic attributes by correcting for tilt and relief displacement.

Stereoplotters use the concept of *stereoscopy,* which is the ability to perceive a photographic image in three dimensions. The process involves using overlapping photographs that depict an area as viewed from two different locations in the air. As each eye sees a different photograph (binocular viewing), the two images merge into a central image that has depth. This is because the closer a point is to the eye, the larger the convergence angle is between the eyes and the point. Thus, when looking at the same feature on two photographs taken from two aerial locations, the highest point would have a larger convergence angle and appear taller.

In the past, the most commonly used steroplotter was the *direct-projection plotter.* This plotter uses diapositive photographs (i.e., prints on glass or mylar) the same size as the original photographs to project enlarged images of the two overlapping photographs onto a tracing table. The two projected images are distinguished from each other by projecting one image through a red filter and the other image through a blue-green filter. The operator then views the images using a pair of glasses with separate red and blue-green lenses, thus allowing each image to be viewed by one eye. The direct-projection plotter has been superseded by the analytical plotter and softcopy processing and is rarely used today. The *analytical stereoplotter* uses a computer to mathematically align images so they line up correctly. It also stores data on a hard drive and can redraw the data to any scale.

Stereoplotters are typically equipped with a "floating-mark" that can be lowered or raised within the three-dimensional view of the model to provide an elevation readout. As part of the orientation process, the floating-mark and its readout are referenced to ground control points in order to map the elevation contours or obtain spot elevations.

The stereoplotter is rapidly being superseded by computerized photogrammetric processes. These processes are often called *softcopy photogrammetry* due to the physical (hard copy) form of the photograph being replaced with a digital version. Soft copy processes use specialized software to digitally manipulate the spatial geometry of the view depicted in the overlapping photographs. Since all manipulation is performed digitally, the elaborate optical-mechanical components of the stereoplotter are eliminated. Thus, soft copy processes have far less equipment requirements. Further, such processes offer the advantages of speed and accuracy with the resulting data more easily integrated into other mapping processes. Soft copy systems also allow the aero-triangulation of data in addition to compilation and editing. In addition, these programs allow the operator to superimpose the resulting topographic map over aerial photography and to view the map in three dimensions.

As an alternative to photogrammetric map compilation, stereopairs of aerial photographs may be used to create a corrected form of photography called an *orthophoto*. An orthophoto shows photographic detail without errors caused by tilt, relief displacement, or scale variation. It can be used to measure true distances since they show an image's correct planimetric position and have uniform scale, and are often used with geographic information systems (GIS). They also display all visible features, whereas conventional line maps only show the digitized features. Orthophotographs are generally less expensive than line maps, which involve extensive manual digitizing.

7. STEREO MODEL ORIENTATION

Before the mapped terrain's elements (e.g., roads, sidewalks, buildings, vegetation) are digitized and compiled, a pair of overlapping photographs must be correctly orientated relative to each other, to the ground coordinate system, and to the camera coordinate system. There are three orientations—inner, relative, and absolute. After these orientations are completed, map elements can be digitized and transformed in real-time to the ground coordinate system.

Inner Orientation

The camera's coordinate system has its origin at the photograph center, and with the x- and y-axes parallel to the sides of the photo. The camera also has fiducial marks located at the four corners of the exposure area and at the middle of each of the four sides to total eight points. These fiducial marks must be digitized, and a linear transformation computed to transform the digitized map elements into the camera's coordinate system.

Relative Orientation

The aerial photograph is similar to a bundle of lines passing through the lens's projection center and through each point appearing on the photo. Six points are selected that appear on both overlapping photos to generate a six-point bundle for each photo. One photo is moved and rotated in 3-D until all pairs of corresponding lines intersect. When all lines intersect, a perfect stereo model is formed and can be viewed through the stereoplotter microscope.

Absolute Orientation

The final (absolute) orientation is the translation rotation and scaling of the stereo model into the ground coordinate system. Surveyed ground points must be identified in the model and have x-, y-, and z-coordinates in the mapping ground coordinate system. These points are usually marked on the ground by a white cross that is clearly visible in the photographic model. The absolute orientation involves a seven-parameter transformation containing three rotations, three translations, and a scale factor.

8. LIDAR MAPPING

Airborne *LiDAR* (Light Detection and Ranging) is a relatively new aerial mapping technology. A form of remote sensing, the LiDAR system is mounted on an aircraft and, during the flight, transmits high-frequency laser pulses toward the earth. The LiDAR sensor records the time difference between the pulse's initial transmission and the reflected laser pulse's return to the aircraft. The laser pulse is projected by a mirror that rapidly rotates from side to side along the flight line.

Airborne LiDAR systems include a geodetic grade airborne GPS receiver that measures the aircraft's position every second and an *inertial measurement unit* (IMU) that determines the aircraft's orientation (e.g., pitch, yaw, and roll) approximately 200 times a second. The integration of the data from these components allows the calculation of precise horizontal and vertical coordinates for each point. Airborne LiDAR typically provides vertical accuracies better than 15 cm RMSE[1] and horizontal accuracies better than 50 cm RSME.

While the LiDAR laser beam is only approximately one micron wide at its source, it can expand wider than two feet during its transmission to the ground. Therefore, the beam can hit several surfaces, such as tree branches, during the descent, and as a result, several different reflections may be recorded from a single pulse.

Some modern LiDAR systems are capable of operating at up to 100 kHz (100,000 pulses per second), resulting in millions of points being defined by three-dimensional geographical coordinates. The number of points is increased because there are often multiple returns for a

---

[1]RMSE, or *root mean squared error*, is a commonly used measure of precision and is calculated as the square root of the mean of the errors.

Field Data Acquisition

particular pulse. (See Fig. 17.3.) Thus, a large "point cloud" of data is produced by LiDAR systems, which can be problematic because it is not always obvious if a particular point was reflected off the ground or off some other above-ground feature. Therefore, creating layers by correctly classifying the points is a significant challenge in using LiDAR data. Various spatial analysis processes have been developed to help classify data points into layers, such as bare earth, vegetation, and structures, based on the trends of the points. Nevertheless, a certain amount of human editing and ground truthing remains necessary to map a bare-earth surface beneath vegetation. When mapping ground beneath vegetation, there may always be some uncertainty. For example, where there are certain sharp ground features hidden by tree cover, these may be misinterpreted as above-ground features.

Figure 17.3 *Multiple Returns from a Single LiDAR Pulse*

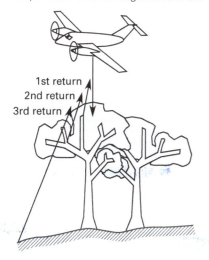

1st return
2nd return
3rd return

Unlike aerial photography, which is passive, LiDAR involves an "active" sensor that can collect data during the day, night, or even under cloud cover. LiDAR missions are often flown at night for better weather and air conditions, and lighter air traffic. However, while LiDAR data acquisition is less weather dependent than aerial photography, weather conditions must still be considered.

The most common use for LiDAR data is *bare-earth digital terrain models* (DTM), although it is also used to map various planimetric features. (See Fig. 17.4.)

Because LiDAR pulses detect above-ground features, prominent vegetation may also be mapped for quantitative analyses (e.g., tree counts, vegetation heights), or for volume analyses (e.g., to monitor agricultural crop yields). LiDAR systems also record an *intensity factor* (the amplitude of the returned wave) as an attribute of each return. The intensity factor can be used to delineate water bodies, as well as various cultural features, and to help distinguish between various classes of data points.

Hydrographic mapping, the mapping of underwater terrain, is a more recent and still-evolving application of airborne LiDAR. Hydrographic mapping is more expensive than topographic mapping and is limited to shallower, less turbid waters. Nevertheless, the use of LiDAR for this purpose can reduce operational costs and increase productivity considerably in comparison to sounding with hydrographic vessels. Results suggest that a precision level of about 0.2 ft may be achieved. Figure 17.5 shows a typical cross section of a spring basin that has been mapped through the use of airborne LiDAR.

Mapping submerged terrain requires a LiDAR system designed for *bathymetry,* the measurement of underwater depth. Most commercial LiDAR systems designed for land applications use a single laser with a wavelength between 800 nm and 1550 nm, and the pulses from such a system tend to be reflected from the water surface.[2] In contrast, a LiDAR system designed for bathymetry uses two lasers, one in the infrared spectrum with a typical wavelength of 1064 nm and another in the green spectrum with a typical wavelength of 532 nm. The infrared laser pulse is reflected by the water's surface, while the green laser pulse penetrates the water and is reflected by the submerged land surface. In this way the water's surface and depth can be measured at the same time.

Not only are two lasers needed instead of one, but the green laser pulse is weakened by absorption and scattering as it passes through the water, so a bathymetric LiDAR system needs considerably more energy than a system designed for land applications alone. However, the infrared LiDAR also works in the topographic mode, so both the submerged land and the bordering upland topography may be surveyed using the same system.

One important consideration with bathymetric LiDAR systems is *refraction.* Light travels more slowly in water than it does in air, and this causes light waves to change direction as they enter water (see Fig. 17.6). In calculating a bathymetric measurement, then, it is necessary to account for the angle of refraction by applying a correction to the azimuth from the aircraft to the submerged ground point.[3] In addition, a correction to the range is needed to account for the slower speed of light along the portion of its path that is in water.

---

[2]Nanometers, abbreviated nm, are billionths of a meter.

[3]The angle of refraction can be calculated with *Snell's law*, which states that the ratio of the sines of the angles of incidence and refraction (θ_1 and θ_2 in Fig. 17.6, respectively) is equivalent to the ratio of the velocities of the light in air and water (v_1 and v_2, respectively). This can be stated algebraically as $\sin \theta_1 / \sin \theta_2 = v_1/v_2$.

Figure 17.4 *Cross Section of LiDAR Data Showing Power Lines and Vegetation Layers*

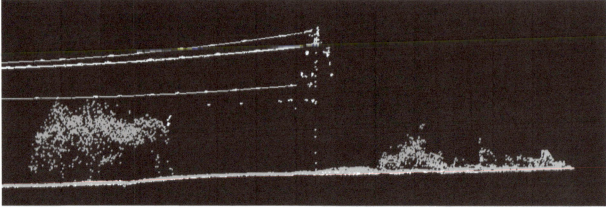

Image courtesy of Terra Remote Sensing.

Figure 17.5 *Typical Cross Section of a Spring Basin Using Airborne LiDAR*

Image courtesy of Suwannee River Water Management District.

Figure 17.6 *Refraction*

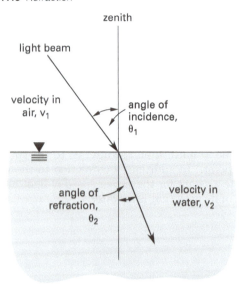

The primary challenge in using airborne LiDAR is in classifying the return pulses. In turbid water, some last return pulses will be reflected from particles suspended in the water column. To avoid mistaking these for pulses reflected from the bathymetric ground, various attributes of the return pulses, such as intensity, must be considered. Making frequent ground truth comparisons will also help with classification.

9. CONTROL FOR AERIAL MAPPING

Traditional photogrammetric processes that use stereo-plotters require *ground control points* to scale and orient each model. Control points are usually marked by aerial targets made of a material whose color contrasts with the background terrain. The optimum size for aerial targets depends on the photograph's scale. Three dimension coordinates are needed for the targets and may be established by conventional or GPS surveying.

In lieu of targets, photo-identification points that can be clearly and precisely identified on the photograph and ground may be used. For example, sharp cultural features such as the intersection of sidewalks and paved roads are ideal photo-identification points.

When mapping large areas involving multiple stereo models, it is not always necessary to establish several targets or photo-identification points on each model. Rather, fewer control points can be established and those points densified by aerotriangulation. *Aerotriangulation* is a process that establishes the geometric relationships between overlapping photographs to determine artificial horizontal and vertical control points. This substantially reduces the requirement for ground surveys and minimizes the cost of photogrammetric mapping. Sophisticated computer software is used to analytically transfer scale and orientation between successive stereo models, and to compute the precise orientation and position of each aerial photo.

Airborne GPS also significantly reduces the need for ground control points since the aircraft's precise position can be determined. However, for mapping accuracy, differential GPS corrections from a nearby GPS ground station must be applied. Therefore, a GPS base station with a known geodetic control point must be operated in the vicinity of the aerial survey. LiDAR surveys do not typically require additional ground control points to supplement the differentially corrected airborne GPS positions. However, for aerial photography, it is generally considered necessary for some ground control points, although the requirements are significantly reduced with the use of airborne GPS.

For any type of aerial mapping, a certain amount of ground verification of the mapping product is important for quality control. For traditional mapping products (hard copy maps), the *National Mapping Accuracy Standards* require that the maps be tested by independent ground surveys to evaluate the accuracy of both the horizontal positions of mapped features and the elevation contours. Those standards require that for maps with scales greater than 1:20,000, no more than 10% of the points tested can have a horizontal error greater than 1/30 inch (0.8 mm). For maps with scales less than 1:20,000, no more than 10% of the points tested can have a horizontal error greater than 1/50 inch (0.5 mm). Regarding vertical accuracy, those standards require that no more than 10% of the points tested can have a vertical error greater than one-half the contour interval.

For modern mapping products, which are primarily digital (see Sec. 17.10) and therefore are essentially files with x-, y-, and z-coordinates for each point, different standards for evaluating the mapping accuracy are necessary since these products have no scale. For these products, a statistical measure of accuracy, such as the RMSE is typically used. As an example, the U.S. Federal Emergency Management Agency (FEMA) specifications for LiDAR flood plain mapping require that a minimum of 20 test points for each major vegetation category (e.g., bare earth and low grass, high grass and crops, brush lands and low trees, fully covered by trees) be checked by ground surveying methods to verify the accuracy of the LiDAR elevation measurements. These specifications require that the RMSE for the total survey not exceed 15 cm.

Therefore, regardless of aerial mapping process used, some ground surveying is usually necessary. Control is required for the actual production of the mapping products and, in addition, independent survey points are necessary for quality control of the process.

10. AERIAL MAPPING PRODUCTS

A number of different types of mapping products can be created using aerial mapping. Traditionally, the products have been in two categories: prints of photographic images and hard copy maps. The maps were generally classified as one of two categories: *Planimetric maps* depict the horizontal position of ground features such as buildings, roads, water courses, and so on. *Topographic maps* have more expansive content and generally depict the ground relief using contours as well as various planimetric features.

The products of most aerial mapping projects are digital. Hard copy maps have generally been replaced by digital elevation models (DEM) or digital terrain models (DTM). These products consist of a file of x-, y-, and z-coordinates, with an identifying attribute for each point mapped.

There are significant advantages to the change from hard copy to digital mapping products. In addition to being generally more economical, digital versions may be plotted at any desired scale. In addition, other georeferenced information, such as boundary surveys, zoning maps, photographic images, and so on, may be easily merged with the mapped data. Therefore, aerial mapping has become an especially useful tool with today's technology.

11. RECENT TRENDS IN AERIAL MAPPING

The last few decades have seen major changes in aerial mapping processes, which have resulted in major changes to survey practice. One such change is the proliferation of digital photography, which has almost completely eliminated the use of film photography for mapping.

Another major change is the development of LiDAR mapping. With its ability to penetrate through vegetation to map the bare earth, LiDAR mapping has almost completely replaced aerial photography for large-scale topographic mapping purposes. (See Chap. 16 for more in-depth coverage of this technology.)

Unmanned aerial vehicles (UAV), also known as drones, have brought about major changes in aerial photography and remote sensing. Thanks to the relatively small cost of UAVs, a large percentage of aerial mapping can be done in-house by traditional surveying firms, as opposed to such services being contracted out to organizations specializing in aerial mapping. Further, UAVs allow relatively close-range aerial photography, precise topographic mapping, and the development of 3-D models of structures, which make it possible to use aerial mapping for surveying tasks traditionally performed using on-the-ground measurements. For example, using a UAV with proper control and procedures, measurements can be made of the spatial relationships between ground points where on-the-ground lines of sight are not available, and construction projects can be monitored at various stages of development.

The rapidly increasing use of these major developments in aerial mapping technology has had a significant impact on surveying practice. The use of these new processes is generally software-dependent; as a result, the knowledge required to properly use LiDAR mapping or a UAV, whether as a pilot or a surveyor creating a georeferenced base map, is heavily dependent on the software package being used. As a result, in addition to becoming skilled in the use of individual software packages that may vary considerably, surveyors also need knowledge of general requirements for the use of these technologies and how to assess the precision of the results.

Chapter 12 provides some guidance regarding accuracy standards for this purpose. UAV technology is regulated by the Federal Aviation Administration (FAA) regulations contained in the Codes of Federal Regulations (CFR), specifically 14 CFR part 107. These regulations require all UAVs weighing 0.55 lbf or more to be registered with the FAA and require all operators of UAVs for commercial purposes to pass an examination on aeronautical knowledge, undergo Transportation Security Administration (TSA) screening, and have a remote pilot's license. 14 CFR part 107 also restricts operation of a UAV to within the visual line of sight of the licensed remote pilot, and restricts operation to altitudes not exceeding 400 ft above ground level, assuming daylight hours with a visibility of 3 statute miles or greater.

Each new surveying technology comes with its own requirements for knowledge. Therefore, the knowledge requirements for the practice of surveying change with the advent of each revolution in technology for the profession.

12. PRACTICE PROBLEMS

1. An aerial photograph is taken from an altitude of 3000 ft using a camera with a focal length of 6 in. The photograph's scale is most nearly

(A) 1 in = 400 ft

(B) 1 in = 500 ft

(C) 1 in = 1200 ft

(D) 1 in = 4800 ft

2. What is most nearly the minimum negative scale needed to cover a standard public land survey section, with one photograph, using a camera with a 6 in focal length and 9 in by 9 in format film?

(A) 1:8000

(B) 1:10,000

(C) 1:12,000

(D) 1:14,000

3. Most nearly, what height above the mean terrain should a photograph be taken to achieve a negative

scale of 1:18,000 using a camera with a focal length of 6 in?

(A) 4000 ft

(B) 6000 ft

(C) 9000 ft

(D) 12,000 ft

4. A 300 ft tall tower is located at the southwestern corner of the southeast quarter of the northeast quarter of a standard public land survey section. Assuming that a vertical aerial photograph, taken with a 6 in focal length camera at an altitude of 9000 ft, is centered on the section, what is most nearly the relief displacement for the top of the tower?

(A) 0.03 in

(B) 0.1 in

(C) 0.3 in

(D) 0.7 in

5. If the relief displacement for the top of a tower located 1000 feet from the center of a vertical photograph was 0.1 in on the photograph and the flying height with a 6 in focal length camera was 6000 ft, what would be the approximate height of the tower?

(A) 345 ft

(B) 445 ft

(C) 545 ft

(D) 645 ft

6. If a vertical photograph is taken at a negative scale of 1:15,000, the expected measurement of the distance on the photograph between two adjacent public land survey section corners would most nearly be

(A) 0.42 in

(B) 2.2 in

(C) 3.3 in

(D) 4.2 in

7. Aerial photography taken from an unmanned aerial vehicle must be taken from at or below what height above ground level?

(A) 300 ft

(B) 400 ft

(C) 500 ft

(D) 500 ft, or 2000 ft w/NAVD

8. An unmanned aerial vehicle must be operated within what distance of the remote pilot controlling the flight?

   (A)   1000 ft

   (B)   1250 ft

   (C)   1250 ft or within the visual line of sight of the pilot, whichever is least

   (D)   within the visual line of sight of the pilot

SOLUTIONS

1. Using a focal length of 6 in (0.5 ft) and a flying height of 3000 ft with Eq. 17.1, the scale is

$$S = \frac{f}{H} = \frac{0.5 \text{ ft}}{3000 \text{ ft}} = \frac{1}{6000}$$

This may be restated as

$$1 \text{ in} = 6000 \text{ in} = \frac{6000 \text{ in}}{12 \frac{\text{in}}{\text{ft}}} = 500 \text{ ft}$$

The answer is (B).

2. Scale of a photograph may be expressed as the ratio of the dimension of an image on the photograph to the dimension of that object on the ground. Since a standard public land survey section is one mile (5280 ft) on each side, the scale may be expressed as

$$9 \text{ in} = (5280 \text{ ft})\left(12 \frac{\text{in}}{\text{ft}}\right) = 63{,}360 \text{ in}$$

This may be restated as 9 in = 63,360 in, or

$$1 \text{ in} = \frac{63{,}360 \text{ in}}{9 \text{ in}}$$
$$= 7040 \text{ in} \quad (1{:}8000)$$

The answer is (A).

3. Using the scale of 1:18,000 and a 6 in (0.5 ft) focal length with Eq. 17.2,

$$H = \frac{f}{S} = \frac{0.5 \text{ ft}}{\dfrac{1 \text{ in}}{18{,}000 \text{ in}}} = 9000 \text{ ft}$$

The answer is (C).

4. Assuming the section is of standard size, the distance from the center of the section to the tower is 1320 ft. The length of the tower's "shadow," D, on the ground may be calculated using Eq. 17.5.

$$D = \frac{hR}{H-h} = \frac{(300 \text{ ft})(1320 \text{ ft})}{9000 \text{ ft} - 300 \text{ ft}} = 45.52 \text{ ft}$$

Then, using Eq. 17.4, the relief displacement on the photograph may be calculated.

$$d = \frac{Df}{H} = \frac{(45.52 \text{ ft})(0.5 \text{ ft})}{9000 \text{ ft}} = 0.0025 \text{ ft} \quad (0.03 \text{ in})$$

The answer is (A).

5. Rearranging Eq. 17.4,

$$D = \frac{dH}{f} = \frac{\left(\dfrac{0.1\text{ in}}{12\ \frac{\text{in}}{\text{ft}}}\right)(6000\text{ ft})}{0.5\text{ ft}} = 100\text{ ft}$$

Rearranging Eq. 17.5,

$$h = \frac{DH}{R+D} = \frac{(100\text{ ft})(6000\text{ ft})}{1000\text{ ft} + 100\text{ ft}}$$
$$= 545.5\text{ ft}\quad(545\text{ ft})$$

The answer is (C).

6. The width of a standard public land section is 5280 ft. The photo scale of 1:15,000 may be restated as

$$1\text{ in} = 15{,}000\text{ in or }\frac{15{,}000\text{ in}}{12\ \frac{\text{in}}{\text{ft}}} = 1250\text{ ft}$$

Therefore, the photo distance is

$$\frac{5280\text{ ft}}{1250\ \frac{\text{ft}}{\text{in}}} = 4.22\text{ in}\quad(4.2\text{ in})$$

The answer is (D).

7. 14 CFR part 107 requires unmanned aerial vehicles operated for commercial purposes to be piloted by a licensed remote pilot at heights not exceeding 400 ft above ground level.

The answer is (B).

8. 14 CFR part 107 requires unmanned aerial vehicles to be operated within the visual line of sight of the licensed remote pilot.

The answer is (D).

18 Hydrographic Surveying

Nomenclature

d	depth	ft	m
DHQ	diurnal high water quality	ft	m
DLQ	diurnal low water quality	ft	m
HHW	mean higher high water for observation period	ft	m
HW	high water	ft	m
LLW	mean lower low water for observation period	ft	m
LW	low water	ft	m
MHHW	19-year mean higher high water	ft	m
MHW	19-year mean high water	ft	m
MLH	mean diurnal low water inequality	ft	m
MLLW	19-year mean low water	ft	m
MLW	19-year mean lower low water	ft	m
MR	19-year mean range	ft	m
MTL	19-year mean tide level	ft	m
R	mean range for observation period	ft	m
t	time	sec	s
v	velocity	ft/sec	m/s

Symbols

θ	angle	deg	deg

Subscripts

c	control station
s	subordinate station

1. INTRODUCTION

Hydrographic surveying measures and records the shape, location, and contour of underwater terrain. It is similar to topographic surveying, but more complex because the terrain being mapped is not directly viewable. Whereas in topographic surveying land elevation is recording height in relation to a fixed datum, hydrographic surveying measures depth below a fixed datum.

Bathymetry, the study of underwater depth, uses contour lines to map floor terrain. These maps use the water surface as a reference plane, but because the surface is constantly changing from tidal actions or meteorological conditions, specialized surveying equipment and practices must be used.

2. HYDROGRAPHIC EQUIPMENT

Water depth and water level can be measured using a wide range of equipment, from simple lead lines and graduated staffs to complex echo sounding systems and acoustical gauges. Equipment choices depend on the purpose of the survey, what equipment is available, the size of the project, the water depth, and a project's budget.

Water Depth Measurement

A *lead line* is the traditional means of measuring water depths. It consists of a lead weight attached to a graduated line. Although the line may have a soft exterior to allow ease of handling, a stranded wire center should be used to avoid problems due to stretching and shrinking. Waterproof, solid-braid cotton rope with a phosphor-bronze wire center is typically used. Lead lines are traditionally marked with a color-coded thread at even units and a standard color at odd units. They are often used in small projects when it would not be cost-effective to use more complex measuring equipment. Graduated staffs and lead lines are also used to confirm least depths over shoals or sunken rocks, to confirm echo sounding in grassy areas, and to determine echo sounding corrections.

In the early 1900s, *echo sounders* (*fathometers*) revolutionized hydrographic surveying and made surveying in deeper water practical. Echo sounders measure, or "sound," the time required for a sound wave to travel from its point of origin and back, and convert the measured time to depth. Echo sounders are typically designed with an electronic component that generates

Field Data Acquisition

an electrical pulse at a specified repetition rate. A *transducer* mounted on or beside the hull of the survey vessel converts the electrical pulse to acoustic energy and then reconverts the returning echo to an electrical pulse to measure the elapsed time. Echo sounders are designed with different acoustical pulse frequencies, durations, and widths depending on how and where they will be used.

Choosing a pulse frequency is often a compromise. *Lower frequencies* in the audible range (below 15 kHz) are used for sounding in deeper waters because low frequency pulses have a lower absorption rate and higher penetrating power. However, low-frequency pulses are less accurate at measuring shoal depths because the pulses carry the same frequency as the noise created by the sounding vessel as it moves through the water. *Ultrasonic frequencies* (20 kHz) are unaffected by noise interference and, therefore, allow more accurate measurements of shoal depths, more detailed profiles of irregular bottoms, and higher directivities of sound waves. However, their higher absorption rates make them ineffective in deep water. They also suffer greater attenuation in turbulent water and in areas with variable water temperature or density.

The *duration of the acoustical pulse* varies. Deep-water echo sounders use longer pulses of 0.001–0.04 sec to ensure the pulse reaches the floor. However, the accuracy of the pulse deteriorates at greater depths and provides a relatively poor definition of the floor. General-purpose echo sounders transmit shorter pulses of less than 0.0002 sec. These shorter pulses do not have sufficient power for reaching deeper waters, but they do provide good definition in shallow waters.

The *transducer beam width* of echo sounders generally varies between 2° and 50°, depending on the equipment's purpose. Beam widths are inversely proportional to the size of the transducer, so that the larger the transducer diameter, the narrower the beam. For the greatest accuracy in echo sounding, the beam should be extremely narrow. However, practical considerations also govern the selection of beam widths, since at some frequencies the transducer would need to be very large to produce a narrow beam signal.

Considering the variables of frequency, duration of acoustical pulse, and transducer beam width, echo sounders are generally produced in three classes: deep-water sounders with low frequencies (less than 15 kHz) and wide beam widths (as much as 50°); medium depth sounders designed to operate in depths less than 200 ft (600 m), with medium frequencies (15–20 kHz) and relatively narrow beam widths; and shallow depth sounders designed to operate in waters less than 30 ft (100 m), with high frequencies (above 20 kHz), short wavelengths, and narrow beam widths for detailed bottom definition.

Two modern innovations in echo sounders are used with more advanced systems: the *multiple-beam fathometer* and *side-scan sonar*. The multiple-beam fathometer is equipped with an array of transducers oriented to varying angles off the vertical axis. This type of system allows for coverage of a wide swath, and is used when 100% coverage of the bottom is required. The side-scan sonar uses a beam oriented slightly below the horizontal axis and can locate underwater features protruding above the floor better than vertically oriented echo sounders, although the accuracy of the depth of objects mapped with side-scan sonar is typically less that that measured by fathometers.

Water Level Measurement

Water level is measured using a wide range of equipment. For short-term observations, a *graduated staff* is often used. It requires little equipment and a minimum amount of installation: it can either be driven into the water floor or secured to a piling. Manually read staffs are also more practical for areas like inter-tidal marshlands where it is desirable to determine a tidal datum, as other gauges (such as float operated gauges) experience problems in areas that are dry for a portion of the tidal cycle.

Most tidal observations lasting longer than a few hours are made using gauges that record the water level either continuously or at fixed intervals. The simplest of these are *float-type gauges*, which are operated by a float that moves in a stilling well with the rise and fall of the water level.

A *stilling well* is a vertical tube that extends below the lowest possible tide. Water enters the well through an intake opening near the bottom of the well and rises to the average level of the water surface outside the well. The size of the opening controls the amount of damping that takes place, and must be large enough to allow sufficient flow of water for tidal measurements, while still dampening the rapid water level fluctuation caused by heavy seas. Stilling wells should always be used with float-type gauges so that short period waves (such as wind waves) are dampened, and only the longer-period tidal waves are allowed to move the float. While early float-type gauges used the vertical action of the water to draw a graph of the water level versus time, more recent gauges convert the motion of the float to digital signals that are recorded by some other device. Because float-type gauges cannot be directly leveled to a benchmark or ground elevation, a tidal staff is usually installed near the gauge. Comparisons are made between the gauge and the staff to determine the constant difference between the two readings.

Acoustic gauges also measure water level. They use sensors that measure the water height by timing sound (acoustic) waves or radar beams that bounce off the water surface. There are a variety of output devices. More modern acoustic gauges have a leveling point on the transducer to level against.

A recent trend in tidal measurement has equipped gauges with *telemetry devices* to allow near-real time monitoring of tides in remote locations. These telemetry devices collect information at a remote location and transmit it to another location. The National Ocean Service (NOS) has equipped most of its permanent tidal stations with telemetry devices that relay the data to their headquarters via satellite.

3. SOUNDING PATTERNS

With most bathymetric surveys, it is impractical to develop 100% coverage of the floor within the survey area. It is also typically impossible (unlike with topographic surveys) to see significant breaks in the terrain that should be mapped for an accurate floor representation. Therefore, a systematic sampling pattern must be used to produce a representative floor model. As a result, bathymetric or depth-measurement surveys are generally made along a series of pre-planned sounding lines. A series of evenly spaced parallel lines is generally considered optimal for a systematic delineation of hydrographic features. The spacing of the sounding lines depends on a number of factors, including the purpose of the survey, the depth of the water, the topographic configuration of the bottom, and the beam width of the fathometer being used.

For best definition, the sounding lines should run perpendicular to the depth contours, although for steep features such as ridges or trenches, the two lines should cross at 45°. In general, cross lines should supplement the sounding lines at angles of 45° to 90°. When floor topography has significant features, such as shoal areas or pinnacles, an additional pattern of closely spaced lines, usually parallel to the axis of the feature being mapped, should be run to adequately develop the feature.

4. CORRECTIONS TO DEPTH MEASUREMENTS

It is necessary to make a number of corrections when using echo sounders so that the soundings reflect the actual depth relative to the selected reference datum. Figure 18.1 illustrates such corrections.

Velocity of Sound Correction

Echo sounders measure depth based on the time a sound pulse takes to travel to the bottom and return. The measured depth, d, is calculated using Eq. 18.1, where v is the velocity of sound in water, and t is the time for the sound pulse to travel to the floor and back.

$$d = \tfrac{1}{2}\,\mathrm{v}t \qquad\qquad 18.1$$

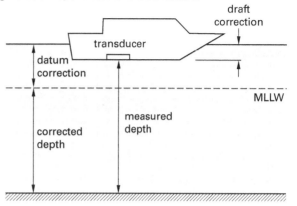

Figure 18.1 *Depth Measurement Corrections*

To perform measurements, most echo sounders are calibrated to a velocity of 800 fathoms per second (4800 ft/ sec (1440 m/s)), which is the approximate average velocity of sound in the sea. However, the actual velocity varies due to the water's differing salinities and temperatures. Therefore, except for surveys in very shallow waters, a velocity correction should be used.

There are two commonly used methods for determining *velocity corrections*. The first method directly compares readings at various depths using a horizontally suspended bar below the vessel that is raised and lowered to various depths. (A less desirable version of this method is to vertically cast a lead line and compare the readings.) The second method calculates the velocity of sound at various depths using temperature and salinity measurements taken from oceanographic sensors. Some echo sounding systems detect the actual velocity of sound and automatically correct the soundings, making these two correction methods unnecessary.

Draft Corrections

Echo sounders measure the depth below the transducer, not the surface of the water; therefore, corrections must be made for the difference. *Dynamic drafts* correct soundings for the height difference between the surface of the water and the face of the transducer. This correction is the algebraic sum of three corrections: static draft, settlement, and squat. *Static draft* is the height difference between the water level and the transducer when the sounding vessel is stationary. Static draft varies with the load aboard the vessel, so this correction needs to be determined for various loads. *Settlement* is the general difference between the elevations of the stationary and moving vessel, while *squat* is the change in trim in the moving vessel—generally, it is a lowering of the stern and a rise of the bow. All factors vary with the speed of the vessel, so they should be determined for various speeds.

Instrument Corrections

Although generally not present in soundings obtained by digital methods, soundings scaled from graphic recorders may require corrections to compensate for

Field Data Acquisition

various errors associated with the depth recorder. Typically, errors occur when setting the index (initial pulse) or with other adjustments to the recorder. Generally, errors can be avoided by properly operating the recorder.

Datum of Reference Correction

Since the depth of water at a specific location in a tidally affected water body can vary with time due to the tides and meteorological conditions, depths measured in hydrographic surveys must be adjusted to some common plane of reference. Techniques and datums used for this purpose are discussed in Sec. 18.6.

5. HORIZONTAL CONTROL OF HYDROGRAPHIC MEASUREMENTS

The horizontal location of the soundings is as important as the accuracy of the depth measurements. Before GPS technology was available, determining the horizontal location was often problematic, since soundings are generally taken from a moving vessel. GPS technology largely overcame this problem using real-time, differentially corrected code observations to provide submeter precision. *Submeter precision* requires a base station at a known point to receive a correction signal. In the United States, correction signals from continuously operating control stations are commonly used. Where such signals are not available, a second GPS receiver may be set on a nearby known point and its observations used for real-time correction.

For small hydrographic projects, it can be more efficient to provide horizontal control using conventional surveying techniques, such as theodolite intersection. *Theodolite intersection* determines the position of the sounding vessel by simultaneously turning angles to the vessel from two known positions on the shore. The resulting position can then be calculated using the bearing-bearing method.

Regardless of the method used for horizontal positioning, time must be associated with both the soundings and the horizontal positions for coordinating between the two. It is additionally important that the sounding vessel's point be located directly over the echo sounder transducer. For example, when using GPS positioning, the GPS antenna should be located directly over the transducer or at a known azimuth and distance offset.

6. VERTICAL CONTROL OF HYDROGRAPHIC MEASUREMENTS

Depth measurements are traditionally referenced from the floor to the surface of the water. Since the surface can vary in height depending on the tide, the surface elevation must be monitored during the course of a hydrographic survey to correct soundings to a common, stable plane of reference. Traditionally, hydrographic soundings in tidally affected waters are reduced to a low-water tidal datum. However, for engineering applications and in non-tidal waters, the soundings can be compared to a geodetic datum. Regardless of the datum used, the basic procedure for correcting soundings involves monitoring the water level during the course of the survey. The monitoring information, combined with the time the depth measurements were taken, can be used to correct the soundings.

A fairly recent process utilizing GPS provides vertical control for hydrographic measurements that bypasses the need to separately measure the water level. GPS provides both horizontal and vertical positions in reference to the ellipsoid for a specific area. If the relationship between the desired vertical reference datum and the ellipsoid is known, GPS can provide the relationship to that desired datum. Such a relationship between the desired vertical datum and the ellipsoid can be determined by static GPS observations at a benchmark, such as a tidal benchmark, where the elevation in reference to the desired vertical datum is known. The resulting difference, combined with the height difference between the fathometer transducer and the phase center of the GPS antenna, can then be used to provide corrections between the observed soundings and the desired vertical datum. One significant advantage of using GPS observations for the vertical datum correction is that it also accounts for the dynamic draft correction, since the GPS measures the instantaneous position of the sounding vessel.

7. TIDAL DATUM CALCULATIONS

Hydrographic soundings are traditionally reduced to a low-water tidal datum, since mariners prefer that hydrographic charts show minimum depths. Therefore, a low-water datum is generally used when a tidal datum is used as a reference. Unfortunately, there is a lack of international uniformity in picking a low-water datum. In the United States, the national charting agency uses *mean lower low water* (MLLW) for hydrographic surveys along all coasts. Other nations use a datum closer to the *lowest possible low water*.

Correct application of corrections to a tidal datum requires some understanding of tidal datum theory. A visible tidal cycle is a composite of numerous constituent cycles of various periods that reflect the relationships between the earth, moon, and sun. The longest of these constituent cycles, that associated with the regression of the moon's nodes, has a period of 18.6 years.[1] To develop a statistically significant tidal value, a tidal datum should include all of the periodic variations in tidal height. Therefore, a tidal datum is usually

---

[1]Regression of the moon's nodes refers to the movement of the intersection of the moon's orbital plane and the plane of the earth's equator, which completes a 360° circuit in 18.6 years.

considered the average of all occurrences of a certain tidal extreme for a period of 19 years. Such a period is called a *tidal epoch*. As shown in Fig. 18.2, *mean high water* (MIIW) is an cxample of tidal datum planes, and is the average height of all the high waters occurring over a period of 19 years.[2] Likewise, *mean low water* (MLW) is defined as the average of all of the low tides over a 19-year tidal epoch.

Figure 18.2 *Tidal Datums*

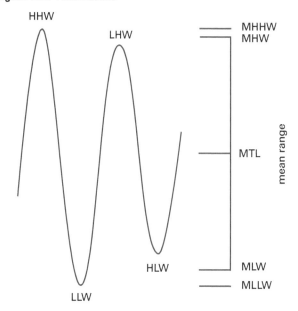

average tidal cycle

Mean tide level (MTL) or *half-tide level* is the plane located halfway between mean high and mean low water, and is used for datum computation purposes. Mean tide level should not be confused with *mean sea level* (MSL), which is the average level of the sea as measured from hourly heights over a tidal epoch. The relationship between mean sea level and mean tide level varies from location to location, depending on the phase and amplitude relationships of the location's various tidal constituents.

Also significant to hydrographic surveying are the mean higher high water and the mean lower low water datum planes. *Mean higher high water* (MHHW) is the average of the higher of the high tides occurring each day. *Mean lower low water* (MLLW) is the average of the lower of the low tides occurring each day. Both of these averages are calculated over a tidal epoch. The height difference between the mean higher high water and mean high water is called *diurnal high water inequality* (DHQ). The difference between mean lower low water and mean low water is called *diurnal low water inequality* (DLQ).

The primary determination of a tidal datum involves the relatively simple determination of the arithmetic mean, or average, of all the occurrences of a certain tidal extreme over a 19-year tidal epoch. In practice, this is usually accomplished by computing the mean values of the various tidal extremes for each calendar month, and then computing the annual mean values by averaging the 12 monthly means for each extreme for each calendar year. Then, the mean values for the tidal epoch are determined by averaging the annual mean values for the 19 years composing the epoch.

Traditionally, all tidal datum values published in the United States are referred to a specific epoch called the *National Tidal Datum Epoch*. A specific 19-year period is used, since apparently non-periodic variation in mean sea level is noted from one 19-year period to the next. A new epoch has historically been adopted every two or three decades when a significant change has occurred. At such times, adjustments are made to all datum elevations. In effect, a quantum jump occurs in the elevations of all tidal datum planes for stations published by the National Ocean Service at those times. The most current epoch of 1983–2001 was adopted in 2003. Previously, the epoch of 1960–1978 was used.

The Range Ratio Method

There are a number of different topographic factors shaping incoming tidal waves that result in significant elevation differences from point to point in a tidal datum—even for points in the same general vicinity. Therefore, a tidal datum must be determined in the immediate area of its intended use. Since it is impractical to observe tides for a full 19-year tidal epoch at every location where a tidal datum is needed, most tidal datum elevations are determined from observations of less than 19 years. The average of the tidal extremes observed over a shorter period can be reduced to a value equivalent to a 19-year mean by a correlation process using a ratio of tidal ranges observed at the station and at a control station with a known 19-year value. This correlation process is known as the range ratio method and is based on the following equations.

Equation 18.2 is used to find the mean range for an observation period, and Eq. 18.3 is used to find the mean tide level for that period.

$$R = \text{HW} - \text{LW} \qquad 18.2$$

$$\text{TL} = \frac{\text{HW} + \text{LW}}{2} \qquad 18.3$$

Equation 18.4 calculates the equivalent 19-year mean range at the subordinate station.

$$\text{MR}_s = \frac{(\text{MR}_c)R_s}{R_c} \qquad 18.4$$

---

[2]18.6 years rounded to the nearest whole year to include a multiple of the annual cycle associated with the declination of the sun.

Field Data
Acquisition

Equation 18.5 calculates the equivalent 19-year mean tide level at the subordinate station.

$$MTL_s = MTL_c + TL_s - TL_c \qquad 18.5$$

The equivalent 19-year mean high water at the subordinate station can be determined by adding one-half of the 19-year mean range to the 19-year mean tide level.

$$MHW_s = MTL_s + \frac{MR_s}{2} \qquad 18.6$$

Likewise, the equivalent 19-year mean low water at the subordinate station can be determined by subtracting one-half of the 19-year mean range from the 19-year mean tide level.

$$MLW_s = MTL_s - \frac{MR_s}{2} \qquad 18.7$$

The values for 19-year mean higher high water and lower low water can be determined using Eq. 18.8 through Eq. 18.10.

$$MHHW_s = MHW_s$$
$$+ \frac{(MHHW_c - MHW_c) \times (HHW_s - HW_s)}{HHW_c - HW_c} \qquad 18.8$$

$$MLLW_s = MLW_s$$
$$- \frac{(MLW_c - MLLW_c) \times (LW_s - LLW_s)}{LW_c - LLW_c} \qquad 18.9$$

$$MLLW_c = MLW_c - DLQ_c \qquad 18.10$$

Example 18.1

Tidal observations were made for one month at a survey site, resulting in the following monthly mean values from a gauge datum.

$$\text{high water (HW)} = 6.21 \text{ ft}$$
$$\text{low water (LW)} = 2.62 \text{ ft}$$

Simultaneous observations were made at a nearby control tidal station that reported the following 19-year mean values from a gauge datum.

$$\text{mean tidal level (MTL)} = 4.55 \text{ ft}$$
$$\text{mean tidal range (MR)} = 3.40 \text{ ft}$$

The monthly mean values on the gauge datum at the control site were

$$\text{high water (HW)} = 5.20 \text{ ft}$$
$$\text{low water (LW)} = 2.00 \text{ ft}$$

What is the calculated value of the survey site's mean high water (MHW)?

Solution

Using Eq. 18.2, the observed subordinate range, R_s, is

$$R_s = HW_s - LW_s = 6.21 \text{ ft} - 2.62 \text{ ft} = 3.59 \text{ ft}$$

The control range, R_c, is

$$R_c = HW_c - LW_c = 5.20 \text{ ft} - 2.00 \text{ ft} = 3.20 \text{ ft}$$

Using Eq. 18.3, the observed tide level, TL_s, is

$$TL_s = \frac{HW_s + LW_s}{2} = \frac{6.21 \text{ ft} + 2.62 \text{ ft}}{2} = 4.415 \text{ ft}$$

The control observed tide level, TL_c, is

$$TL_c = \frac{HW_c + LW_c}{2} = \frac{5.20 \text{ ft} + 2.00 \text{ ft}}{2} = 3.60 \text{ ft}$$

Using Eq. 18.4, the subordinate station mean range is

$$MR_s = \frac{(MR_c)R_s}{R_c} = \frac{(3.40 \text{ ft})(3.59 \text{ ft})}{3.20 \text{ ft}} = 3.81 \text{ ft}$$

Using Eq. 18.5, the subordinate mean tide level is

$$MTL_s = MTL_c + TL_s - TL_c$$
$$= 4.55 \text{ ft} + 4.415 \text{ ft} - 3.60 \text{ ft}$$
$$= 5.365 \text{ ft}$$

Therefore, using Eq. 18.6, the survey site's mean high water level is

$$MHW_s = MTL_s + \frac{MR_s}{2} = 5.365 \text{ ft} + \frac{3.81 \text{ ft}}{2} = 7.27 \text{ ft}$$

Datum Calculations

The process for datum calculations uses a strong network of control tidal stations that have been established in many of the coastal areas of the United States, as well as in a number of other countries. Typically, these networks have base primary stations that have been operated for a full 19-year tidal epoch. Such networks are then typically densified with a larger number of secondary and tertiary stations, where observations are conducted for a shorter period simultaneously with the observations at the primary stations. The data at such densified stations are corrected to their equivalent 19-year mean values using the previously described range ratio method. At each station, a series of benchmarks are established to perpetuate the resulting data; that is, the elevations are preserved by setting benchmarks and leveling to them. Then, even if the gauge is removed, the elevations are preserved.

When a hydrographic survey uses a tidal datum as its reference datum, any existing local tidal stations must first be located. In the United States, the National Ocean Service is the national repository for such data. If the survey is in the immediate area of an existing control station, a water level staff or gauge may be installed at the station and leveled to the benchmarks for that station. Water level observations made during the survey are used to reduce the soundings to the reference data.

If the survey is not in the immediate area of an existing tidal station, a new tidal datum must be established by installing a gauge both at the most appropriate existing control station and in the survey area. Simultaneous observations taken at the gauges allow the equivalent 19-year mean values to be calculated for the survey area. The observation period necessary for a valid determination will vary depending on the distance between gauges. It is typically recommended that such simultaneous observations be conducted for a minimum of one month to capture the dominant 28-day lunar cycle. However, if the control station is within a few miles and in the same hydrographic regime, as few as three repeated tidal cycles may be sufficient. For greater distances, at least one year of observations may be necessary.

Tidal datums may also serve as a basis for mapping the limits of the water body being surveyed. Most jurisdictions use tidal datum lines to delineate the boundary between the publicly owned submerged lands and the bordering uplands, which are subject to private ownership. Additionally, tidal datum lines are used as a basis for offshore maritime boundaries, making them an essential part of hydrographic surveys.

8. AIRBORNE LiDAR BATHYMETRY

Airborne LiDAR bathymetry is a newly evolving technology for hydrographic mapping of relatively shallow waters.[3] In waters where suitable, this technology can provide a precise, high-density mapping of the bathymetric ground, as well as topography of the surrounding uplands, far more rapidly and with more density than with sounding from hydrographic vessels.

Most commercial LiDAR systems for land applications use lasers with wavelengths between 800 and 1550 nm.[4] The pulses from such systems tend to be reflected from the water surface. In contrast, LiDAR systems designed for bathymetry use two systems, one in the infrared spectrum with a typical wavelength of 1064 nm, and another in the green spectrum with a typical wavelength of 532 nm. The infrared wavelength signal is reflected by the water surface and determines the distance to the water surface. The green wavelength signal penetrates the water and is reflected from the submerged land surface. Since longer-wavelength LiDAR may also work in the topographic mode, both the submerged land and bordering upland topography may be surveyed using the same system. Figure 18.3 shows a typical cross section of a LiDAR survey.

Figure 18.3 *Typical Cross Section of Bathymetric LiDAR Survey of a Spring Basin*

Image courtesy of Suwannee River Water Management District.

In areas where it is suitable, this technology offers excellent resolution and eliminates the need for all corrections associated with sounding from surface vessels except adjustments to the datum being used. With LiDAR bathymetry, each resulting bathymetric ground point has both horizontal and vertical coordinates referenced to the ellipsoid. Therefore, as with GPS-controlled soundings from hydrographic vessels, the relationship between the desired vertical datum of reference and the ellipsoid must be known for the specific area being surveyed. This may be the relationship to a tidal datum if one such as mean lower low water is used, or to the geoid if orthometric elevations are used. If a tidal datum is used for the underwater surface and orthometric elevations are used for the bordering upland, differing corrections would be used.

One important consideration with bathymetric LiDAR systems is refraction. Light waves bend when they enter water due to the difference in the speed of light in air and water (Fig. 18.4). Therefore, to calculate a precise position for a bathymetric measurement, it is necessary to apply a correction to the azimuth from the aircraft to

---

[3] Systems with greater power are being developed that may allow bathymetric measurements of deeper waters. In addition, since a limiting factor is reflection from particles suspended in the water column, the ongoing development of processes for improved filtering and detection of valid bathymetric ground returns may also increase the ability of such systems for bathymetry.
[4] Nanometers (billionths of a meter)

the submerged ground point to account for the angle of refraction based on Snell's law.[5] In addition, a correction to the range is necessary due to the different speeds of light between the portions of the path in air and in water. Since the green laser pulse is subject to absorption and scattering as it passes through the water, bathymetric LiDAR systems also need higher energy than systems for land applications.

Figure 18.4 *Speed of Light in Air and Water*

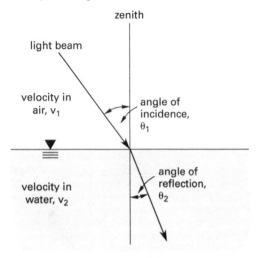

The primary challenge in using airborne LiDAR is in the classification of the return pulses. Due to turbidity in the water, some last return pulses may result from suspended particles in the water column, as opposed to the bathymetric ground, and may therefore be mistaken for ground points. As a result, various attributes of the return pulses, such as intensity values, should be considered to identify which return pulses are actually from the bathymetric ground. In addition, frequent ground truth comparisons should be used to help in the classification process. Results to date suggest that a precision level of about 0.2 ft, more or less, may currently be achieved by this technology.

9. PRACTICE PROBLEMS

1. The Marianas Trench has a maximum depth of 35,800 ft. Assuming the standard velocity of sound in sea water, most nearly how long would a fathometer sound pulse take to travel to the floor and back?

(A) 3.5 sec

(B) 7.5 sec

(C) 15 sec

(D) 16 sec

2. Bar checks determine that the velocity of sound at a survey site is 4600 ft/sec and the draft correction for the sounding vessel is 2.5 ft. The uncorrected sounding is 59.8 ft. Accounting for sound velocity and draft, what is most nearly the corrected sounding?

(A) 57.3 ft

(B) 58.6 ft

(C) 59.8 ft

(D) 61.2 ft

3. Table A provides depth measurements for a channel cross section. Table B provides tidal height measurements for the same day at a tide station near the channel. What is most nearly the measured depth above the mean lower low water at 100 ft distance?

Table A: depth measurements

distance (ft)	time	depth (ft)
0	4:00	0.5
20	5:00	2.0
40	6:00	10.0
60	7:00	20.0
80	8:00	50.0
100	9:00	50.0
120	10:00	51.0
140	11:00	35.0
160	12:00	15.0
180	13:00	5.0
200	14:00	5.0

---

[5] Snell's law states that the ratio of the sines of the angles of incidence (θ_1) and refraction (θ_2) is equivalent to the ratio of the velocity of the light in the air (v_1) and water (v_2), which may be stated algebraically as

$$\frac{\sin\theta_1}{\sin\theta_2} = \frac{v_1}{v_2}$$

Table B: tidal heights

time	height with MLLW (ft)
4:00	2.80
5:00	5.29
6:00	7.38
7:00	8.82
8:00	9.57
9:00	9.17
10:00	7.75
11:00	5.94
12:00	4.28
13:00	2.84
14:00	1.68

(A)   9.2 ft

(B)   40.8 ft

(C)   50.0 ft

(D)   59.2 ft

Use the following information for Prob. 4 through Prob. 6.

Tidal observations were made for one month at a survey site, resulting in the following mean values from the gauge datum.

higher high water (HHW)	7.45 ft
high water (HW)	7.11 ft
low water (LW)	1.72 ft
lower low water (LLW)	1.55 ft

Simultaneous observations were made at a nearby control tide station with the following known mean values.

mean tide level (MTL)	4.75 ft
mean tidal range (MR)	5.96 ft
mean diurnal high water inequality (DHQ)	0.35 ft
mean diurnal low water inequality (DLQ)	0.19 ft

The monthly mean values on the gauge datum at the control site were

higher high water (HHW)	8.26 ft
high water (HW)	7.91 ft
low water (LW)	1.89 ft
lower low water (LLW)	1.70 ft

4. What is most nearly the mean high water (MHW) at the survey site?

(A)   3.26 ft

(B)   4.26 ft

(C)   5.34 ft

(D)   6.93 ft

5. What is most nearly the mean low water (MLW) at the survey site?

(A)   1.43 ft

(B)   1.55 ft

(C)   1.60 ft

(D)   1.68 ft

6. What is most nearly the mean lower low water (MLLW) at the survey site?

(A)   1.33 ft

(B)   1.43 ft

(C)   1.58 ft

(D)   1.77 ft

Field Data
Acquisition

SOLUTIONS

1. Rearrange Eq. 18.1 to find how long it would take a fathometer sound pulse to travel to the ocean floor and back.

$$d = \frac{1}{2} \, \text{v}t$$

$$t = \frac{2d}{\text{v}}$$

$$= \frac{(2)(35{,}800 \text{ ft})}{4800 \, \dfrac{\text{ft}}{\text{sec}}}$$

$$= 14.9 \text{ sec} \quad (15 \text{ sec})$$

The answer is (C).

2. With the standard velocity of sound, 4800 ft/sec, used by the fathometer, and a recorded depth of 59.8 ft, the time it takes to measure the depth may be calculated by rearranging Eq. 18.1 as follows.

$$d = \frac{1}{2} \, \text{v}t$$

$$t = \frac{2d}{\text{v}}$$

$$= \frac{(2)(59.8 \text{ ft})}{4800 \, \dfrac{\text{ft}}{\text{sec}}}$$

$$= 0.0249 \text{ sec}$$

With that time and the actual velocity of sound, 4600 ft/sec, the actual depth below the transducer may then be calculated using Eq. 18.1.

$$d = \frac{1}{2}\text{v}t = \left(\frac{1}{2}\right)\left(4600 \, \frac{\text{ft}}{\text{sec}}\right)(0.0249 \text{ sec}) = 57.3 \text{ ft}$$

Depth from the water surface may then be calculated by adding the draft correction.

$$\begin{aligned}
\text{corrected depth} &= \text{depth below the tranducer} \\
&\quad + \text{draft correction} \\
&= 57.3 \text{ ft} + 2.5 \text{ ft} \\
&= 59.8 \text{ ft}
\end{aligned}$$

The answer is (C).

3. At the sounding made at 100 ft, the uncorrected depth is 50.0 ft. At the time of that sounding (9:00), the tidal height above MLLW is 9.17 ft. Therefore, the corrected depth is

$$50 \text{ ft} - 9.17 \text{ ft} = 40.83 \text{ ft} \quad (40.8 \text{ ft})$$

The answer is (B).

4. Using Eq. 18.2, the observed subordinate range, R_s, is

$$\begin{aligned}
R_s &= \text{HW}_s - \text{LW}_s \\
&\quad - 7.11 \text{ ft} - 1.72 \text{ ft} \\
&= 5.39 \text{ ft}
\end{aligned}$$

The observed control range, R_c, is

$$\begin{aligned}
R_c &= \text{HW}_c - \text{LW}_c \\
&= 7.91 \text{ ft} - 1.89 \text{ ft} \\
&= 6.02 \text{ ft}
\end{aligned}$$

From Eq. 18.3, the observed subordinate tide level, TL_s, is

$$\begin{aligned}
\text{TL}_s &= \frac{\text{HW}_s + \text{LW}_s}{2} \\
&= \frac{7.11 \text{ ft} + 1.72 \text{ ft}}{2} \\
&= 4.415 \text{ ft}
\end{aligned}$$

The observed control tide level, TL_c, is

$$\begin{aligned}
\text{TL}_c &= \frac{\text{HW}_c + \text{LW}_c}{2} \\
&= \frac{7.91 \text{ ft} + 1.89 \text{ ft}}{2} \\
&= 4.90 \text{ ft}
\end{aligned}$$

Using Eq. 18.4, the subordinate station mean range is

$$\begin{aligned}
\text{MR}_s &= \frac{(\text{MR}_c) R_s}{R_c} \\
&= \frac{(5.96 \text{ ft})(5.39 \text{ ft})}{6.02 \text{ ft}} \\
&= 5.336 \text{ ft}
\end{aligned}$$

Using Eq. 18.5, the subordinate mean tide level is

$$\begin{aligned}
\text{MTL}_s &= \text{MTL}_c + \text{TL}_s - \text{TL}_c \\
&= 4.75 \text{ ft} + 4.415 \text{ ft} - 4.90 \text{ ft} \\
&= 4.265 \text{ ft}
\end{aligned}$$

Using Eq. 18.6, the subordinate mean high water level is

$$\begin{aligned}
\text{MHW}_s &= \text{MTL}_s + \frac{\text{MR}_s}{2} \\
&= 4.265 \text{ ft} + \frac{5.336 \text{ ft}}{2} \\
&= 6.993 \text{ ft} \quad (6.93 \text{ ft})
\end{aligned}$$

The answer is (D).

Field Data
Acquisition

5. Using Eq. 18.7, the subordinate mean low water is

$$MLW_s = MTL_s - \frac{MR_s}{2}$$
$$= 4.265 \text{ ft} - \frac{5.336 \text{ ft}}{2}$$
$$= 1.597 \text{ ft} \quad (1.60 \text{ ft})$$

The answer is (C).

6. Using Eq. 18.7, mean low water at the control station is

$$MLW_c = MTL_c - \frac{MR_c}{2}$$
$$= 4.75 \text{ ft} - \frac{5.96 \text{ ft}}{2}$$
$$= 1.77 \text{ ft}$$

From Eq. 18.10, the control mean lower low water is

$$MLLW_c = MLW_c - DLQ_c = 1.77 \text{ ft} - 0.19 \text{ ft} = 1.58 \text{ ft}$$

Using Eq. 18.10 and the subordinate mean low water from Prob. 5, the subordinate mean lower low water is

$$MLLW_s = MLW_s - \frac{(MLW_c - MLLW_c)(LW_s - LLW_s)}{LW_c - LLW_c}$$
$$= 1.597 \text{ ft} - \frac{(1.77 \text{ ft} - 1.58 \text{ ft})(1.72 \text{ ft} - 1.55 \text{ ft})}{(1.89 \text{ ft} - 1.70 \text{ ft})}$$
$$= 1.427 \text{ ft} \quad (1.43 \text{ ft})$$

The answer is (B).

Topic III: Plane Survey Calculations

Plane Survey
Calculations

19 Traverses

Nomenclature

A	azimuth	deg	deg
d	distance	ft	m
E	departure	ft	m
N	latitude	ft	m

Symbols

θ	angle	deg	deg

Subscripts

t	total

1. TRAVERSES

A *traverse* is a series of straight lines connecting successive instrument stations in a survey. The relative position of each station in the traverse is determined by measuring line lengths and angles between consecutive lines. Typically, *total stations* are used for angular and distance measurements. For traversing, most angles are measured using total stations in a clockwise direction. Either interior angles or exterior angles may be measured.[1]

Traversing is a quick and convenient method for establishing horizontal control in an area. It is particularly useful for small projects and in locations where GPS is not effective, such as areas with overhead obstructions like buildings or trees. Typical uses of traversing include surveying property boundaries, locating and laying out construction projects, and establishing control points for aerial mapping or laser scanning projects. Traverses are usually classified as being either open or closed.

2. TYPES OF TRAVERSES

A *closed traverse* starts and ends at the same point (see Fig. 19.1). Because a closed traverse is a closed polygon, its interior angles can be checked for accuracy and mathematically adjusted, and its distances can be checked for blunders and random errors. However, even a closed traverse will not detect systematic errors related to distance, such as EDM scale errors, so precautions must be taken by other means to eliminate such errors.

An *open traverse* begins at one point and ends at a different point. (See Fig. 19.1.) Because an open traverse is not a closed figure, no mathematical check can be made on its angles and distances. Procedures such as redundant measurements can eliminate blunders, but even when such procedures are used, the lack of closure makes the measurements of an open traverse suspect. Therefore, open traverses should only be used when absolutely necessary.

3. CONTROL POINTS

A traverse only indicates relative positions unless it includes a *baseline* (a line with a known direction) and *control points* (locations with known coordinates). Control points also allow systematic errors to be detected for a traverse and allow closure checks to be made for open traverses. Figure 19.2 shows the open and closed traverses from Fig. 19.1 with control points.

4. TRAVERSE CALCULATIONS

Traverse calculations are derived from basic trigonometric and algebraic principles related to right triangles.

Forward Calculation

The forward calculation of a traverse involves three steps to determine coordinates for each point in the traverse.

---

[1]Historically, compasses were used to measure line direction.

Figure 19.1 *Open and Closed Traverses*

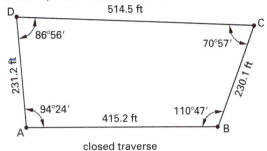

closed traverse

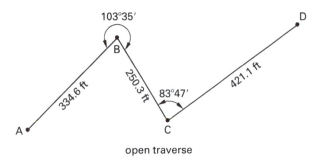

open traverse

Figure 19.2 *Traverses Including Control Points*

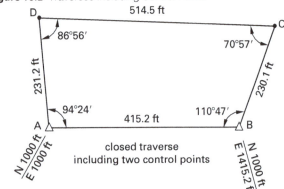

closed traverse
including two control points

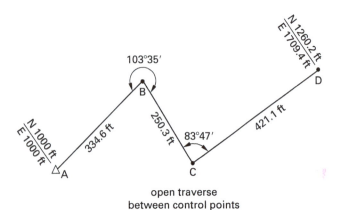

open traverse
between control points

step 1: Determine the azimuth of each segment of the traverse. For this process, an azimuth for the initial course must be known or assumed. Once that azimuth is determined, the azimuth for each succeeding course may be determined by adding 180° to the azimuth of the previous course, then adding the angle to the right (the clockwise angle) to the resulting sum. If the resulting value is greater than 360°, subtract 360°.

step 2: Calculate the latitude (ΔN) and departure (ΔE) for each segment of the traverse by use of the azimuth (AZ) and the distance (length) of the segment in Eq. 19.1 and Eq. 19.2. (See Fig. 19.3.)

$$\Delta N = d \cos A \qquad 19.1$$

$$\Delta E = d \sin A \qquad 19.2$$

Figure 19.3 *Latitude (ΔN) and Departure (ΔE)*

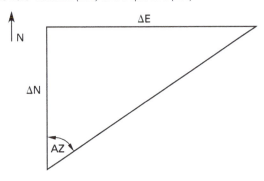

step 3: Once the latitude and departure for a segment of a traverse is known, the northing coordinate (N) and easting coordinate (E) for the end point of that segment may be calculated by adding the ΔN to the previous northing coordinate and the ΔE to the previous easting coordinate.

Inverse Calculations

The azimuth (A) and distance (d) between two points with known coordinates may be calculated using Eq. 19.3 and Eq. 19.4, using the latitude (ΔN) and departure (ΔE) for the segment between the points.

$$A = \arctan\left(\frac{\Delta E}{\Delta N}\right) \qquad 19.3$$

$$d = \sqrt{\Delta N^2 + \Delta E^2} \qquad 19.4$$

Example 19.1

Calculate the coordinates for points B and C for the traverse shown.

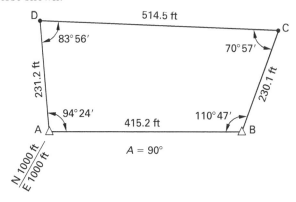

Solution

Because the azimuth of line AB, AZ_{AB}, is known, the latitude, ΔN_{AB}, and departure, ΔE_{AB}, can be calculated using Eq. 19.1 and Eq. 19.2, respectively.

$$\Delta N_{AB} = d_{AB}\cos A_{AB} = (415.2 \text{ ft})\cos(90°)$$
$$= 0.0 \text{ ft}$$

$$\Delta E_{AB} = d_{AB}\sin A_{AB} = (415.2 \text{ ft})\sin(90°)$$
$$= 415.2 \text{ ft}$$

Since Point A has a northing, N_A, of 1000 ft and an easting, E_A, of 1000 ft, the coordinates for point B can be calculated as

$$N_B = N_A + \Delta N_{AB} = 1000 \text{ ft} + 0.0 \text{ ft}$$
$$= 1000.0 \text{ ft}$$

$$E_B = E_A + \Delta E_{AB} = 1000 \text{ ft} + 415.2 \text{ ft}$$
$$= 1415.2 \text{ ft}$$

Since the azimuth for line AB is given as 90°, the azimuth for line BC may be determined by adding 180° to that azimuth and then adding the angle to the right. Since the result is greater than 360°, 360° is subtracted from the resulting sum.

$$A_{BC} = 90° + 180° + 110°47' - 360°$$
$$= 20°47'$$

The latitude and departure for line BC can then be calculated using Eq. 19.1 and Eq. 19.2, respectively.

$$\Delta N_{BC} = d_{BC}\cos A_{BC} = (230.1 \text{ ft})\cos(20°47')$$
$$= 215.1 \text{ ft}$$

$$\Delta E_{BC} = d_{BC}\sin A_{BC} = (230.1 \text{ ft})\sin(20°47')$$
$$= 81.6 \text{ ft}$$

The coordinates for point C may then be calculated by adding ΔN to the previous northern coordinate and ΔE to the previous easting coordinate.

$$N_C = N_B + \Delta N_{BC} = 1000 \text{ ft} + 215.1 \text{ ft}$$
$$= 1215.1 \text{ ft}$$

$$E_C = E_B + \Delta E_{BC} = 1415.2 \text{ ft} + 81.6 \text{ ft}$$
$$= 1496.8 \text{ ft}$$

This process can then be repeated for lines CD and DA, as illustrated in the table shown. It is helpful to use a table to keep track of calculations. (All distances are in feet.)

pt	dist	angle	az	ΔN	ΔE	N	E
A						1000.0	1000.0
	415.2		90°00'	0.0	415.2		
B		110°47'				1000.0	1415.2
	230.1		20°47'	215.1	81.6		
C		70°57'				1215.1	1496.8
	514.5		271°42'	15.3	−514.3		
D		83°56'				1230.7	982.5
	231.2		175°40'	−230.5	17.5		
A		94°24'				1000.2	1000.0
sum	1391.0	360°04'		0.2	0.0		

5. ANGULAR CLOSURE ERROR

For a closed traverse, the angular closure error should be checked. First, the interior angles of the traverse polygon should be summed. This is illustrated in Ex. 19.1.

Once the sum of the interior angles is known, the sum is compared with the theoretical sum of angles calculated from Eq. 19.5, where n is the number of sides of the traverse polygon.

$$\text{theoretical sum of angles} = (n-2)(180°) \quad \textit{19.5}$$

From Eq. 19.5, it may be seen that if no angular errors exist for the traverse, the sum of the interior angles in a triangle will be 180°, the sum of the interior angles in a quadrilateral will be 360°, and so forth. The *angular closure error* for a traverse is the difference between the actual sum and the theoretical sum (Eq. 19.6).

$$\text{angular closure error} = \text{sum of measured angles} \\ -\text{theoretical sum of} \quad \textit{19.6} \\ \text{measured angles}$$

The angular closure error is used as a correction and is distributed evenly to the angles unless a least squares adjustment is being made or an angular blunder is suspected. The error distribution is called *balancing the angles.* The correction per angle may be calculated using Eq. 19.7, where n is the number of angles in the traverse.

$$\frac{\text{correction}}{\text{per angle}} = -\left(\frac{\text{angular closure error}}{n}\right) \quad 19.7$$

Example 19.2

Using the closed traverse shown in Ex. 19.1, perform an angular closure check, then balance the angles.

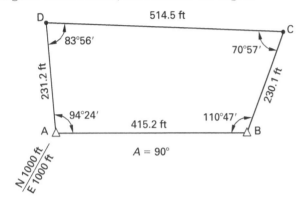

Solution

Find the theoretical sum of the angles using Eq. 19.5.

$$\begin{aligned}
\text{theoretical sum of angles} &= (n-2)(180°) \\
&= (4-2)(180°) \\
&= 360°
\end{aligned}$$

Sum the interior angles of the traverse.

$$\begin{aligned}
\text{sum of angles} &= \angle A + \angle B + \angle C + \angle D \\
&= 94°24' + 110°47' + 70°57' + 83°56' \\
&= 360°04'
\end{aligned}$$

Find the angular closure error using Eq. 19.6.

$$\begin{aligned}
\text{angular closure error} &= \text{sum of measured angles} \\
&\quad - \text{theoretical sum of angles} \\
&= 360°04' - 360° \\
&= +4'
\end{aligned}$$

Find the correction per angle using Eq. 19.7.

$$\begin{aligned}
\text{correction per angle} &= -\left(\frac{\text{angular closure error}}{n}\right) \\
&= -\left(\frac{+4'}{4}\right) \\
&= -1
\end{aligned}$$

Balance the angles of the traverse.

angle	measured	correction	balanced
A	94°24'	−1'	94°23'
B	110°47'	−1'	110°46'
C	70°57'	−1'	70°56'
D	83°56'	−1'	83°55'
	360°04'	−4'	360°00'

6. POSITIONAL CLOSURE

A *positional closure* check must be done for a closed traverse or an open traverse that closes on a point with a known spatial relationship to the beginning point. The positional closure or error is determined after the angles have been balanced by comparing the coordinates of the beginning point with the closing coordinates for that point, as shown in Fig. 19.4. If the differences in northing and easting are not known for the closing point, the sum of the latitudes and departures for the traverse will provide the same values. The *error of closure* for the traverse can be found using the Pythagorean theorem, as shown in Eq. 19.8.

$$\text{error of closure} = \sqrt{\left(\sum \Delta N\right)^2 + \left(\sum \Delta E\right)^2} \quad 19.8$$

Figure 19.4 *Positional Closure*

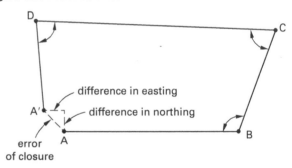

The error of closure is often expressed as a ratio. The *ratio of precision*, also known as the *ratio of error*, can be determined using the ratio of the traverse perimeter to the error of closure.

$$\text{ratio of precision} = 1 : \frac{\text{traverse perimeter}}{\text{error of closure}} \quad 19.9$$

Example 19.3

Find the error of closure and the ratio of precision for the closed traverse shown.

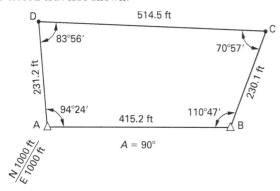

Solution

Use the closed traverse from Ex. 19.1 and Ex. 19.2, balance the angles using Eq. 19.6 and Eq. 19.7, and create a table of data with angular corrections applied. (All distances are in feet.)

Using Eq. 19.8, the error of closure is

$$
\begin{aligned}
\text{error of closure} &= \sqrt{\left(\sum \Delta \text{N}\right)^2 + \left(\sum \Delta \text{E}\right)^2} \\
&= \sqrt{(0.0 \text{ ft})^2 + (-0.2 \text{ ft})^2} \\
&= 0.2 \text{ ft}
\end{aligned}
$$

Using Eq. 19.9, the ratio of precision is

$$
\begin{aligned}
\text{ratio of precision} &= 1 : \frac{\text{traverse perimeter}}{\text{error of closure}} \\
&= 1 : \frac{1391.0}{0.2} \\
&= 1 : 6955
\end{aligned}
$$

See *Example 19.3 Solution.*

7. ADJUSTMENT OF TRAVERSES

After the error of closure has been calculated and blunders have been eliminated, each course of the traverse should be adjusted based on the random error. There are several approaches to the process of *error distribution*, also known as *adjusting the traverse.*

Least Squares Adjustment of Traverses

The most statistically valid method for balancing a traverse is the *least squares adjustment method.* This method allows observations to be weighted based on the level of precision required for the instrument and the technique used to obtain the measurements.

With the widespread use of computers in surveying, the least squares adjustment method is becoming the norm.[2] The current specifications for boundary surveys using the American Land Title Association/American Congress on Surveying and Mapping (ALTA/ACSM) standards require least squares adjustment to quantify the precision of a survey, unless an independent higher-order survey is used to check results. However, because the use of a computer is required, the procedure for least squares adjustment is outside the scope of this book.

Compass Rule Adjustment of Traverses

Traditionally, the most widely used method for adjusting a traverse is the *compass rule*, also known as the *Bowditch method.*[3] The compass rule is still widely used for traverses where the angular accuracy and linear accuracy are about the same.

With the compass rule, the differences in the sums of the latitudes and departures are distributed over each course based on the ratio of the length, d, of that course to the total length, d_t, of the traverse. Adjusted latitudes, ΔN, are calculated using Eq. 19.10, and adjusted departures, ΔE, are calculated using Eq. 19.11. ΔN_t is the sum of the latitudes, and ΔE_t is the sum of the departures.

$$
\text{adjusted } \Delta \text{N}_i = \Delta \text{N}_i + \left(\frac{d_i}{d_t}\right)(-\Delta \text{N}_t) \qquad \textit{19.10}
$$

$$
\text{adjusted } \Delta \text{E}_i = \Delta \text{E}_i + \left(\frac{d_i}{d_t}\right)(-\Delta \text{E}_t) \qquad \textit{19.11}
$$

Once all of the adjusted ΔN and ΔE values are calculated, the adjusted N and E coordinates may be calculated.

As an example, referring to the illustration for Ex. 19.3, the length of segment BC is 230.1 ft, the total of all the legs of the traverse is 1391.0 ft, and the total of all of the ΔE values is 0.2 ft. Equation 19.11 can be used to calculate the adjusted ΔE for the line between B and C as follows.

$$
\begin{aligned}
\text{adjusted } \Delta \text{E}_{\text{BC}} &= \Delta \text{E}_i + \left(\frac{d_i}{d_t}\right)(-\Delta \text{E}_t) \\
&= 81.6 \text{ ft} + \left(\frac{230.1 \text{ ft}}{1391.0 \text{ ft}}\right)(-0.2 \text{ ft}) \\
&= 81.56 \text{ ft}
\end{aligned}
$$

After all of the adjusted latitudes and departures have been calculated, they are used to adjust the northing and easting coordinates for each point in the traverse.

---

[2] Prior to the use of computers in surveying, this method was seldom used because of its complexity.
[3] Another balancing process, the *transit method*, was used frequently prior to the widespread use of EDM instruments for traverses with much greater angular accuracy than linear accuracy.

Example 19.3 Solution

point	distance	angle	balanced angle	A	ΔN	ΔE	N	E
A							1000.0	1000.0
	415.2			90°00′	0.0	415.2		
B		110°47′	110°46				1000.0	1415.2
	230.1			20°46′	215.2	81.6		
C		70°57′	70°56′				1215.2	1496.8
	514.5			271°42′	15.3	−514.3		
D		83°56′	83°55′				1230.5	982.5
	231.2			175°37′	−230.5	17.7		
A		94°24′	94°23′				1000.0	1000.2
sum	1391.0	360°04′	360°00′		0.0	0.2	0.0	−0.2

Example 19.4

Adjust the closed traverse from Ex. 19.3 using the compass rule. Note that since there is no error in the northing coordinate, no adjustment is needed.

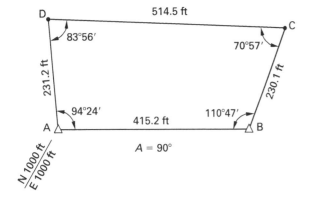

Solution

Note that after balancing the angles and adjusting the traverse, the coordinates for Point A at the end of the traverse are the same as the beginning values for that point.

See *Example 19.4 Solution.*

8. DEFLECTION ANGLE TRAVERSES

A *deflection angle* is an angle between a line and the extension of the preceding line, as shown in Fig. 19.5. It may be turned either right or left from the extension, but the direction of turning must be recorded with the angular measurement. In a deflection angle traverse, the straight lines between the points of change in direction are known as *tangents,* and the points of change in direction are known as *points of intersection.*

Figure 19.5 *Deflection Angle Traverse*

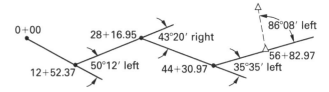

Deflection angle traverses are often used for roadway surveys because the deflection angle at the point of intersection of two tangents along the centerline of a highway is equal to the central angle of the circular arc that is inserted to connect two tangents.

The deflection angle traverse shown in Fig. 19.6 begins at point A. The azimuth of AB is found by adding the deflection angle at A to the azimuth XA. Azimuths of the other lines of the traverse are found by adding right deflection angles to the forward azimuth of the preceding line and subtracting left deflection angles from the forward azimuth of the preceding line. At the ending point of the traverse (Point E), the azimuth of the line EY is used to check the angular closure of the traverse.

Figure 19.6 *Deflection Angle Traverse and Calculations*

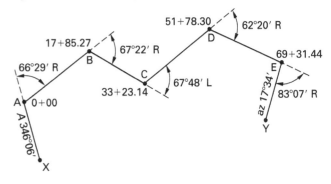

Example 19.4 Solution

							adjusted values		
point	distance	A	ΔN	ΔE	ΔN	ΔE	N	E	
A							1000.0	1000.0	
	415.2	90°00′	0.0	415.2	0.0	415.1			
B							1000.0	1415.1	
	230.1	20°46′	215.2	81.6	215.2	81.6			
C							1215.2	1496.7	
	514.5	271°42′	15.3	−514.3	15.3	−514.4			
D							1230.5	982.3	
	231.2	175°37′	−230.5	17.7	−230.5	17.7			
A							1000.0	1000.0	
		sum	0.0	0.2	0.0	0.0			

9. CONNECTING TRAVERSES

Open traverses are often run as *connecting traverses* that start on one established control point and end on another. By including control points for the beginning and ending points in an open traverse, *x*- and *y*-coordinates can be calculated, the direction can be found, and closure checks can be made.

Figure 19.7 shows a connecting traverse between triangulation station WAAF and triangulation station PRICE. Traverse stations A, B, and C are established as part of the connecting traverse.

10. INTERSECTIONS OF TRAVERSE LINES

There are three commonly used trigonometric methods for determining the point of intersection of two traverse lines: the bearing-bearing method, the bearing-distance method, and the distance-distance method. Selection of the method to use depends on the measurements available. Traverse calculations and the law of sines (see Eq. 19.12) or the law of cosines (see Eq. 19.13) are used for these methods.

$$\frac{d_{BC}}{\sin\theta_A} = \frac{d_{AC}}{\sin\theta_B} = \frac{d_{AB}}{\sin\theta_C} \qquad 19.12$$

$$\cos\theta_A = \frac{d_{AC}^2 + d_{AB}^2 - d_{BC}^2}{2d_{AC}d_{AB}} \qquad 19.13$$

Bearing-Bearing Method (Triangulation)

If the coordinates of the ending points of one side of a triangle are known and the bearings of the other two sides are known, the coordinates of the point of intersection of the other two sides can be found using the *bearing-bearing method*. This approach is also known as *triangulation* and has been widely used in large-scale geodetic networks.

Figure 19.6 *Connecting Traverse*

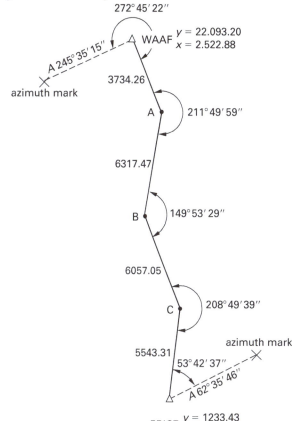

Example 19.5

Find the coordinates of point C at the intersection of lines AC and BC in the illustration shown using the bearing-bearing method.

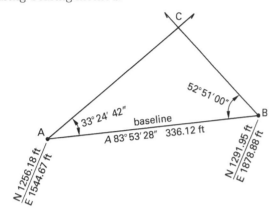

Solution

Find the azimuth AC.

$$A_{AC} = A_{AB} - \theta_A$$
$$= 83°53'27'' - 33°24'42''$$
$$= 50°28'46''$$

Determine angle C.

$$\theta_C = 180° - (\theta_A + \theta_B) = 93°44'18''$$

Find the distance AC using the law of sines, Eq. 19.12.

$$\frac{d_{AC}}{\sin\theta_B} = \frac{d_{BC}}{\sin\theta_A} = \frac{d_{AB}}{\sin\theta_C}$$

$$d_{AC} = \frac{d_{AB}\sin\theta_B}{\sin\theta_C}$$

$$= \frac{(336.12 \text{ ft})\sin 52°51'00''}{\sin 93°44'18''}$$

$$= 268.48 \text{ ft}$$

Then find the northing and easting coordinate for Point C.

$$N_C = N_A + d_{AC}\cos A_{AC}$$
$$= 1256.18 \text{ ft} + (268.48 \text{ ft})\cos 50°28'46''$$
$$= 1427.03 \text{ ft}$$

$$E_C = E_A + d_{AC}\sin A_{AC}$$
$$= 1544.67 \text{ ft} + (268.48 \text{ ft})\sin 50°28'46''$$
$$= 1751.77 \text{ ft}$$

Bearing-Distance Method

If the coordinates of the end points of one side of a triangle are known, and the bearing of one of the other two sides and the distance for the third side are known, the coordinates of the point of intersection of the other two sides can be found using the *bearing-distance method*.

Example 19.6

Find the coordinates of the point of intersection of lines AC and BC in the illustration shown using the bearing-distance method.

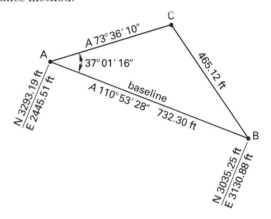

Solution

Determine angle C using the law of sines, Eq. 19.12.

$$\frac{d_{AC}}{\sin\theta_B} = \frac{d_{BC}}{\sin\theta_A} = \frac{d_{AB}}{\sin\theta_C}$$

$$\theta_C = \arcsin\frac{d_{AB}\sin\theta_A}{d_{BC}}$$

$$= \arcsin\frac{(732.30 \text{ ft})\sin 37°01'16''}{465.12 \text{ ft}}$$

$$= 71°26'17''$$

With a bearing-distance situation, two solutions are possible and are 180° apart. An arc with a radius of 465.12 ft swung from point B would intersect a line from point A with an azimuth of 73°36'10'' at two points. By inspection, angle C is greater than 71°. Therefore, the correct solution for angle C is $180° - 71°26'17'' = 108°33'43''$.

Determine angle B.

$$\theta_B = 180° - \theta_A - \theta_C$$
$$= 180° - 108°33'43'' - 37°01'16''$$
$$= 34°25'01''$$

Find the distance AC using the law of sines, Eq. 19.12.

$$\frac{d_{AC}}{\sin\theta_B} = \frac{d_{BC}}{\sin\theta_A} = \frac{d_{AB}}{\sin\theta_C}$$

$$d_{AC} = \frac{d_{BC}\sin\theta_B}{\sin\theta_A}$$

$$= \frac{(465.12 \text{ ft})\sin 34°25'01''}{\sin 37°01'16''}$$

$$= 436.62 \text{ ft}$$

Then find the northing and easting coordinates for Point C.

$$N_C = N_A + d_{AC} \cos A_{AC}$$
$$= 3293.19 \text{ ft} + (436.62 \text{ ft})\cos 73°36'10''$$
$$= 3416.45 \text{ ft}$$

$$E_C = E_A + d_{AC} \sin A_{AC}$$
$$= 2445.51 \text{ ft} + (436.62 \text{ ft})\sin 73°36'10''$$
$$= 2864.37 \text{ ft}$$

Distance-Distance Method (Trilateration)

If the coordinates of the end points of one side of a triangle and the distances of the other two sides are known, the coordinates of the point of intersection of the other two sides can be found using the *distance-distance method*. This method is also known as *trilateration*.

Example 19.7

Find the coordinates of the point of intersection of lines AC and BC in the illustration shown using the distance-distance method.

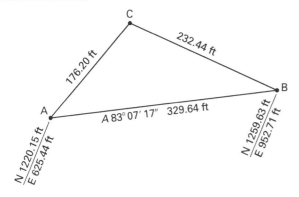

Solution

Use the law of cosines, Eq. 19.13, to determine angle A.

$$\cos \theta_A = \frac{d_{AC}^2 + d_{AB}^2 - d_{BC}^2}{2 d_{AC} d_{AB}}$$

$$\theta_A = \arccos \frac{d_{AC}^2 + d_{AB}^2 - d_{BC}^2}{2 d_{AC} d_{AB}}$$

$$= \arccos \frac{(176.20 \text{ ft})^2 + (329.64 \text{ ft})^2 - (232.44 \text{ ft})^2}{(2)(176.20 \text{ ft})(329.64 \text{ ft})}$$

$$= 42°28'29''$$

Then calculate A_{AC}.

$$A_{AC} = A_{AB} - \theta_A$$
$$= 83°07'17'' - 42°28'29''$$
$$= 40°38'48''$$

Then find the northing coordinate and the easting coordinates of Point C.

$$N_C = N_A + d_{AC} \cos A_{AC}$$
$$= 1220.15 \text{ ft} + (176.20 \text{ ft})\cos 40°38'48''$$
$$= 1353.84 \text{ ft}$$

$$E_C = E_A + d_{AC} \sin A_{AC}$$
$$= 625.44 \text{ ft} + (176.20 \text{ ft})\sin 40°38'48''$$
$$= 740.22 \text{ ft}$$

11. PRACTICE PROBLEMS

1. The interior angles shown were measured for a five-sided closed traverse.

point	measured angle
A	67°06'30''
B	216°19'00''
C	65°12'30''
D	95°18'30''
E	96°02'00''

To balance the angles, the correction that should be added to each angle is most nearly

(A)   18''

(B)   21''

(C)   54''

(D)   1'30''

2. In the illustration shown, the azimuth of line AB is 15°22'.

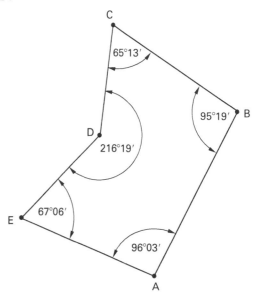

The azimuth of line BC is most nearly

(A) 110°41′

(B) 95°19′

(C) 275°19′

(D) 290°41′

3. Triangle ABC is shown.

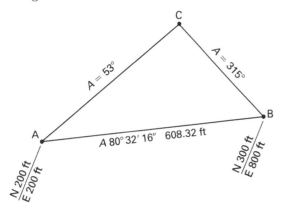

Based on the bearing-bearing method, the coordinates for point C are most nearly

(A) N = 450.2 ft; E = 540.4 ft

(B) N = 475.1 ft; E = 502.8 ft

(C) N = 490.5 ft; E = 612.5 ft

(D) N = 500.8 ft; E = 599.2 ft

4. Triangle ABC is shown.

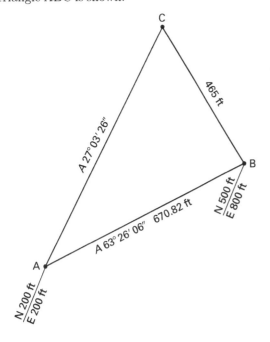

Based on the bearing-distance method, the coordinates for point C are most nearly

(A) N = 473.4 ft; E = 335.7 ft

(B) N = 484.9 ft; E = 344.8 ft

(C) N = 690.3 ft; E = 462.4 ft

(D) N = 895.3 ft; E = 555.2 ft

5. Triangle ABC is shown.

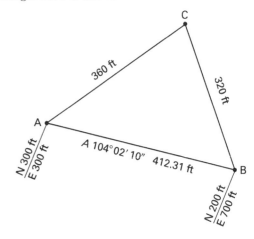

Based on the distance-distance method, the coordinates for point C are most nearly

(A) N = 503.6 ft; E = 597.3 ft

(B) N = 512.9 ft; E = 571.2 ft

(C) N = 562.2 ft; E = 569.2 ft

(D) N = 527.2 ft; E = 588.1 ft

6. The starting coordinates for a closed traverse with a total perimeter of 14,575.1 ft are 10,000.0 (N) and 10,000.0 ft (E). Closing coordinates for the traverse are 10,001.0 (N) and 9,999.4 ft (E). The error of closure and ratio of precision are most nearly

(A) 1.0 ft; 1:10,544

(B) 1.2 ft, 1:12,457

(C) 1.6 ft, 1:11,234

(D) 1.7 ft, 1:9,876

7. A point in a traverse has coordinates of 10,000.0 (N) and 10,000.0 (E). The azimuth from the previous point to the occupied point is 10°, the angle to the right to the next point is 23°17′, and the measured distance to the next point is 572.2 ft. The coordinates for the next point in the traverse are most nearly

(A) 9684.3 ft (N), 10,480.8 ft (E)

(B) 9829.3 ft (N), 10,521.4 ft (E)

(C) 9521.6 ft (N), 9685.9 ft (E)

(D) 10,521.4 ft (N), 9829.3 ft (E)

8. Point A has coordinates of 1276.3 ft (N) and 1533.4 ft (E). Point B has coordinates of 2256.0 ft (N) and 1644.0 ft (E). The azimuth and distance from Point A to B is most nearly

(A) 6°26′27″, 985.9 ft

(B) 20°19′52″, 1121.2 ft

(C) 109°40′08″, 1052.9 ft

(D) 173°33′33″, 874.2 ft

SOLUTIONS

1. Because there are five angles in the traverse, use Eq. 19.5 to calculate the theoretical sum of the angles.

$$\text{theoretical sum of angles} = (n-2)(180°)$$
$$= (5-2)(180°)$$
$$= 540°$$

The sum of the measured angles is

$$\text{sum of angles} = \theta_A + \theta_B + \theta_C + \theta_D + \theta_E$$
$$= 67°06′30″ + 216°19′00″ + 65°12′30″$$
$$+ 95°18′30″ + 96°02′00″$$
$$= 539°58′30″$$

Therefore, the angular closure error is

$$\text{error} = \text{sum of measured angles}$$
$$- \text{theoretical sum of angles}$$
$$= 539°58′30″ - 540°$$
$$= -1′30″$$

Using Eq. 19.7, divide the angular closure error by the number of angles to find the correction for each angle.

$$\text{correction per angle} = -\left(\frac{\text{angular closure error}}{n}\right)$$
$$= -\left(\frac{-1′30″}{5}\right)$$
$$= +18″$$

The answer is (A).

2. Since the azimuth for line AB is given as 15°22′, the azimuth for line BC may be determined by adding 180° to that azimuth and then adding the angle to the right.

$$A_{BC} = 15°22′ + 180° + 95°19′ = 290°41′$$

The answer is (D).

Plane Survey Calculations

3.

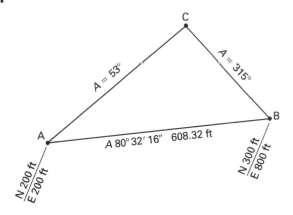

The coordinates of point C in the illustration can be determined by the bearing-bearing method.

Determine angle A.

$$\theta_A = A_{AB} - A_{AC} = 27°32'16''$$

Determine angle B.

$$\theta_B = A_{BC} - A_{BA} = 54°27'44''$$

Determine angle C.

$$\theta_C = A_{CB} - A_{CA} = 98°$$

Find the distance AC using the law of sines, Eq. 19.12.

$$\frac{d_{BC}}{\sin\theta_A} = \frac{d_{AC}}{\sin\theta_B} = \frac{d_{AB}}{\sin\theta_C}$$

$$d_{AC} = \frac{d_{AB}\sin\theta_B}{\sin\theta_C}$$

$$= \frac{(608.32 \text{ ft})\sin 54°27'44''}{\sin 98°}$$

$$= 499.87 \text{ ft}$$

Then find the northing and easting coordinates for point C.

$$N_C = N_A + d_{AC}\cos A_{AC}$$

$$= 200 \text{ ft} + (499.87 \text{ ft})\cos 53°$$

$$= 500.8 \text{ ft}$$

$$E_C = E_A + d_{AC}\sin A_{AC}$$

$$= 200 \text{ ft} + (499.87 \text{ ft})\sin 53°$$

$$= 599.2 \text{ ft}$$

The answer is (D).

4.

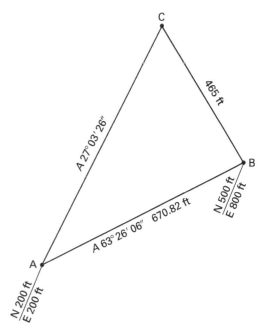

The coordinates of point C in the illustration can be determined by the bearing-distance method.

Find the measurement of angle A.

$$\theta_A = A_{AB} - A_{AC}$$

$$= 63°26'06'' - 27°03'26''$$

$$= 36°22'40''$$

Use the law of sines (Eq. 19.12) to find angle C.

$$\frac{d_{BC}}{\sin\theta_A} = \frac{d_{AC}}{\sin\theta_B} = \frac{d_{AB}}{\sin\theta_C}$$

$$\theta_C = \arcsin\frac{d_{AB}\sin\theta_A}{d_{BC}}$$

$$= \arcsin\left(\frac{(670.82 \text{ ft})\sin 36°22'40''}{465 \text{ ft}}\right)$$

$$= 58°49'46''$$

Subtract the measurements of angle A and angle C from 180° to get the measurement of angle B.

$$\theta_B = 180° - 36°22'40'' - 58°49'46'' = 84°47'34''$$

Use the law of sines to find the distance of line AC.

$$\frac{d_{BC}}{\sin\theta_A} = \frac{d_{AC}}{\sin\theta_B} = \frac{d_{AB}}{\sin\theta_C}$$

$$\begin{aligned} d_{AC} &= \frac{d_{BC}\sin\theta_B}{\sin\theta_A} \\ &= \frac{(465\ \text{ft})\sin 84°47'34''}{\sin 36°22'40''} \\ &= 780.8\ \text{ft} \end{aligned}$$

Then find the northing coordinate and the easting coordinate.

$$\begin{aligned} N_C &= N_A + d_{AC}\cos A_{AC} \\ &= 200\ \text{ft} + (780.8\ \text{ft})\cos 27°03'26'' \\ &= 895.3\ \text{ft} \end{aligned}$$

$$\begin{aligned} E_C &= E_A + d_{AC}\sin A_{AC} \\ &= 200\ \text{ft} + (780.8\ \text{ft})\sin 27°03'26'' \\ &= 555.2\ \text{ft} \end{aligned}$$

The answer is (D).

5.

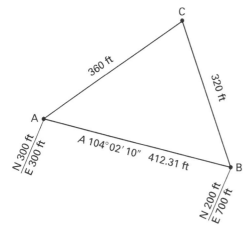

The coordinates of point C in the illustration can be determined by the distance-distance method.

Use the law of cosines (Eq. 19.13) to find the cosine of angle A.

$$\begin{aligned} \cos\theta_A &= \frac{d_{AC}^2 + d_{AB}^2 - d_{BC}^2}{2d_{AC}d_{AB}} \\ &= \frac{(360\ \text{ft})^2 + (412.31\ \text{ft})^2 - (320\ \text{ft})^2}{(2)(360\ \text{ft})(412.31\ \text{ft})} \\ &= 0.6642775 \end{aligned}$$

Calculate the measurement of angle A.

$$\theta_A = \arccos 0.6642775 = 48°22'23''$$

Find the azimuth of line AC.

$$\begin{aligned} A_{AC} &= A_{AB} - \theta_A \\ &= 180° - 75°57'50'' - 48°22'23'' \\ &= 55°39'47'' \end{aligned}$$

Find the northing coordinate and the easting coordinate.

$$\begin{aligned} N_C &= N_A + d_{AC}\cos\theta_{AC} \\ &= 300\ \text{ft} + (360\ \text{ft})\cos 55°39'47'' \\ &= 503.06\ \text{ft} \quad (503.1\ \text{ft}) \end{aligned}$$

$$\begin{aligned} E_C &= E_A + d_{AC}\sin\theta_{AC} \\ &= 300\ \text{ft} + (360\ \text{ft})\sin 55°39'47'' \\ &= 597.26\ \text{ft} \quad (597.3\ \text{ft}) \end{aligned}$$

The answer is (A).

6. The difference in northing is $10{,}000.0 - 10{,}001.0 = -1.0$ ft.

The difference in easting is $10{,}000.0 - 9{,}999.40 = 0.6$ ft.

$$\begin{aligned} \text{error of closure} &= \sqrt{\left(\sum\Delta N\right)^2 + \left(\sum\Delta E\right)^2} \\ &= \sqrt{(-1.0\ \text{ft})^2 + (0.6\ \text{ft})^2} \\ &= 1.17\ \text{ft} \quad (1.2\ \text{ft}) \end{aligned}$$

$$\begin{aligned} \text{ratio of precision} &= 1{:}\frac{\text{traverse perimeter}}{\text{error of closure}} \\ &= 1{:}\frac{14{,}575.1\ \text{ft}}{1.17\ \text{ft}} \\ &= 1{:}12{,}457 \end{aligned}$$

The answer is (B).

7. The azimuth to the next point is determined by adding 180° to the azimuth of the previous course, then adding the angle to the right (clockwise angle) to the resulting sum. If the resulting value is greater than 360°, then subtract 360°.

$$A = 10° + 180° + 23°17' = 213°17'$$

Use Eq. 19.10 and Eq. 19.11 to find the northing coordinate and the easting coordinate.

$$\begin{aligned} N &= N_{BS} + d_{AC}\cos A_{AC} \\ &= 10{,}000.0\ \text{ft} + (572.2\ \text{ft})\cos 213°17' \\ &= 9521.6\ \text{ft} \end{aligned}$$

$$E = E_{BS} + d_{AC} \sin A_{AC}$$
$$= 10{,}000.0 \text{ ft} + (572.2 \text{ ft}) \sin 213°17'$$
$$= 9685.9 \text{ ft}$$

The answer is (C).

8. Use Eq. 19.3 to find the azimuth from Point A to B.

$$A = \arctan\left(\frac{\Delta E}{\Delta N}\right)$$
$$= \arctan\left(\frac{1644.0 \text{ ft} - 1533.4 \text{ ft}}{2256.0 \text{ ft} - 1276.3 \text{ ft}}\right)$$
$$= 6°26'27''$$

Use Eq. 19.4 to find the distance from point A to point B.

$$d = \sqrt{\Delta N^2 + \Delta E^2}$$
$$= \sqrt{(979.7)^2 + (110.6)^2}$$
$$= 985.9 \text{ ft}$$

The answer is (A).

20 Area of a Traverse

Nomenclature

A	area	ft^2	m^2
D	regular interval	ft	m
E	departure	ft	m
LC	long chord	ft	m
M	middle ordinate	ft	m
N	latitude	ft	m
r	radius	ft	m
T	tie distance	ft	m

Symbols

θ	angle	deg	deg
ϕ	central angle	deg	deg

Subscripts

s	sector
t	triangle

1. METHODS FOR COMPUTATION OF AREA

The area within a traverse can be computed by the *double meridian distance* (DMD) *method*, by the coordinate method, or by use of geometric or trigonometric formulas.

2. DMD METHOD

Prior to the widespread use of coordinates, advanced calculators, and computers, the DMD method was widely used, since it provides a quick way of calculating area using the balanced latitude (ΔN) and departure (ΔE) for the segments of the traverse. This method sets up a series of trapezoids and triangles, both inside and outside of the traverse. It calculates each of these areas and determines the area of the traverse from them.

The *double meridian distance* (DMD) is simply twice the meridian distance. The DMD is used instead of the meridian distance (MD) to simplify the arithmetic. If the MD were used, division by two would be required several times. Using DMD, division by two is required only once.

The area of a trapezoid is one-half the sum of the bases times the altitude (i.e., the average of the bases times the altitude). In the DMD method, the meridian distance for each course of the traverse serves as the average of the bases of a trapezoid. The DMDs of the courses are obtained from the departures of the courses. Thus, the only data needed are latitudes and departures.

The *meridian distance* (MD) of a course is the right-angle distance from the midpoint of the course to a reference meridian. MDs are illustrated in Fig. 20.1. Since east and west departures are used, algebraic signs must be considered. To simplify the use of plus and minus values for departure, the entire traverse should be placed in the northeast quadrant. This can be done by taking the reference meridian through the most westerly point in the traverse. That point is determined by inspection.

Figure 20.1 *Meridian Distance*

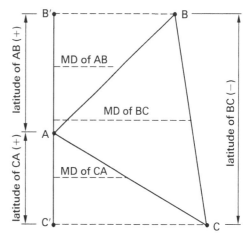

In Fig. 20.2, the MD of EA = $\frac{1}{2}E_{EA}$. The DMD of EA = E_{EA}. The MD of AB = MD of EA + $\frac{1}{2}E_{EA}$ + $\frac{1}{2}E_{AB}$. The DMD of AB = DMD of EA + E_{EA} + E_{AB}.

Figure 20.2 *DMD Rules*

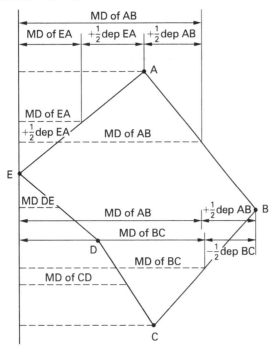

The MD of BC = MD of AB + ½E_{AB} − ½E_{BC}. The DMD of BC = DMD of AB + E_{AB} − E.

From these examples and from further examination, the following rules can be derived for computing DMDs.

1. The DMD of the first course is equal to the departure of the first course.

2. The DMD of any course is equal to the DMD of the preceding course plus the departure of the preceding course plus the departure of the course itself.

3. The DMD of the last course is equal to the departure of the last course with opposite sign. ("Opposite sign" means that for a westerly course, the DMD will be positive.)

The area of a traverse can be found by multiplying the DMD of each course by the latitude of that course, with north latitudes producing positive areas and south latitudes producing negative areas, adding the areas algebraically, and dividing by two. The algebraic sign of DMDs is always positive.

Referring back to Fig. 20.1,

$$A_{ABC} = A_{B'BCC'} - A_{AB'B} - A_{ACC'}$$

If MDs are positive, north latitudes are positive, and south latitudes are negative, then

$$A_{B'BCC'} = (\text{MD of BC})N_{BC} \quad [\text{negative}]$$
$$A_{AB'B} = (\text{MD of AB})N_{AB} \quad [\text{positive}]$$
$$A_{ACC'} = (\text{MD of CA})N_{CA} \quad [\text{positive}]$$
$$A_{ABC} = \text{algebraic sum of B'BCC', AB'B, ACC'}$$

The sign of the sum can be either plus or minus. If DMDs are used, the area will be double the area determined by using MDs, and the area of the traverse can be determined by dividing the double area by two.

Example 20.1

Table for Example 20.1 shows the tabulation of computations for the area of a traverse ABCDEA by the DMD method. Latitudes and departures shown are balanced. Find the DMD.

Solution

The first step is to determine the most westerly traverse point. In lieu of a sketch showing the traverse, it is found using this method.

1. Because AB has a northwest direction, B is west of A. Looking at the departure column, it is 507.97 ft west of A.

2. C is 243.72 ft west of B, so B is not the most westerly point.

3. Courses CD, DE, and EA have east departures; therefore, C is the most westerly point.

With C the most westerly point, the first DMD computed is for the course CD. Remembering that the DMD for the first course is the departure of the first course, and also remembering the definition of the DMD for any course,

$$E_{CD} = +373.77 \text{ ft} = \text{DMD of CD}$$
$$E_{CD} = +373.77 \text{ ft}$$
$$E_{DE} = +232.27 \text{ ft}$$
$$\qquad +979.81 \text{ ft} = \text{DMD of DE}$$
$$E_{DE} = +232.27 \text{ ft}$$
$$E_{EA} = +145.65 \text{ ft}$$
$$\qquad +1357.73 \text{ ft} = \text{DMD of EA}$$

$$E_{EA} = +145.65 \text{ ft}$$
$$E_{AB} = -507.97 \text{ ft}$$
$$\qquad +995.41 \text{ ft} = \text{DMD of AB}$$
$$E_{AB} = -507.97 \text{ ft}$$
$$E_{BC} = -243.72 \text{ ft}$$
$$\qquad +243.72 \text{ ft} = \text{DMD of BC}$$

Plane Survey Calculations

Table for Example 20.1

| point | bearing | distance | latitude | | departure | | DMD | area | |
			north	south	east	west		plus	minus
AB	N65°04′W	560.27	236.11			507.97	995.41	235,026	
BC	S30°14′W	484.14		418.39		243.72	243.72		101,970
CD	S84°33′E	375.42		35.71	373.77		373.77		13,347
DE	S48°13′E	311.44		207.56	232.27		979.81		203,369
EA	N18°53′E	449.83	425.55		145.65		1357.73	577,782	
								812,808	318,686

Note that the DMD of BC is the same as its departure except that it has a positive sign. The area of the traverse ABCDEA is found from the double area sums in the table.

$$A = \left(\frac{1}{2}\right)(812{,}808 \text{ ft}^2 - 318{,}686 \text{ ft}^2)$$
$$= 247{,}061 \text{ ft}^2$$

If necessary, this area can be converted to acres by dividing 247,061 ft^2 by 43,560.

$$A = \frac{247{,}061 \text{ ft}^2}{43{,}560 \dfrac{\text{ft}^2}{\text{ac}}}$$
$$= 5.672 \text{ ac}$$

3. COORDINATE METHOD

After the coordinates of the corners of a tract of land are determined, the area of the tract can be computed by the *coordinate method*. This method is also known as the "criss-cross" or "shoelace" method.

The coordinate method can be represented by Eq. 20.1.

$$A = \frac{1}{2}\sum (E_1N_2 - E_2N_1)$$
$$+ (E_2N_3 - E_3N_2) + \dots \qquad 20.1$$
$$+ (E_{n-1}N_n - E_nN_{n-1})$$
$$+ (E_nN_1 - E_1N_n)$$

Example 20.2

Given the traverse 1-2-3-4-5-1 illustrated in Fig. 20.3, find the area inside the traverse by the coordinate method.

Figure 20.3 *Coordinates of Traverse Points*

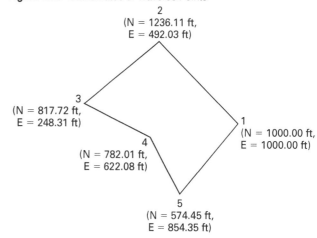

Solution

From Eq. 20.1,

$$A = \frac{1}{2}\sum (E_1N_2 - E_2N_1) + (E_2N_3 - E_3N_2) + \dots$$
$$+ (E_{n-1}N_n - E_nN_{n-1}) + (E_nN_1 - E_1N_n)$$

$$= \frac{1}{2}\sum \begin{pmatrix} (1000.0 \text{ ft})(1236.11 \text{ ft}) \\ -(492.03 \text{ ft})(1000.00 \text{ ft}) \\ +(492.03 \text{ ft})(817.72 \text{ ft}) \\ -(248.31 \text{ ft})(1236.11 \text{ ft}) \\ +(248.31 \text{ ft})(782.07 \text{ ft}) \\ -(622.08 \text{ ft})(817.72 \text{ ft}) \\ +(622.08 \text{ ft})(574.45 \text{ ft}) \\ -(854.35 \text{ ft})(782.01 \text{ ft}) \\ +(854.35 \text{ ft})(1000.0 \text{ ft}) \\ -(1000.00 \text{ ft})(574.45 \text{ ft}) \end{pmatrix}$$

$$= \left(\frac{1}{2}\right)(494{,}122 \text{ ft}^2)$$
$$= 247{,}061 \text{ ft}^2$$

$$A = 247{,}061 \text{ ft}^2 = \frac{247{,}061 \text{ ft}^2}{43{,}560 \; \frac{\text{ft}^2}{\text{ac}}} = 5.7 \text{ ac}$$

4. TRIANGLE METHOD

When small traverses do not warrant computations of latitudes and departures or coordinates, their areas can be determined by using formulas for the area of a triangle.

$$A_t = \tfrac{1}{2}ab\sin\theta_{\mathrm{C}} \qquad\qquad 20.2$$

(Where a and b are any two sides, and θ_{C} is the angle included between them.)

In Fig. 20.4(a), a tract of land has been divided into two triangles. Two sides and an included angle have been measured in each triangle. The areas of the triangles can be computed by using Eq. 20.3. Their sum is the area of the tract. In Fig. 20.4(b), the property line has become covered with brush so that the four triangles have been formed from a central point. Angles at the central point have been measured for each triangle, and distances from the central point to each corner have also been measured.

Areas can again be determined by using Eq. 20.2.

Equation 20.3 is also applicable in determining areas.

$$A_t = \sqrt{(s)(s-a)(s-b)(s-c)} \qquad\qquad 20.3$$

s is one-half of the perimeter of the triangle, and a, b, and c are the sides of the triangle.

Example 20.3

Find the area of a triangle with sides 32 ft, 46 ft, and 68 ft long.

Solution

$$s = \left(\frac{1}{2}\right)(32 \text{ ft} + 46 \text{ ft} + 68 \text{ ft}) = 73 \text{ ft}$$

$$A_t = \sqrt{\begin{array}{l}(73 \text{ ft})(73 \text{ ft} - 32 \text{ ft}) \\ \times (73 \text{ ft} - 46 \text{ ft})(73 \text{ ft} - 68 \text{ ft})\end{array}}$$

$$= 636 \text{ ft}^2$$

5. AREA ALONG AN IRREGULAR BOUNDARY

When a tract of land is bounded on one side by an irregular boundary, such as a stream or lake, the traverse can be composed of straight lines so that closure can be computed. Points along the irregular side can be tied to one of the sides of the traverse by right-angle offset measurements. The area between the irregular side and the traverse line is approximated by dividing the area into

Figure 20.4 *Area by Triangles*

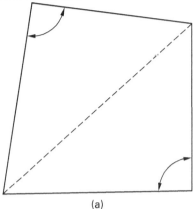

(a)

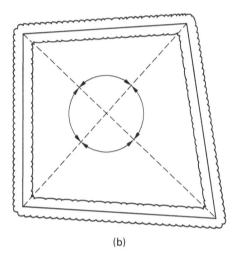

(b)

trapezoids and triangles formed by the ties to the breaks in the irregular side. This irregular area is then added to the traverse area. The irregular area can be computed by applying the trapezoidal rule or Simpson's one-third rule.

6. THE TRAPEZOIDAL RULE

In using the *trapezoidal rule*, it is assumed that the irregular boundary is made up of a series of straight lines. When the ties are taken close enough, a curved line connecting the ends of any two ties is very nearly a straight line, and no significant error is introduced.

The trapezoidal rule applies only to the part of the area where the ties are at regular intervals and form trapezoids. Triangles and trapezoids that do not have altitudes of the regular interval are computed separately and are added to the area found by applying the trapezoidal rule.

The rule is given by Eq. 20.4.

$$A_t = D\left(\frac{T_1}{2} + T_2 + T_3 + T_4 + \cdots + \frac{T_n}{2}\right) \qquad 20.4$$

D is the regular interval, and T is the tie distance.

Example 20.4

Using the trapezoidal rule, find the area of the following illustration.

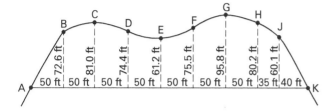

Solution

$$A = (50 \text{ ft})\left(\frac{72.6 \text{ ft}}{2}\right)$$

$$+ (50 \text{ ft})\left(\frac{\frac{72.6 \text{ ft}}{2} + 81.0 \text{ ft} + 74.4 \text{ ft} + 61.2 \text{ ft}}{+75.5 \text{ ft} + 95.8 \text{ ft} + \frac{80.2 \text{ ft}}{2}}\right)$$

$$+ (35 \text{ ft})\left(\frac{80.2 \text{ ft} + 60.1 \text{ ft}}{2}\right) + (40 \text{ ft})\left(\frac{60.1 \text{ ft}}{2}\right)$$

$$= 28{,}687 \text{ ft}^2$$

7. AREA OF A SEGMENT OF A CIRCLE

Land along highways, streets, and railroads often has a circular arc for a boundary. A traverse of straight lines can be run by using the *long chord* (LC) of the circular arc as one of the sides of the traverse. The area of the tract can be found by adding the area of the segment formed by the chord and the arc to the area within the traverse. It is usually practical to measure the chord length and the middle ordinate length. Using these two lengths and formulas derived for computing circular curves, the area of the segment can be found.

ϕ is the central angle, M is the middle ordinate, LC is the long chord, and r is the radius.

$$A_s = \frac{\phi \pi r^2}{360°} \qquad\qquad 20.5$$

$$A_t = \frac{r(r \sin\phi)}{2}$$

$$= \frac{r^2 \sin\phi}{2} \qquad\qquad 20.6$$

The radius is not known, but it can be determined if the long chord and the middle ordinate are known. These lengths can be measured. Two formulas used to compute r are

$$\tan\frac{\phi}{4} = \frac{2M}{LC} \qquad\qquad 20.7$$

$$r = \frac{LC}{2\sin\dfrac{\phi}{2}} \qquad\qquad 20.8$$

In this case, A represents the area of the segment, A_s represents the area of the sector, and A_t represents the area of the triangle.

$$A = A_s - A_t$$

$$= \frac{\phi \pi r^2}{360°} - \frac{r^2 \sin\phi}{2}$$

$$= r^2\left(\frac{\phi\pi}{360°} - \frac{\sin\phi}{2}\right) \qquad\qquad 20.9$$

Example 20.5

The tract of land shown in Fig. 20.5 consists of the area within the traverse ABCDEA plus the area in the segment bounded by side DE and arc DE. The area of the segment is equal to the area of the sector DOE minus the area of the triangle DOE.

Find the area in the segment bounded by side DE and arc DE in Fig. 20.5 if the long chord is 325.48 ft and the middle ordinate is 42.16 ft.

Figure 20.5 *Circular Segment Areas*

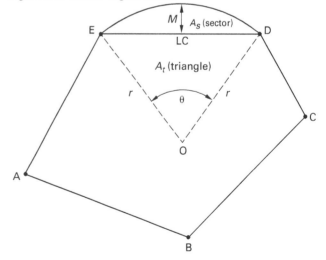

Solution

$$\tan\frac{\phi}{4} = \frac{(2)(42.16\text{ ft})}{325.48\text{ ft}}$$

$$\phi = 58.0958°$$

$$r = \frac{LC}{2\sin\dfrac{\phi}{2}}$$

$$= \frac{325.48\text{ ft}}{2\sin\dfrac{58.0958°}{2}}$$

$$= 335.17\text{ ft}$$

$$A = r^2\left(\frac{\pi\phi}{360°} - \frac{\sin\phi}{2}\right)$$

$$= (335.17\text{ ft})^2\left(\frac{(58.0958°)\pi}{360°} - \frac{\sin 58.0958°}{2}\right)$$

$$= 9270\text{ ft}^2$$

8. SPECIAL FORMULA

The formula for the area of a triangle is given in Sec. 20.4, $A_t = \frac{1}{2}\,ab\sin\theta_C$, can be used in deriving another useful formula for the area of a triangle. Use the law of sines and substitute $b = a\sin\theta_B/\sin\theta_A$ in the formula $A_t = \frac{1}{2}\,ab\sin\theta_C$ to get

$$A_t = \frac{a^2\sin\theta_B\sin\theta_C}{2\sin\theta_A} \qquad \textit{20.10}$$

Also,

$$A_t = \frac{b^2\sin\theta_C\sin\theta_A}{2\sin\theta_B}$$

$$= \frac{c^2\sin\theta_A\sin\theta_B}{2\sin\theta_C} \qquad \textit{20.11}$$

Example 20.6

Compute the area of the tract 1-2-3-4-5-6-1 shown in *Illustration for Example 20.6*.

	bearing	distance
1-2	N02°27′50″W	761.49
2-3	N87°35′37″E	1076.62
3-4	S02°24′23″E	290.00
4-5	S09°49′21″W	826.10
5-6	S30°30′21″W	68.00
6-1	N67°56′54″W	949.09

Solution

Calculate the areas as shown in *Tzable for Example 20.6*.

$$\frac{1{,}981{,}768\text{ ft}^2}{2} = 990{,}884\text{ ft}^2$$

$$A_{1\text{-}2\text{-}3\text{-}4\text{-}5\text{-}6\text{-}1} = 990{,}884\text{ ft}^2$$

$$A_s = r^2\left(\frac{\phi\pi}{360} - \frac{\sin\phi}{2}\right)$$

$$= 24{,}425\text{ ft}^2$$

$$A_t = 990{,}884\text{ ft}^2 + 24{,}425\text{ ft}^2$$

$$= 1{,}015{,}309\text{ ft}^2$$

Example 20.7

A 3 ac triangular tract is to be cut off the northeast corner of a larger tract with bearings as shown. Find the lengths of the sides of the triangle.

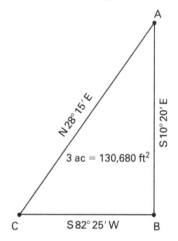

Solution

$$\theta_A = 38°35', \quad \theta_B = 87°15', \quad \text{and} \quad \theta_C = 54°10'.$$

$$A_t = \frac{a^2\sin\theta_B\sin\theta_C}{2\sin\theta_A}$$

$$a^2 = \frac{A(2\sin\theta_A)}{\sin\theta_B\sin\theta_C}$$

Also,

$$A_t = \frac{b^2\sin\theta_C\sin\theta_A}{2\sin\theta_B}$$

$$b^2 = \frac{A(2\sin\theta_B)}{\sin\theta_C\sin\theta_A}$$

Illustration for Example 20.6

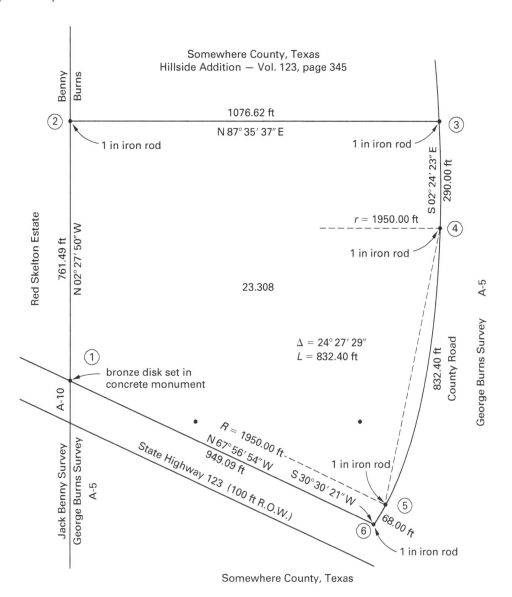

Table for Example 20.6

	bearing	distance	latitude	departure	DMD	area
1-2	N02°27′50″W	761.49	+760.79	−32.74	32.74	+24,908
2-3	N87°35′37″E	1076.62	+45.20	+1075.67	1075.67	+48,620
3-4	S02°24′23″E	290.00	−289.74	+12.18	2163.52	−626,858
4-5	S09°49′21″W	826.10	−813.99	−140.93	2034.77	−1,656,282
5-6	S30°30′21″W	68.00	−58.59	−34.52	1859.32	−108,938
6-1	N67°56′54″W	949.09	+356.33	−879.66	945.14	+336,782
						−1,981,768

Also,

$$A_t = \frac{c^2 \sin \theta_A \sin \theta_B}{2 \sin \theta_C}$$

$$c^2 = \frac{A(2 \sin \theta_C)}{\sin \theta_A \sin \theta_B}$$

Therefore,

$$a = \sqrt{\frac{A(2 \sin \theta_A)}{\sin \theta_B \sin \theta_C}} = \sqrt{\frac{(130{,}680 \text{ ft}^2)(2 \sin 38°35')}{(\sin 87°15')(\sin 54°10')}}$$

$$= 448.6 \text{ ft}$$

$$b = \sqrt{\frac{A(2 \sin \theta_B)}{\sin \theta_C \sin a}} = \sqrt{\frac{(130{,}680 \text{ ft}^2)(2 \sin 87°15')}{(2 \sin 54°10')(\sin 38°35')}}$$

$$= 718.5 \text{ ft}$$

$$c = \sqrt{\frac{A(2 \sin C)}{\sin a \sin B}} = \sqrt{\frac{(130{,}680 \text{ ft}^2)(2 \sin 54°10')}{(\sin 38°35')(\sin 87°13')}}$$

$$= 583.2 \text{ ft}$$

In summary,

$$AB = 583.2 \text{ ft}$$
$$BC = 448.6 \text{ ft}$$
$$CA = 718.5 \text{ ft}$$

9. PRACTICE PROBLEMS

(All dimensions and distances are in feet.)

1. Calculate the area of the traverse ABCDEA by the coordinate method.

point	coordinates N	E
A	1000.00	1000.00
B	1493.12	1265.66
C	1761.09	862.37
D	1390.05	804.65
E	1170.41	583.74

2. Calculate the area of the city lot shown by using the formula $A = \frac{1}{2}ab \sin \theta_C$.

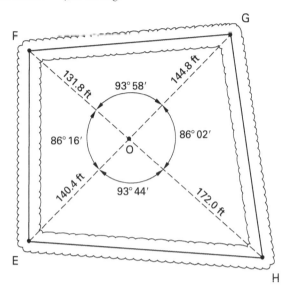

3. Calculate the area of the city lot shown using the formula $A = \frac{1}{2}ab \sin \theta_C$.

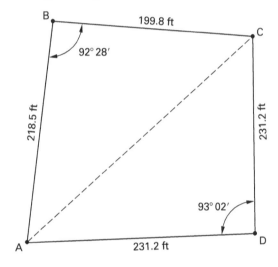

4. Find the area of a triangle with sides 12 ft, 14 ft, and 20 ft using the formula

$$A = \sqrt{s(s-a)(s-b)(s-c)}$$

5. Calculate the area along the irregular boundary shown by using the trapezoidal rule.

Ties: B = 48.1 ft, C = 52.6 ft, D = 46.8 ft, E = 39.9 ft, F = 43.7 ft, G = 58.0 ft, H = 51.6 ft, and J = 40.0 ft.

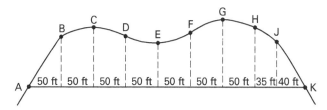

6. Calculate the area of a segment with a long chord of 491.67 ft and a middle ordinate of 98.23 ft.

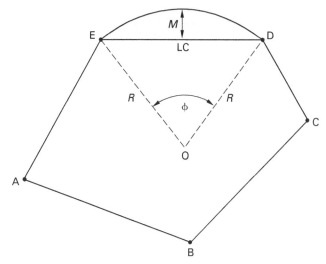

7. Calculate the area of the traverse ABCDEFA by the DMD method.

point	bearing	length	latitude N	latitude S	departure E	departure W
A						
	north	500.00	500		0	
B						
	N45°00′W	848.6	600			600
C						
	S69°27′W	854.4		300		800
D						
	S11°19′W	1019.8		1000		200
E						
	S79°42′E	1118.0		200	1100	
F						
	N51°20′E	640.3	400		500	
A						

8. Calculate the area of the traverse ABCDEA by the DMD method.

point	bearing	length	latitude N	latitude S	departure E	departure W
A						
	N28°19′E	560.27	493.12		265.66	
B						
	N56°23′W	484.18	267.97			403.29
C						
	S08°50′W	375.42		371.04		57.72
D						
	S45°10′W	311.44		219.64		220.91
E						
	S67°45′E	449.83		170.41	416.26	

SOLUTIONS

1.

$$A = \frac{1}{2}\sum (E_1N_2 - E_2N_1) + (E_2N_3 - E_3N_2) + \dots$$
$$+ (E_{n-1}N_n - E_nN_{n-1}) + (E_nN_1 - E_1N_n)$$

$$= \frac{1}{2}\sum \begin{pmatrix} (1000.00 \text{ ft})(1493.12 \text{ ft}) \\ - (1265.66 \text{ ft})(1000.00 \text{ ft}) \\ + (1265.66 \text{ ft})(1761.09 \text{ ft}) \\ - (862.37 \text{ ft})(1493.12 \text{ ft}) \\ + (862.37 \text{ ft})(1390.05 \text{ ft}) \\ - (804.65 \text{ ft})(1761.09 \text{ ft}) \\ + (804.65 \text{ ft})(1170.41 \text{ ft}) \\ - (583.74 \text{ ft})(1390.05 \text{ ft}) \\ + (583.74 \text{ ft})(1000.00 \text{ ft}) \\ - (1000.00 \text{ ft})(1170.41 \text{ ft}) \end{pmatrix}$$

$$= \frac{494{,}128 \text{ ft}^2}{(2)\left(43{,}560 \dfrac{\text{ft}^2}{\text{ac}}\right)}$$

$$= 5.67 \text{ ac}$$

2.

$$A = \frac{1}{2}ab\sin\theta_C$$

$$= \left(\frac{1}{2}\right)(131.8 \text{ ft})(144.8 \text{ ft})\sin 93°58' = 9{,}519 \text{ ft}^2$$

$$= \left(\frac{1}{2}\right)(144.8 \text{ ft})(172.0 \text{ ft})\sin 86°02' = 12{,}423 \text{ ft}^2$$

$$= \left(\frac{1}{2}\right)(172.0 \text{ ft})(140.4 \text{ ft})\sin 93°44' = 12{,}049 \text{ ft}^2$$

$$= \left(\frac{1}{2}\right)(140.4 \text{ ft})(131.8 \text{ ft})\sin 86°16' = 9{,}233 \text{ ft}^2$$

$$= 43{,}224 \text{ ft}^2$$

3.

$$A = \frac{1}{2}ab\sin\theta_C$$

$$= \left(\frac{1}{2}\right)(218.5 \text{ ft})(199.8 \text{ ft})\sin 92°28' = 21{,}808 \text{ ft}^2$$

$$= \left(\frac{1}{2}\right)(231.2 \text{ ft})(231.2 \text{ ft})\sin 93°02' = 26{,}689 \text{ ft}^2$$

$$= 48{,}497 \text{ ft}^2$$

4.

$$A = \sqrt{s(s-a)(s-b)(s-c)}$$

$$= \sqrt{\begin{array}{c}(23 \text{ ft})(23 \text{ ft} - 12 \text{ ft})(23 \text{ ft} - 14 \text{ ft}) \\ \times (23 \text{ ft} - 20 \text{ ft})\end{array}}$$

$$= 83 \text{ ft}^2$$

5.

$$A = \left(\frac{1}{2}\right)(50 \text{ ft})(48.1 \text{ ft})$$

$$+ (50 \text{ ft})\left(\begin{array}{c}\dfrac{48.1 \text{ ft}}{2} + 52.6 \text{ ft} + 46.8 \text{ ft} + 39.9 \text{ ft} \\ + 43.7 \text{ ft} + 58.0 \text{ ft} + \dfrac{51.6 \text{ ft}}{2}\end{array}\right)$$

$$+ (35 \text{ ft})\left(\frac{51.6 \text{ ft} + 40.0 \text{ ft}}{2}\right) + \left(\frac{1}{2}\right)(40.0 \text{ ft})(40.0 \text{ ft})$$

$$= 18{,}148 \text{ ft}^2$$

6.

$$\tan\frac{\phi}{4} = \frac{2M}{C} = \frac{(2)(98.23 \text{ ft})}{491.67 \text{ ft}}$$

$$\phi = 87.122°$$

$$r = \frac{C}{2\sin\dfrac{\phi}{2}} = \frac{491.67 \text{ ft}}{2\sin\left(\dfrac{87.122°}{2}\right)}$$

$$= 356.73 \text{ ft}$$

$$A = r^2\left(\frac{\pi\phi}{360°} - \frac{\sin\phi}{2}\right)$$

$$= (356.73 \text{ ft})^2\left(\frac{\pi(87.122°)}{360°} - \frac{\sin 87.122°}{2}\right)$$

$$= 33{,}203 \text{ ft}^2$$

7. See *Solution to Problem 7* table.

$$A = \frac{3{,}460{,}000 \text{ ft}^2}{(2)\left(43{,}560 \dfrac{\text{ft}^2}{\text{ac}}\right)} = \frac{1{,}730{,}000 \text{ ft}^2}{43{,}560 \dfrac{\text{ft}^2}{\text{ac}}}$$

$$= 39.715 \text{ ac}$$

8. See *Solution to Problem 8* table.

$$A = \frac{494{,}129 \text{ ft}^2}{(2)\left(43{,}560 \dfrac{\text{ft}^2}{\text{ac}}\right)} = \frac{247{,}065 \text{ ft}^2}{43{,}560 \dfrac{\text{ft}^2}{\text{ac}}}$$

$$= 5.672 \text{ ac}$$

Solution to Problem 7

point	bearing	length	latitude N	latitude S	departure E	departure W	DMD	area plus	area minus
A									
	north	500.00	500		0		3200	1,600,000	
B									
	N45°00'W	848.6	600			600	2600	1,560,000	
C									
	S69°27'W	854.4		300		800	1200		360,000
D									
	S11°19'W	1019.8		1000		200	200		200,000
E									
	S79°42'E	1118.0		200	1100		1100		220,000
F									
	N51°20'E	640.3	400		500		2700	1,080,000	
A								4,240,000	780,000
								780,000	
								3,460,000	

Solution to Problem 8

point	bearing	length	latitude N	latitude S	departure E	departure W	DMD	area plus	area minus
A									
	N28°19'E	560.27	493.12		265.66		1098.18	541,535	
B									
	N56°23'W	484.18	267.97			403.29	960.55	257,399	
C									
	S08°50'W	375.42		371.04		57.72	499.54		185,349
D									
	S45°10'W	311.44		219.64		220.91	220.91		48,521
E									
	S67°45'E	449.83		170.41	416.26		416.26		70,935
								798,934	304,805
								304,805	
								494,129	

21 Horizontal Curves

Nomenclature

A	area	ft^2	m^2
A	azimuth	deg	deg
C	chord length	ft	m
D	degree of curve	deg	deg
E	external distance	ft	m
I	deflection angle	deg	deg
L	length of curve	ft	m
LC	length of long chord	ft	m
M	middle ordinate	ft	m
PC	point of curvature	–	–
PT	point of tangency	–	–
R	radius	ft	m
t	tangent	deg	deg
T	tangent distance	ft	m

Symbol

ϕ	angle	deg	deg

1. SIMPLE CURVES

Highways consist of a series of straight sections joined by curved sections. The straight sections are known as *tangents*. The curves are most often circular arcs, known as *simple curves*, but may be spiral curves. Spiral curves are encountered more often on railroads. Initial locations of highways usually consist of straight lines. Curves are later inserted to connect two intersecting

tangents. Many curves of different radii, or degrees of curve, may be selected for any given intersection of tangents.

Figure 21.1 shows that a choice of circular arcs, or curves, can be made after the tangent locations have been made. The curves are usually classified as to their degree of curve, the angle subtended by a portion of the curve 100 ft long. In selecting the degree of curve, consideration is given to design speed, topographic features, economy, and other variables.

Figure 21.1 *Curves Connecting Tangent Lines*

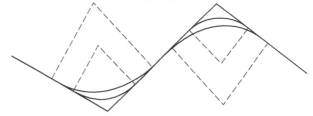

2. GEOMETRY

All the formulas for computations involving circular curves depend on certain principles of geometry and trigonometry.

An *inscribed angle* is an angle that has its vertex on a circle and that has chords for its sides, as shown in Fig. 21.2(a).

An inscribed angle is measured by one-half its intercepted arc, as shown in Fig. 21.2(a).

An angle formed by a tangent and a chord is measured by one-half its intercepted arc, as shown in Fig. 21.2(b).

The radius of a circle is perpendicular to a tangent at the point of tangency, as shown in Fig. 21.3(a).

The perpendicular bisector of a chord passes through the center of the circle, as shown in Fig. 21.3(b).

2 parameters & the stationing of a critical point are needed for a simple curve

Plane Survey Calculations

Figure 21.2 *Tangent and Arc Geometry*

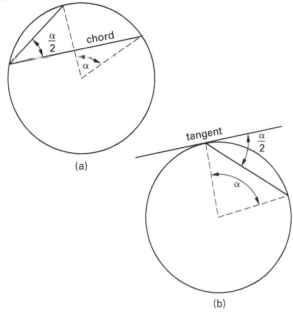

(a)

(b)

Figure 21.3 *Chord Geometry*

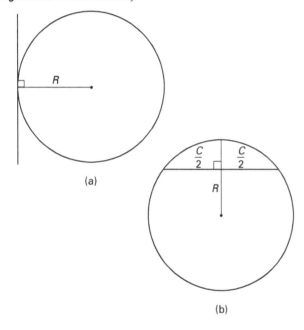

(a)

(b)

3. DEFINITIONS AND SYMBOLS

The following symbols are used in curve computations.

- *C (length of chord):* chord length
- *D (degree of curve):* central angle that subtends a 100 ft arc (arc basis)
- *E (external distance):* distance from PI to middle of curve
- *I (deflection angle):* central angle (angle at PI, or angle at center)

- *L (length of curve):* distance from PC to PT along the arc
- LC *(length of long chord):* distance from PC to PT; chord length for angle ϕ
- *M (middle ordinate):* length of ordinate from middle of long chord to middle of curve
- PC *(point of curvature):* beginning of curve intersect
- PI: point of intersection of back and forward tangents
- PT *(point of tangency):* end of curve
- *R (radius):* a straight line from the center of a circle to the circumference
- *T (tangent distance):* distance from PI to PC, or distance from PI to PT

4. DEFLECTION ANGLE EQUALS CENTRAL ANGLE

In Fig. 21.4, the sum of the interior angles of the polygon O-PC-PI-PT is 360°. The angles at the PC and PT each equal 90°. The sum of the interior angle at the PI and the deflection angle is 180°. The sum of the interior angle at the PI and the central angle is 180°. Therefore, the deflection angle equals the central angle.

Figure 21.4 *Circular Arc*

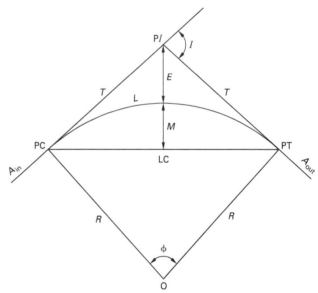

5. HORIZONTAL CURVE FORMULAS

In computing various components of a circular curve, certain formulas derived from trigonometry are useful and necessary. In addition to the usual six trigonometric functions, two others are used in curve computations: *versed sine* (vers) and *external secant* (exsec). Most surveying textbooks include tables for these functions.

$$\text{vers}\,\phi = 1 - \cos\phi \qquad 21.1$$

$$\text{exsec}\,\phi = \sec\phi - 1 \qquad 21.2$$

Other formulas commonly encountered are

$$T = R\tan\frac{I}{2} \qquad 21.3$$

$$L = \frac{\pi R I}{180°} \qquad 21.4$$

$$LC = 2R\sin\frac{I}{2} \qquad 21.5$$

$$M = R\left(1 - \cos\frac{I}{2}\right) = R\,\text{vers}\,\frac{I}{2} \qquad 21.6$$

$$E = R\left(\sec\frac{I}{2} - 1\right) = R\,\text{exsec}\,\frac{I}{2} \qquad 21.7$$

$$R = \frac{LC}{2\sin\dfrac{I}{2}} \qquad 21.8$$

$$I = A_{\text{out}} - A_{\text{in}} \qquad 21.9$$

$$A_{\text{sector}} = \frac{RL}{2} = \frac{\pi R^2 I}{2} \qquad 21.10$$

$$A_{\text{segment}} = \frac{\pi R^2 I}{360} = \frac{R^2\sin I}{2} \qquad 21.11$$

6. DEGREE OF CURVE

There are two definitions of *degree of curve*. The *arc definition* is used for highways and streets. The *chord definition* is used for railroads. (See Fig. 21.5). By the arc definition, degree of curve, D, is the central angle that subtends a 100 ft arc. By the chord definition, degree of curve, D, is the central angle that subtends a 100 ft chord.

Using the arc definition,

$$R = \frac{(360°)(100\text{ ft})}{2\pi D}$$
$$= \frac{5729.58}{D}\quad\left[\begin{array}{l}\text{for }D = 1°,\\ R = 5729.58\text{ ft}\end{array}\right] \qquad 21.12$$

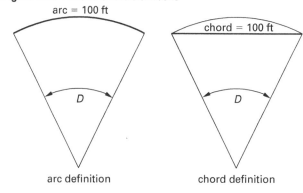

Figure 21.5 *Arc and Chord Definitions*

Using the chord definition,

$$R = \frac{50}{\sin\dfrac{D}{2}}\quad\left[\begin{array}{l}\text{for }D = 1°,\\ R = 5729.65\text{ ft}\end{array}\right] \qquad 21.13$$

When using the arc definition for curve computations with a 100 ft tape to lay out the curve in the field, measurements are actually chord lengths of 100 ft, and the arc length is somewhat greater. For curves up to 4°, the difference in arc length and chord length is negligible. For instance, the chord length for a 100 ft arc on a 4° curve is 99.980 ft.

7. CURVE LAYOUT

Because of their long radii, most curves cannot be laid out by swinging an arc from the center of the circle. They must be laid out by a series of straight lines (chords). This is traditionally done by use of transit and tape.

8. DEFLECTION ANGLE METHOD

The *deflection angle method* is based on the fact that the angle between a tangent and a chord, or between two chords that form an inscribed angle, is one-half the intercepted arc (see Fig. 21.2). In Fig. 21.6, the angle formed by the tangent at the PC and a chord from the PC to a point 100 ft along the arc is equal to one half the degree of curve. Likewise, the angle formed by this chord and a chord from the PC to a point 100 ft farther along the arc is also equal to one-half the degree of curve. These angles are known as *deflection angles*. The deflection angle from the PC to the PT is one-half the central angle I, which provides an important check in computing deflection angles.

Figure 21.6 *Laying Out a Curve*

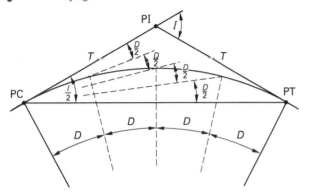

In laying out the curve, the transit is set up at the PC, the PT, or some other point on the curve, and deflection angles are turned for 100 ft arcs along the curve as 100 ft arcs are marked off by a 100 ft tape. For degree of curve up to 4°, the difference in length between chord and arc is slight. For sharper curves, this discrepancy can be corrected by laying out a chord slightly less than 100 ft. This length can be found from Eq. 21.5. On route surveys, stationing is carried continuously along tangent and curve. Thus, the PC and PT will seldom fall on a full station. The first full station will not likely be 100 ft from the PC, and the deflection angle to the first full station will not be $D/2$ but a fraction of $D/2$. When stakes are required on closer intervals than full stations, such as 50 ft, the true length of the 50 ft chord (known as a *subchord*) will be less than 50 ft, and this length can be found by use of Eq. 21.5. (The central angle that subtends a 50 ft arc is $D/2$).

9. LENGTH OF CURVE

The *length of curve*, L (arc definition), is the distance along the arc from the PC to the PT. As any two arcs are proportional to their central angles,

$$L = (100 \text{ ft})\frac{I}{D} \qquad 21.14$$

10. FIELD PROCEDURE IN STAKING A SIMPLE CURVE

These steps can be used when staking out a curve from the PC or PT.

step 1: Measure the deflection angle.

step 2: Select D by considering the design criteria.

step 3: Calculate the tangent distance, T, from the PI to the PC.

Tangent distance for any degree of curve can be found by dividing the tangent distance for a 1° curve by D.

step 4: Measure the tangent distance T from the PI to the PT and set the tacked hub. Measure T from the PI to the PC and set the tacked hub.

step 5: Calculate PC station by subtracting T from the PI station.

step 6: Calculate the length of curve L. Calculate the PT station by adding L to the PC station.

step 7: Calculate the deflection angles at the PC for each station, checking to see that the deflection angle to the PT is $I/2$.

step 8: (a) Set up a transit at the PC and take a foresight on the PI with the telescope normal and with the A vernier set on 0°00′. Turn the deflection angle for each station as the corresponding arc length is marked off by chaining. Make the check at the PT for angle and distance.

(b) Set up the transit at the PT and take a backsight reading on the PC with the telescope normal and with the A vernier set on 0°00′. Turn the deflection angle for each station as the corresponding arc length is marked off by chaining, starting at the PC. Deflection angles are the same as those calculated for the staking curve from the PC. Make the check by sighting on the PI with the deflection angle to the PT.

11. CIRCULAR CURVE COMPUTATIONS

Example 21.1

Compute the parameters of a simple curve.

$$PI = \text{sta } 25+05$$
$$I = 20°$$
$$D = 2°$$

Solution

$$L = (100 \text{ ft})\frac{I}{D} = (100 \text{ ft})\left(\frac{20°}{2°}\right) = 1000 \text{ ft}$$

From Eq. 21.12,

$$R = \frac{5729.58 \text{ ft}}{D} = \frac{5729.58 \text{ ft}}{2} = 2864.79 \text{ ft}$$

$$T = \frac{T_{1° \text{ curve}}}{D} = \frac{1010.28 \text{ ft}}{2} = 505 \text{ ft}$$

Or from Eq. 21.3,

$$T = R\tan\frac{I}{2} = (2864.79 \text{ ft})\tan\frac{20°}{2} = 505 \text{ ft}$$

From Eq. 21.7,

$$E = R\left(\sec \frac{I}{2} - 1\right) = (2864.79 \text{ ft})\left(\sec \frac{20°}{2} - 1\right) = 44.19 \text{ ft}$$

From Eq. 21.6,

$$M = R\left(1 - \cos \frac{I}{2}\right) = (2864.79 \text{ ft})\left(1 - \cos \frac{20°}{2}\right) = 43.52 \text{ ft}$$

$$\begin{aligned}
\text{PI} &= \text{sta } 25{+}05 \\
\text{T} &= \text{sta} - 5{+}05 \\
\text{PC} &= \text{sta } 20{+}00 \\
\text{L} &= +10{+}00 \\
\text{PT} &= \text{sta } 30{+}00
\end{aligned}$$

The deflection angles are

point	station	deflection angle
PC	20+00	
	21+00 = $D/2$	
	22+00 = deflection angle of station	$21{+}00 + D/2 = 2°00'$
	23+00	$22{+}00 + D/2 = 3°00'$
	24+00	$23{+}00 + D/2 = 4°00'$
	25+00	$24{+}00 + D/2 = 5°00'$
	26+00	$25{+}00 + D/2 = 6°00'$
	27+00	$26{+}00 + D/2 = 7°00'$
	28+00	$27{+}00 + D/2 = 8°00'$
	29+00	$28{+}00 + D/2 = 9°00'$
PT	30+00	$29{+}00 + D/2 = 10°00'$

Check: deflection angle to the PT = $I/2 = 10°00'$

12. SHIFTING FORWARD TANGENT

Route locations often require changes in curve locations. One such case involves shifting the forward tangent to a new location parallel to the original tangent and keeping the back tangent in its original location. This produces a change in both the PC and PT stations.

Example 21.2

The forward tangent of the highway curve shown is to be shifted outward so that it will be parallel to and 100 ft from the original tangent. Curve data for the original curve are shown in *Illustration for Example 21.2*. The degree of curve is to remain unchanged. Calculate the PC and PT stations for the new curve.

Solution

$$\begin{aligned}
\text{PI}_1 - \text{PI}_2 &= \frac{100 \text{ ft}}{\sin 60°} = 115.47 \text{ ft} \\
\text{PI}_2 &= 28{+}97.00 + 115.47 \text{ ft} = 30{+}12.47 \\
\text{PC}_2 &= 30{+}12.47 - 551.33 \text{ ft} = 24{+}61.14 \\
\text{PT}_2 &= 24{+}61.14 + 1000.00 \text{ ft} = 34{+}61.14
\end{aligned}$$

13. EASEMENT CURVES

Where simple curves are used, the tangent changes to a curved line at the PC. This means that a vehicle on a tangent arriving at the PC changes direction to a curved path instantaneously. At high speeds, this is impossible. What actually happens in an automobile is that the driver adjusts the steering wheel to make a gradual transition from a straight line to a curved line.

For railroad cars traveling at high speed, the problem is more acute than for automobiles. The rigidity and length of a railroad car cause a sharp thrust on the rails by the wheel flanges. To alleviate this situation, curves that provide a gradual transition from tangent to circular curve (and back again) are inserted between the tangent and the circular curve. These transition curves are known as *easement curves*. These curves also provide a place to increase superelevation from zero to the maximum required for the circular curve. The spiral is a curve that fulfills the requirements for the transition.

14. COMPOUND CURVES

A *compound curve* consists of two or more simple curves with different radii joined together at a common tangent point. Their centers are on the same side of the curve.

Compound curves are not generally used for highways except in mountainous country, because an abrupt change in degree of curve causes a serious hazard even at moderate speeds. They are sometimes used for curvilinear streets in residential subdivisions, however. In Fig. 21.7, the subscript 1 is used for the curve of longer radius, and the subscript 2 is used for the curve of shorter radius. The point of common tangency is called the *point of common curvature*, PCC. The short tangents for the two curves are designated as t_1 and t_2.

Illustration for Example 21.2

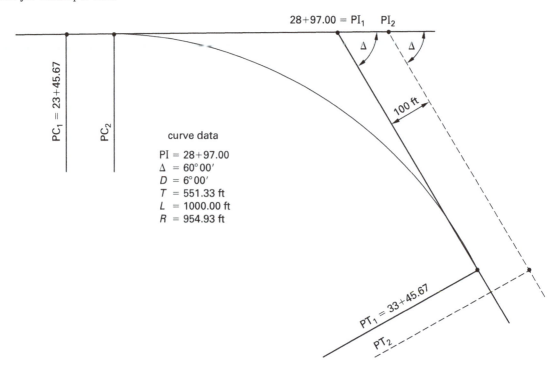

curve data

PI = 28+97.00
Δ = 60°00′
D = 6°00′
T = 551.33 ft
L = 1000.00 ft
R = 954.93 ft

Figure 21.7 *Compound Curve*

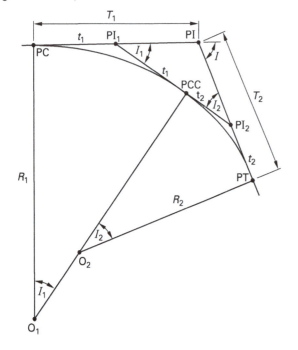

There are seven major parameters of a compound curve: I, I_1, I_2, R_1, R_2, T_1, and T_2. Four of these must be known before calculations can be made. Usually, I is measured, and R_1, R_2, and either Δ_1 or Δ_2 are given. If I, I_1, R_1, and R_2 are given, Eq. 21.15 through Eq. 21.18 can be used.

$$I = I_1 + I_2 \qquad 21.15$$

$$I_2 = I - I_1 \qquad 21.16$$

$$t_1 = R_1 \tan \frac{I_1}{2} \qquad 21.17$$

$$t_2 = R_2 \tan \frac{I_2}{2} \qquad 21.18$$

The triangle PI-PI$_1$-PI$_2$ can be solved by the law of sines. The sine of the angle at the PI equals the sine of I because they are related angles.

$$\text{PI} - \text{PI}_1 = \frac{\sin I_2(t_1 + t_2)}{\sin I} \qquad 21.19$$

$$\text{PI} - \text{PI}_2 = \frac{\sin I_1(t_1 + t_2)}{\sin I} \qquad 21.20$$

$$T_1 = t_1 + \text{PI} - \text{PI}_1 \qquad 21.21$$

$$T_2 = t_2 + \text{PI} - \text{PI}_2 \qquad 21.22$$

To stake out the curve, the PC and PT are located as is done for a simple curve. The PCC is located by establishing the common tangent from either PI$_1$ or PI$_2$. Deflection angles for the two curves are computed separately. The first curve is staked from the PC, and the second curve is staked from the PCC using the common tangent for orientation with the vernier set on 0°00′.

Example 21.3

PC, PCC, and PT stations, deflection angles, and chord lengths are to be calculated from the following information.

$$PI = \text{sta } 15+56.32$$
$$I = 68°00'$$
$$I_1 = 35°00'$$
$$R_1 = 600 \text{ ft}$$
$$R_2 = 400 \text{ ft}$$

Solution

$$I_2 = I - I_1 = 68°00' - 35°00'$$
$$= 33°00'$$

$$t_1 = R_1 \tan\frac{I_1}{2} = (600 \text{ ft})(\tan 17°30')$$
$$= 189.18 \text{ ft}$$

$$t_2 = R_2 \tan\frac{I_2}{2} = (400 \text{ ft})(\tan 16°30')$$
$$= 118.49 \text{ ft}$$

$$PI - PI_1 = \frac{\sin I_2(t_1 + t_2)}{\sin I} = \frac{(\sin 33°)\left(\begin{array}{c}189.18 \text{ ft}\\ +118.49 \text{ ft}\end{array}\right)}{\sin 68°}$$
$$= 180.73 \text{ ft}$$

$$PI - PI_2 = \frac{\sin I_1(t_1 + t_2)}{\sin I} = \frac{(\sin 35°)\left(\begin{array}{c}189.18 \text{ ft}\\ +118.49 \text{ ft}\end{array}\right)}{\sin 68°}$$
$$= 190.33 \text{ ft}$$

$$T_1 = t_1 + PI - PI_1 = 189.18 \text{ ft} + 180.73 \text{ ft}$$
$$= 369.91 \text{ ft}$$
$$T_2 = t_2 + PI - PI_2 = 118.49 \text{ ft} + 190.33 \text{ ft}$$
$$= 308.82 \text{ ft}$$

$$L_1 = \frac{2\pi R_1 I_1}{360°} = \frac{2\pi(600 \text{ ft})(35°)}{360°}$$
$$= 366.52 \text{ ft}$$
$$L_2 = \frac{2\pi R_2 I_2}{360°} = \frac{2\pi(400 \text{ ft})(33°)}{360°}$$
$$= 230.38 \text{ ft}$$

$PI = 15+56.32$

$T_1 = -3+69.91$

$PC = 11+86.41$

$L_1 = +3+66.52$

$PCC = 15+52.93$

$L_2 = +2+30.38$

$PT = 17+83.31$

The deflection angles are

point	station	deflection angles
PC	11+86.41	
	12+00	$\left(\frac{13.59}{366.52}\right)\left(\frac{35}{2}\right) = 0.6489° = 0°39'$
	13+00	$\left(\frac{113.59}{366.52}\right)\left(\frac{35}{2}\right) = 5.4235° = 5°25'$
	14+00	$\left(\frac{213.59}{366.52}\right)\left(\frac{35}{2}\right) = 10.1981° = 10°12'$
	15+00	$\left(\frac{313.59}{366.52}\right)\left(\frac{35}{2}\right) = 14.9728° = 14°59'$
PCC	15+52.93	$\left(\frac{366.52}{366.52}\right)\left(\frac{35}{2}\right) = 17.5000° = 17°30'$
	16+00	$\left(\frac{47.07}{230.38}\right)\left(\frac{33}{2}\right) = 3.3712° = 3°22'$
	17+00	$\left(\frac{147.07}{230.38}\right)\left(\frac{33}{2}\right) = 10.5332° = 10°32'$
PT	17+83.31	$\left(\frac{230.38}{230.38}\right)\left(\frac{33}{2}\right) = 16.5000° = 16°30'$

The chord lengths are

$$C = (1200 \text{ ft}) \sin 0.6489° = 13.59 \text{ ft}$$
$$C = (1200 \text{ ft}) \sin 4.7746° = 99.88 \text{ ft}$$
$$C = (1200 \text{ ft}) \sin 2.5272° = 52.91 \text{ ft}$$
$$C = (800 \text{ ft}) \sin 3.3712° = 47.04 \text{ ft}$$
$$C = (800 \text{ ft}) \sin 7.1620° = 99.74 \text{ ft}$$
$$C = (800 \text{ ft}) \sin 5.9668° = 83.16 \text{ ft}$$

The field notes showing the results of these calculations are shown in *Solution for Example 21.3.*

Solution for Example 21.3

point	station	deflection angle	chord	calculated bearing	curve data
PT	17+83.31	16°30′			
			83.16′		
	17+00.00	10°32′			
			99.74′		
	16+00.00	3°22.3′			$I = 68°00′$
PI	15+56.32		47.04′		$R_1 = 600$ ft
PCC	15+52.93	17°30′			$I_1 = 35°00′$
			52.91′		$R_2 = 400$ ft
	15+00.00	14°58.7′			$I_2 = 33°00′$
			99.88′		$T_1 = 369.91$ ft
	14+00.00	10°12.1′			$T_2 = 308.82$ ft
			99.88′		$L_1 = 366.52$ ft
	13+00.00	5°25.5′	$L_2 = 230.38$ ft		$L_2 = 230.38$ ft
			99.88′		
	12+00.00	0°38.9′			
				13.59′	
	11+86.41	0°00′			

Plane Survey Calculations

15. REVERSE CURVES

Reverse curves are compound curves with deflections in opposite directions (Fig. 21.8(a)). The curves may have equal or unequal radii and/or deflection angles. Each of the curves in a reverse curve is treated in a similar manner as a horizontal curve.

In some situations, good roadway design practice requires a short tangent section between reverse curves (Fig. 21.8(b)). This allows for a reversal of the superelevation of the roadway surface when transitioning between curves with opposite deflections.

Figure 21.8 *Reverse Curves*

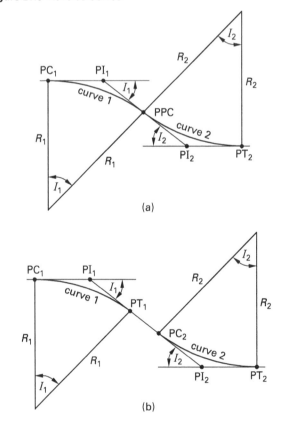

16. PRACTICE PROBLEMS

1. Provide the missing word or words in each sentence.

(a) Highway curves are most often _____ arcs, known as simple curves.

(b) An inscribed angle is an angle that has its vertex on a _____ and that has _____ for its sides.

(c) An inscribed angle is measured by _____ its intercepted arc.

(d) An angle formed by a tangent and a chord is measured by _____ its intercepted arc.

(e) The radius of a circle is _____ to a tangent at the point of tangency.

(f) A perpendicular bisector of a chord passes through the _____ of the circle.

(g) By the arc definition, the degree of curve is the central angle that subtends a 100 ft _____.

(h) By the chord definition, the degree of curve is the central angle that subtends a 100 ft _____.

(i) By the arc definition, the radius of a 1° curve is _____ ft.

(j) The deflection angle for a full station for a 1° curve is _____.

2. Place all symbols pertinent to a circular curve on the figure shown.

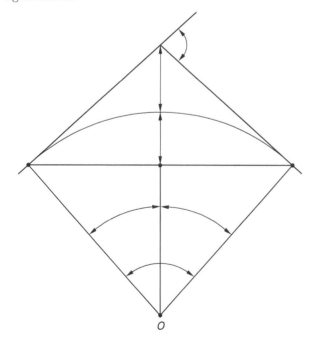

3. Calculate the PC and PT stations and the deflection angles for each full station of the simple highway curve with the data given. Round the tangent distance to the nearest foot.

$$PI = sta\, 25{+}01$$
$$I = 10°$$
$$D = 1°$$

4. Calculate the PC and PT stations and the deflection angles for each full station of the simple highway curve with the data given. Express the length to two decimal places.

$$PI = sta\, 45{+}11.75$$
$$I = 30°$$
$$D = 3°$$

5. Prepare field notes to be used in staking the centerline of a simple horizontal curve for a highway with the data given.

$$PI = sta\, 29 + 62.78$$
$$I = 40\ deg.\ 21\ min.$$
$$D = 5\ deg.\ 15\ min.$$
$$back\ tangent\ bearing = N56°12'\,W$$

6. A preliminary highway location has been made by locating tangents. Deflection angles have been measured at each PI, and the station number of each PI has been established by measuring along the tangents. Circular curves have not been located, but the degree of curve has been established for each curve. The beginning point is at sta 0+00. Using the data given, make necessary calculations to establish stations for PCs and PTs of the curves and for the end of the line.

PI	original station	I	D
1	10+35.27	13°34′	R 1°30′
2	36+15.44	15°18′	L 2°30′
3	52+98.40	18°05′	R 3°00′
end	61+32.77	end of line	

7. In locating a highway, the PI of two tangents falls in a lake and is inaccessible. Point A on the back tangent has been established at sta 26+52.61. Point B has been established on the forward tangent and is visible from point A. Angle A has been measured and found to be 23°13′; angle B has been measured and found to be 19°55′. The length of

AB has been found to be 434.87 ft. Find the stations for a 3° curve.

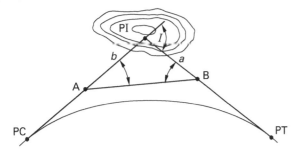

8. The forward tangent of the highway curve shown is to be shifted outward so that it will be parallel to and 50 ft from the original tangent. Data for the original curve are shown in the figure. The degree of curve is to be unchanged. Find the PC and PT stations for the new curve.

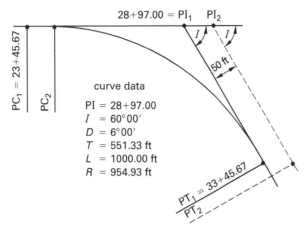

9. Find L for the deflection angle and degree of curve indicated.

(a) $I = 32°18'$

$D = 2°30'$

(b) $I = 41°27'$

$D = 3°15'$

10. Find T for the deflection angle and degree of curve indicated.

(a) $I = 41°51'$

$D = 1°45'$

(b) $I = 39°14'$

$D = 2°15'$

11. Find E for the deflection angle and degree of curve indicated.

(a) $I = 31°30'$

$D = 1°30'$

(b) $I = 42°21'$

$D = 2°45'$

12. Find D for the nearest full degree for the deflection angle and approximate tangent distance indicated.

(a) $I = 32°56'$

$T = 600$ ft

(b) $I = 40°10'$

$T = 1000$ ft

13. Use trigonometric equations to solve the following problems.

(a) Find R for $D = 2°$.

(b) Find D for $R = 1909.86$ ft.

(c) Find T for $I = 34°44'$ and $R = 800$ ft.

(d) Find E for $I = 37°20'$ and $R = 650$ ft.

(e) Find M for $I = 42°51'$ and $R = 800$ ft.

(f) Find LC for $I = 32°55'$ and $R = 850$ ft.

(g) Find the chord length for $D = 8°$, $R = 716.20$ ft, and arc $= 50$ ft.

14. Calculate PC, PCC, and PT stations and deflection angles for full stations for the compound curve with the information given.

PI = sta 14+78.32

$I = 68°00'$

$I_1 = 36°00'$

$R_1 = 400$ ft

$R_2 = 300$ ft

15. Prepare field notes to be used in staking the centerline of the compound curve on full stations, given the data shown.

PI = sta 12+65.35

$I = 70°00'$

$I_1 = 36°00'$

$R_1 = 900$ ft

$R_2 = 600$ ft

16. Calculate the area of the traverse shown to the nearest tenth of an acre.

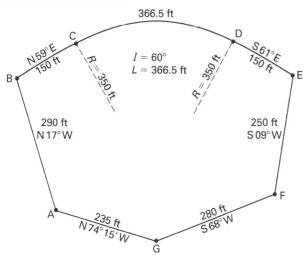

SOLUTIONS

1.

(a) circular	(b) circle; chords
(c) one-half	(d) one-half
(e) perpendicular	(f) center
(g) arc	(h) chord
(i) 5729.58	(j) $0°30'$

2.

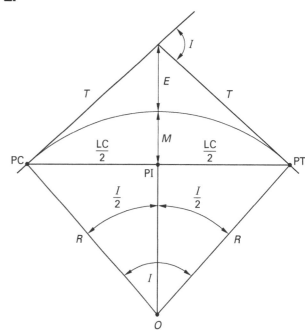

See Sec. 21.3 for definitions of these symbols.

3.

$$L = (100 \text{ ft})\left(\frac{I}{D}\right) = (100 \text{ ft})\left(\frac{10°}{1°}\right) = 1000 \text{ ft}$$

$T = 5{+}01$
$D = 5729.58 \text{ ft}$

The PC and PT stations are

$$
\begin{aligned}
\text{PI} &= 25{+}01 \\
T &= 5{+}01 \\
\text{PC} &= 20{+}00 \\
L &= 10{+}00 \\
\text{PT} &= 30{+}00
\end{aligned}
$$

Plane Survey Calculations

The deflection angles are

point	station	deflection angle
PC	20+00	0°00′
	21+00	0°30′
	22+00	1°00′
	23+00	1°30′
	24+00	2°00′
	25+00	2°30′
	26+00	3°00′
	27+00	3°30′
	28+00	4°00′
	29+00	4°30′
PT	30+00	5°00′

4.

$$L = (100 \text{ ft})\left(\frac{I}{D}\right) = (100 \text{ ft})\left(\frac{30°}{3°}\right) = 1000 \text{ ft}$$

$T = 511.75 \text{ ft}$
$R = 1909.86 \text{ ft}$

The PC and PT stations are

$$PI = 45{+}11.75$$
$$T = 5{+}11.75$$
$$PC = 40{+}00$$
$$L = 10{+}00$$
$$PT = 50{+}00$$

The deflection angles are as follows.

point	station	deflection angle
PC	40+00	0°00′
	41+00	1°30′
	42+00	3°00′
	43+00	4°30′
	44+00	6°00′
	45+00	7°30′
	46+00	9°00′
	47+00	10°30′
	48+00	12°00′
	49+00	13°30′
PT	50+00	15°00′

5.

point	station	deflection angle	calculated chord bearing	curve data
PT	33+30.35	20°10.5″	S83°27′W	
	33+00	19°22.5′		
	32+00	16°45′		
	31+00	14°07.5′		$I = 40°21'$
				$D = 5°15'$
				$T = 401.00 \text{ ft}$
	30+00	11°30′		$L = 768.57 \text{ ft}$
	29+00	8°52.5′		$R = 1091.35 \text{ ft}$
	28+00	6°15′		
	27+00	3°37.5′		
	26+00	1°00′		
PC	25+61.78	0°00′	N56°12′W	

6.

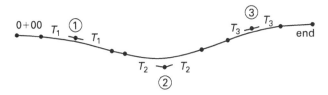

PI	original station	T	L	PC station	PT station	end
1	10+35.27	454.35 ft	904.44 ft	5+80.92	14+85.36	
2	36+15.44	307.83 ft	612.00 ft	33+03.35	39+15.35	
3	52+98.40	303.92 ft	602.78 ft	49+86.56	55+89.34	
end						61+19.79

7.

$$I = 23°13' + 19°55' = 43°08'$$

$$L = (100 \text{ ft})\frac{I}{D} = (100 \text{ ft})\left(\frac{43°08'}{3°}\right)$$
$$= 1437.78 \text{ ft}$$

$$T = R\tan\frac{I}{2}$$
$$= \left(\frac{5729.58}{3}\right)\left(\tan\frac{43°08'}{2}\right) = 754.88 \text{ ft}$$

$$a = \frac{(434.87 \text{ ft})(\sin 23°13')}{\sin(180° - 43°08')} = 250.74 \text{ ft}$$

$$b = \frac{(434.87 \text{ ft})(\sin 19°55')}{\sin(180° - 43°08')} = 216.67 \text{ ft}$$

$$PC = (26{+}52.61) + 216.67 \text{ ft} - 754.88 \text{ ft} = 21{+}14.40$$
$$PT = (21{+}14.40) + 1437.78 \text{ ft} = 35{+}52.18$$

8.

$$\mathrm{PI}_1 - \mathrm{PI}_2 = \frac{50 \text{ ft}}{\sin 60°} = 57.74 \text{ ft}$$
$$\mathrm{PC} = (23{+}45.67) + 57.74 = 24{+}03.41$$
$$\mathrm{PT} = (24{+}03.41) + 1000.00 = 34{+}03.41$$

9.

(a)

$$L = (100 \text{ ft})\left(\frac{I}{D}\right)$$
$$= (100 \text{ ft})\left(\frac{32°18'}{2°30'}\right)$$
$$= 1292.00 \text{ ft}$$

(b)

$$L = (100 \text{ ft})\left(\frac{I}{D}\right)$$
$$= (100 \text{ ft})\left(\frac{41°27'}{3°15'}\right)$$
$$= 1275.38 \text{ ft}$$

10.

(a)

$$T = R \tan \frac{I}{2}$$
$$= \left(\frac{5729.58 \text{ ft}}{1°45'}\right)\left(\tan \frac{41°51'}{2}\right)$$
$$= 1251.87 \text{ ft}$$

(b)

$$T = R \tan \frac{I}{2}$$
$$= \left(\frac{5729.58 \text{ ft}}{2°15'}\right)\left(\tan \frac{39°14'}{2}\right)$$
$$= 907.60 \text{ ft}$$

11.

(a)

$$E = \frac{223.51 \text{ ft}}{1°30'} = 149.01 \text{ ft}$$

(b)

$$E = \frac{414.86 \text{ ft}}{2°45'} = 150.86 \text{ ft}$$

12.

(a)

$$D = \left(\frac{5729.58 \text{ ft}}{600 \text{ ft}}\right)\left(\tan \frac{32°56'}{2}\right) = 3°$$

(b)

$$D = \left(\frac{5729.58 \text{ ft}}{1000 \text{ ft}}\right)\left(\tan \frac{40°10'}{2}\right) = 2°$$

13.

(a)

$$R = \frac{5729.58 \text{ ft}}{2} = 2864.79 \text{ ft}$$

(b)

$$D = \frac{5729.58 \text{ ft}}{1909.86 \text{ ft}} = 3°$$

(c)

$$T = R \tan \frac{I}{2} = (800 \text{ ft})\left(\tan \frac{34°44'}{2}\right)$$
$$= 250.19 \text{ ft}$$

(d)

$$E = (650 \text{ ft})\left(\sec \frac{37°20'}{2} - 1\right)$$
$$= (650 \text{ ft})\left(\frac{1}{\cos \frac{37°20'}{2}} - 1\right)$$
$$= 36.09 \text{ ft}$$

(e)

$$M = R\left(1 - \cos \frac{I}{2}\right)$$
$$= (800 \text{ ft})\left(1 - \cos \frac{42°51'}{2}\right)$$
$$= 55.28 \text{ ft}$$

Plane Survey Calculations

(f)

$$LC = (2)(850 \text{ ft})\left(\sin\frac{32°55'}{2}\right)$$
$$= 481.64 \text{ ft}$$

(g)

$$\text{chord} = (2)(716.20 \text{ ft})\left(\sin\frac{4°}{2}\right)$$
$$= 49.99 \text{ ft}$$

14.

$$I_2 = 68° - 36°$$
$$= 32°00'$$
$$t_1 = (400 \text{ ft})(\tan 18°)$$
$$= 129.97 \text{ ft}$$
$$t_2 = (300 \text{ ft})(\tan 16°)$$
$$= 86.02 \text{ ft}$$
$$PI - PI_1 = \frac{(215.99 \text{ ft})(\sin 32°)}{\sin 68°}$$
$$= 123.45 \text{ ft}$$
$$PI - PI_2 = \frac{(215.99 \text{ ft})(\sin 36°)}{\sin 68°}$$
$$= 136.93 \text{ ft}$$

$$T_1 = 129.97 + 123.45$$
$$= 253.42 \text{ ft}$$
$$T_2 = 86.02 + 136.93$$
$$= 222.95 \text{ ft}$$
$$L_1 = \left(\frac{36}{360}\right)(2\pi)(400)$$
$$= 251.33 \text{ ft}$$
$$L_2 = \left(\frac{32}{360}\right)(2\pi)(300)$$
$$= 167.55 \text{ ft}$$
$$PI = 14+78.32$$

$$T_1 = 2+53.42$$
$$PC = 12+24.90$$
$$L_1 = 2+51.33$$
$$PCC = 14+76.23$$
$$L_2 = 1+67.55$$
$$PT = 16+43.78$$

point	station		deflection angle
PC	12+24.90		0°00'
	13+00	$\left(\frac{75.10 \text{ ft}}{251.33 \text{ ft}}\right)\left(\frac{36°}{2}\right) = 5.3786°$	5°23'
	14+00	$\left(\frac{175.10 \text{ ft}}{251.33 \text{ ft}}\right)\left(\frac{36°}{2}\right) = 12.5405°$	12°32'
PCC	14+76.23	$\left(\frac{251.33 \text{ ft}}{251.33 \text{ ft}}\right)\left(\frac{36°}{2}\right) = 18.0000°$	18°00'
	15+00	$\left(\frac{23.77 \text{ ft}}{167.55 \text{ ft}}\right)\left(\frac{32°}{2}\right) = 2.2699°$	2°16'
	16+00	$\left(\frac{123.77 \text{ ft}}{167.55 \text{ ft}}\right)\left(\frac{32°}{2}\right) = 11.8193°$	11°49'
PT	16+43.78	$\left(\frac{167.55 \text{ ft}}{167.55 \text{ ft}}\right)\left(\frac{32°}{2}\right) = 16.0000°$	16°00'

15.

point	station	deflection angle	curve data
PT	16+11.28	17°00'	
	16+00	16°28'	
	15+00	11°41'	
			$I = 70°$
	13+00	2°08'	$R_1 = 900 \text{ ft}$
			$R_2 = 600 \text{ ft}$
PCC	12+55.23	18°00'	$I_1 = 36°$
			$I_2 = 34°$
	12+00	16°15'	$T_1 = 575.61 \text{ ft}$
			$T_2 = 481.10 \text{ ft}$
	11+00	13°04'	$L_1 = 565.49 \text{ ft}$
			$L_2 = 356.05 \text{ ft}$
	10+00	9°53'	
	9+00	6°42'	
	8+00	3°31'	
	7+00	0°20'	
PC	6+89.74		

Table for Solution 16

line	bearing	distance (ft)	latitude	departure (ft)	DMD	area (ft^2)
AB	N17°W	290	+277.3	−84.8	84.8	+23,515
BC	N59°E	150	+77.3	+128.6	128.6	+9941
CD	N89°E	350	+6.1	+349.9	607.1	+3703
DE	S61°E	150	−72.7	+131.2	1088.2	−79,112
EF	S09°W	250	−246.9	−39.1	1180.3	−291,416
FG	S68°W	280	−104.9	−259.6	881.6	−92,480
GA	N74°15′ W	235	+63.8	−226.2	395.8	+25,252
			0.0	0.0		400,597
						400,597/2 = 200,298

16. Bearings, distances, and other values for the traverse are given in *Table for Solution 16*.

$$A_{\text{segment}} = \frac{\phi \pi r^2}{360°} - \frac{r^2 \sin\phi}{2}$$

$$= \frac{(60°)(3.1416)(350\text{ ft}^2)^2}{360°} - \frac{(350\text{ ft}^2)^2 \sin(60°)}{2}$$

$$= 11,097\text{ ft}^2$$

The total area is

$$200,298\text{ ft}^2 + 11,097\text{ ft}^2 = 211,395\text{ ft}^2$$

$$A_{\text{total}} = \frac{211,396\text{ ft}^2}{43,560\ \dfrac{\text{ft}^2}{\text{ac}}} = 4.9\text{ ac}$$

22 Vertical Alignment

Nomenclature

E	elevation	ft	m
g	gradient		
L	length	sta	sta
r	rate of change	%/sta	%/sta

1. GRADE (GRADIENT)

In highway construction, the slope of the line that is the profile of the centerline is known as the *grade* or *gradient*, *g*. The grade of a highway is calculated in the same way that the slope of a line is calculated. Horizontal distances are usually expressed in stations; vertical distances are expressed in feet.

Example 22.1

Determine the gradient of a highway that has a centerline elevation of 444.50 ft at sta 20+75.00 and a centerline elevation of 472.20 ft at sta 32+25.00.

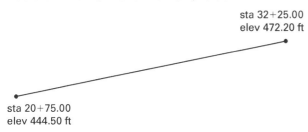

sta 32+25.00
elev 472.20 ft

sta 20+75.00
elev 444.50 ft

Solution

$$g = \frac{472.20 \text{ ft} - 444.50 \text{ ft}}{3225.00 \text{ ft} - 2075.00 \text{ ft}} = \frac{27.70 \text{ ft}}{1150.00 \text{ ft}}$$
$$= +0.02409 \text{ ft/ft}$$

If the numerator is expressed in feet and the denominator is expressed in stations, the decimal point in the gradient will move two places to the right. The gradient can then be expressed as a percent. In this form, gradient expresses change in elevation per station. For Ex. 22.1,

$$g = \frac{27.70 \text{ ft}}{11.50 \text{ sta}} = 2.409 \text{ ft/sta} \quad (+2.409\%)$$

2. POINTS OF INTERSECTION

Vertical alignment for a highway is located similarly to horizontal alignment. Straight lines are located from point to point, and vertical curves are inserted. The points of intersecting gradients are known as *points of intersection*. These lines, after vertical curves have been inserted, are the centerline profile of the highway. Usually the profile is the finish elevation (pavement) profile, but it may be the subgrade (earthwork) profile.

3. TANGENT ELEVATIONS

After points of intersection have been located and connected by tangents (straight lines), elevations of each station on the tangent need to be determined before finding elevations on the vertical curve. The gradient is the change in elevation per station. If the station number and elevation of each PI is known, the elevation at each station can be calculated. The gradient should be calculated as a percentage with three decimal places. Finish elevations should be calculated to two decimal places in feet (hundredths of a foot).

Example 22.2

From the information shown, calculate the gradient of each tangent and the elevation at each full station on the tangents.

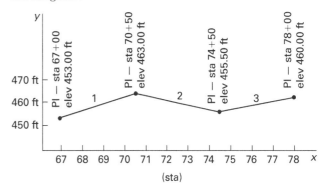

Plane Survey Calculations

Solution

$$g_1 = \frac{463.00 \text{ ft} - 453.00 \text{ ft}}{70.50 \text{ sta} - 67.00 \text{ sta}}$$
$$- 2.857 \text{ ft/sta} \quad (+2.857\%)$$

$$g_2 = \frac{455.50 \text{ ft} - 463.00 \text{ ft}}{74.50 \text{ sta} - 70.50 \text{ sta}}$$
$$= -1.875 \text{ ft/sta} \quad (-1.875\%)$$

$$g_3 = \frac{460.00 \text{ ft} - 455.50 \text{ ft}}{78.00 \text{ sta} - 74.50 \text{ sta}}$$
$$= 1.286 \text{ ft/sta} \quad (+1.286\%)$$

Elevation Calculations

point		calculation	elevation (ft)
PI	67+00		453.00
	68+00	$453.00 + (1)(2.857)$	455.86
	69+00	$453.00 + (2)(2.857)$	458.71
	70+00	$453.00 + (3)(2.857)$	461.57
PI	70+50	$453.00 + (3.5)(2.857)$	463.00
	71+00	$463.00 - (0.5)(1.875)$	462.06
	72+00	$463.00 - (1.5)(1.875)$	460.19
	73+00	$463.00 - (2.5)(1.875)$	458.31
	74+00	$463.00 - (3.5)(1.875)$	456.44
PI	74+50	$463.00 - (4.0)(1.875)$	455.50
	75+00	$455.50 + (0.5)(1.286)$	456.14
	76+00	$455.50 + (1.5)(1.286)$	457.43
	77+00	$455.50 + (2.5)(1.286)$	458.72
	78+00	$455.50 + (3.5)(1.286)$	460.00

4. VERTICAL CURVES

Just as horizontal curves connect two tangents in horizontal alignment, vertical curves connect two tangents in vertical alignment. However, a horizontal curve is usually an arc of a circle, and a vertical curve is usually a parabola. Vertical curves may be *sag curves* or *crest curves* (Fig. 22.1). They are usually symmetrical about the point of intersection of the two tangents. As a result, the downward and upward tangents are equal in length, even though the high or low points of the curve may not correspond with the station of the point of intersection of the tangents.

For parabolic equal tangent vertical curves, the rate of change, r, for the curve may be determined by Eq. 22.1, where the gradients of the two intersecting grades are g_1

Figure 22.1 *Vertical Curves*

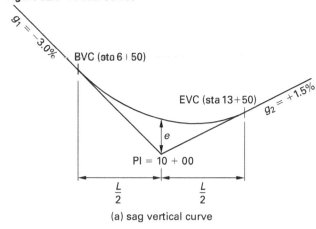

(a) sag vertical curve

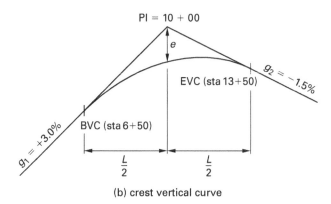

(b) crest vertical curve

and g_2 and the total length of the curve, L, is expressed in stations, where 100 ft = one station (or where 100 meters = one station).

$$r = \frac{g_2 - g_1}{L} \qquad \text{22.1}$$

For parabolic equal tangent vertical curves, when the total length of curve and the station of the PI is known, the station of the beginning of vertical curve (BVC) and end of vertical curve (EVC) may be determined by simply subtracting or adding half the total length of the curve from or to the station of the PI. When the elevation of the PI is known, the elevation of the BVC may be determined by Eq. 22.2.

$$E_{\text{BVC}} = E_{\text{PI}} - g_1\left(\frac{L}{2}\right) \qquad \text{22.2}$$

Finish elevations along a vertical curve may be calculated by the basic equation for a parabola, Eq. 22.3, where x is the distance in stations from the BVC and y is in the units of the elevation of the BVC.

$$y = \left(\frac{r}{2}\right)x^2 + xg_1 + E_{\text{BVC}} \qquad \text{22.3}$$

Other useful equations for vertical curves are as follows:

$$E_{EVC} = E_{PI} - g_2\left(\frac{L}{2}\right) \qquad 22.4$$

$$g_1 = \frac{E_{PI} - E_{BVC}}{\frac{L}{2}} \qquad 22.5$$

$$g_2 = \frac{E_{PI} - E_{EVC}}{\frac{L}{2}} \qquad 22.6$$

$$sta\,BVC = sta\,PI - \frac{L}{2} \qquad 22.7$$

$$sta\,EVC = sta\,PI + \frac{L}{2} \qquad 22.8$$

$$sta\,EVC = sta\,BVC + L \qquad 22.9$$

Example 22.3

A -1.50% grade meets a $+2.25\%$ grade at sta 36+50, elev 452.00 ft. A vertical curve of length 600 ft (six stations) will be used. The elevation at the BVC is 456.50 ft. Calculate the finish elevations for each full station from the BVC to the EVC.

Solution

From the information provided, the station of the BVC is (36+50) − (3+00), or (33+50). The station of the EVC is (36+50) + (3+00), or (39+50).

Using Eq. 22.1,

$$r = \frac{g_2 - g_1}{L} = \frac{(2.25\% + 1.50\%)}{6\,sta} = 0.625$$

Using Eq. 22.3, for station 34+00,

$$y = \left(\frac{r}{2}\right)x^2 + xg_1 + E_{BVC}$$
$$= \left(\frac{0.625}{2}\right)(34.0 - 33.5)^2 + (34.0 - 33.5)(-1.50) + 456.50\,ft$$
$$= 455.83\,ft$$

Elevations for other stations can be calculated similarly.

point	station	finish elevation (ft)
PC	33+50	456.50
	34+00	455.83
	35+00	454.95
	36+00	454.70
PI	36+50	454.81
	37+00	455.07
	38+00	456.07
	39+00	457.70
PT	39+50	458.75

5. TURNING POINT ON VERTICAL CURVE

Unless the incoming and outgoing grades are the same, the highest or lowest point on a vertical parabolic curve is not vertically below (or above) the PI. Yet the location of that point is critical for drainage purposes. This point is called the *turning point*. The distance x from the PC to the turning point can be found from Eq. 22.10.

$$x = \frac{-g_1}{r} \qquad 22.10$$

Example 22.4

A $+1.500\%$ grade meets a -2.500% grade at sta 12+50. Determine the distance from the PC to the turning point if a 600 ft vertical curve is used.

Solution

$$r = \frac{g_2 - g_1}{L}$$
$$= \frac{(-2.50\% - 1.50\%)}{6\,sta}$$
$$= -0.66667\,\%/sta$$

$$x = \frac{-g_1}{r}$$
$$= \frac{-1.50\%}{-0.66667\,\frac{\%}{sta}}$$
$$= 2.25\,sta \quad (225\,ft)$$

6. PRACTICE PROBLEMS

1. Determine the gradients between the points on the highway profiles shown.

Example:

$$PI = 5+50$$
$$E = 452.00\,ft$$
$$PI = 8+50$$
$$E = 455.00\,ft$$
$$PI = 11+00$$
$$E = 453.00\,ft$$

Solution:

$$g_1 = \frac{455.00\,ft - 452.00\,ft}{8.50\,sta - 5.50\,sta}$$
$$= 1.00\,ft/sta \quad (+1.000\%)$$
$$g_2 = \frac{455.00\,ft - 453.00\,ft}{11.00\,sta - 8.50\,sta}$$
$$= -0.800\,ft/sta \quad (-0.800\%)$$

(a)

$$PI = 20+70$$
$$E = 504.00 \text{ ft}$$
$$PI = 23+50$$
$$E = 498.00 \text{ ft}$$
$$PI = 26+60$$
$$E = 503.00 \text{ ft}$$

(b)

$$PI = 40+00$$
$$E = 461.00 \text{ ft}$$
$$PI = 46+00$$
$$E = 459.00 \text{ ft}$$
$$PI = 52+00$$
$$E = 465.00 \text{ ft}$$

(c)

$$PI = 55+00$$
$$E = 474.00 \text{ ft}$$
$$PI = 59+00$$
$$E = 469.00 \text{ ft}$$
$$PI = 64+00$$
$$E = 477.50 \text{ ft}$$
$$PI = 67+00$$
$$E = 477.50 \text{ ft}$$

(d)

$$PI = 29+25$$
$$E = 445.00 \text{ ft}$$
$$PI = 32+50$$
$$E = 432.00 \text{ ft}$$
$$PI = 37+75$$
$$E = 432.00 \text{ ft}$$
$$PI = 41+00$$
$$E = 437.70 \text{ ft}$$

2. Calculate the gradient for each tangent of the highway profile shown, and the elevation of each full station on the tangents.

point	station	tangent elevation (ft)
PI	25+00	466.00
PI	31+00	458.00
PI	35+50	472.00
PI	39+00	465.00
PI	43+00	472.00

3. A +1.25% grade meets a −2.75% grade at sta 18+00, at an elevation of 270.19 ft. The vertical curve is 200 ft long. Calculate the elevation on the curve at half station intervals, beginning at the BVC.

4. Determine the station and elevation of the highest point on the vertical curve in Prob. 3.

SOLUTIONS

1.

(a)

$$g_1 = \frac{504.00 \text{ ft} - 498.00 \text{ ft}}{23.50 \text{ sta} - 20.70 \text{ sta}}$$
$$= -2.143 \text{ ft/sta} \quad (-2.143\%)$$

$$g_2 = \frac{503.00 \text{ ft} - 498.00 \text{ ft}}{26.60 \text{ sta} - 23.50 \text{ sta}}$$
$$= 1.613 \text{ ft/sta} \quad (+1.613\%)$$

(b)

$$g_1 = \frac{461.00 \text{ ft} - 459.00 \text{ ft}}{46.00 \text{ sta} - 40.00 \text{ sta}}$$
$$= -0.333 \text{ ft/sta} \quad (-0.333\%)$$

$$g_2 = \frac{465.00 \text{ ft} - 459.00 \text{ ft}}{52.00 \text{ sta} - 46.00 \text{ sta}}$$
$$= 1.000 \text{ ft/sta} \quad (+1.000\%)$$

(c)

$$g_1 = \frac{474.00 \text{ ft} - 469.00 \text{ ft}}{59.00 \text{ sta} - 55.00 \text{ sta}}$$
$$= -1.250 \text{ ft/sta} \quad (-1.250\%)$$

$$g_2 = \frac{477.50 \text{ ft} - 469.00 \text{ ft}}{64.00 \text{ sta} - 59.00 \text{ sta}}$$
$$= 1.700 \text{ ft/sta} \quad (+1.700\%)$$

$$g_3 = \frac{477.50 \text{ ft} - 477.50 \text{ ft}}{67.00 \text{ sta} - 64.00 \text{ sta}}$$
$$= 0 \text{ ft/sta} \quad (0\%)$$

(d)

$$g_1 = \frac{445.00 \text{ ft} - 432.00 \text{ ft}}{32.50 \text{ sta} - 29.25 \text{ sta}}$$
$$= -4.000 \text{ ft/sta} \quad (-4.000\%)$$

$$g_2 = \frac{432.00 \text{ ft} - 432.00 \text{ ft}}{37.75 \text{ sta} - 32.50 \text{ sta}}$$
$$= 0 \text{ ft/sta} \quad (0\%)$$

$$g_3 = \frac{437.70 \text{ ft} - 432.00 \text{ ft}}{41.00 \text{ sta} - 37.75 \text{ sta}}$$
$$= 1.754 \text{ ft/sta} \quad (+1.754\%)$$

2.

point	station	tangent elevation (ft)
PI	25+00	466.00
	26+00	464.67
	27+00	463.33
$g_1 = -1.333\%$	28+00	462.00
	29+00	460.67
	30+00	459.33
PI	31+00	458.00
	32+00	461.11
	33+00	464.22
$g_2 = +3.111\%$	34+00	467.33
	35+00	470.44
PI	35+50	472.00
	36+00	471.00
$g_3 = -2.000\%$	37+00	469.00
	38+00	467.00
PI	39+00	465.00
	40+00	466.75
$g_4 = +1.750\%$	41+00	468.50
	42+00	470.25
PI	43+00	472.00

3. The station at the BVC is

$$\text{sta BVC} = 18+00 - \frac{L}{2} = 17+00$$

From Eq. 22.1,

$$r = \frac{g_2 - g_1}{L}$$
$$= \frac{-2.75\% - 1.25\%}{2 \text{ sta}}$$
$$= -2.00\%/\text{sta}$$

$$E_{\text{BVC}} = E_{\text{PI}} - g_1\left(\frac{L}{2}\right)$$
$$= (270.19 \text{ ft} - 1.25\%)\left(\frac{2 \text{ sta}}{2}\right)$$
$$= 268.94 \text{ ft}$$

Therefore, the elevation of any point on this curve may be calculated using a modification of Eq. 22.3.

$$y = \frac{r}{2}x^2 + g_1x + E_{\text{BVC}} = -x^2 + 1.25x + 268.94 \text{ ft}$$

sta 17+50: $y = -(0.5 \text{ ft})^2 + (1.25)(0.5 \text{ ft}) + 268.94 \text{ ft}$
$\qquad = 269.32 \text{ ft}$

sta 18+00: $y = -(1 \text{ ft})^2 + (1.25)(1 \text{ ft}) + 268.94 \text{ ft}$
$\qquad = 269.19 \text{ ft}$

sta 18+50: $y = -(1.5 \text{ ft})^2 + (1.25)(1.5 \text{ ft}) + 268.94 \text{ ft}$
$\qquad = 268.56 \text{ ft}$

sta 19+00: $y = -(2.0 \text{ ft})^2 + (1.25)(2.0 \text{ ft}) + 268.94 \text{ ft}$
$\qquad = 267.44 \text{ ft}$

4. From Eq. 22.10,

$$x = -\frac{g_1}{r} = -\frac{1.25\%}{-2.00 \ \frac{\%}{\text{sta}}} = 0.625 \text{ sta}$$

$$\text{sta}_{\text{high point}} = \text{sta}_{\text{BVC}} + 0.625 = 17 + 62.5$$

$$E = -x^2 + 1.25x + 268.94 \text{ ft}$$
$$= -(0.625 \text{ ft})^2 + (1.25)(0.625 \text{ ft}) + 268.94 \text{ ft}$$
$$= 269.33 \text{ ft}$$

Plane Survey Calculations

23 Map Projections and Plane Coordinate Systems

Nomenclature

a	length of semimajor axis	ft	m
A	azimuth	deg	deg
b	length of semiminor axis	ft	m
C	correction	various	various
d	distance	ft	m
D	observed distance	ft	m
f	flattening factor	–	–
h	ellipsoid height	ft	m
H	orthometric height	ft	m
H	elevation above sea level or geoid	ft	m
k	scale factor	–	–
N	geoidal separation	ft	m
R	radius	ft	m
S	distance on the ellipsoid	ft	m

Symbols

θ	angle	deg	deg
ϕ	latitude	deg	deg

1. INTRODUCTION TO MAP PROJECTIONS

Most land surveying processes are ultimately used to create maps of the features surveyed. Therefore, understanding how to create precise maps based on measurements from the earth's curved face is essential. As early as the sixth century B.C., Greek scholars were proposing methods to project the earth's curvature onto a plane surface.

Many scholars have contributed to the current understanding of map projection, but possibly the greatest contribution was made by Claudius Ptolemaeus (c. AD 100–170), also known as Ptolemy. As the last great geographer of the Roman Empire, he is often called the father of modern surveying and geography. Ptolemy is believed to have invented the terms *latitude* and *longitude* and used them to record location. In his most famous geographical work, the *Geography*, he devised how to project the earth's curved face onto a plane surface.

2. THE FIGURE OF THE EARTH

The earth is often considered spherical. However, though it approximates a sphere, it flattens near the polar regions, bulges nearer the equator, and has an irregular surface even over the oceans. One approach to describing this irregular surface is to use an equipotential surface (i.e., a surface closely approximating the mean sea level where the gravity potential is equal).

This surface, which can also be considered as the level of the undisturbed sea, is called the *geoid*. The uneven distribution of the earth's mass makes the geoid irregular. It has no complete mathematical expression, although models for selected areas of the geoid have been constructed.

Since it is impractical to describe the geoid mathematically, a more simple figure is used for coordinate calculation purposes. That figure is an *ellipsoid of revolution*, one that would result if an ellipse is spun about its minor axis as shown in Fig. 23.1. Ellipsoids are typically described by two factors: the flattening factor and the semi-major axis. The *flattening factor*, f, is the ratio of the difference between the length of the semimajor axis, a, and the length of the semiminor axis, b, to the semimajor axis. It can be calculated using Eq. 23.1.

$$f = \frac{a - b}{a} \qquad 23.1$$

Prior to the existence of artificial satellites, the *Clarke 1866 Ellipsoid* was used as a basis for geodetic coordinates. It worked well for North America, Central America, and Greenland, and was used as the basis for datums (i.e., reference frames used to make position measurements) such as the North American Datum of 1927 (NAD 27). In South America, the *South America 1969 Ellipsoid* served the same purpose.

With the advent of satellites, it was necessary to create an ellipsoid to fit the entire earth. As a result, an earth-centered ellipsoid identified as the *Geodetic Reference System of 1980* (GRS 80) was developed. GRS 80 has a semi-major axis length of 6,378,137 m and a flattening factor, f, of 1/298.257222101. The United States Department of Defense uses the *World Geodetic System*

Figure 23.1 *Ellipse*

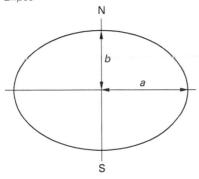

of 1984 (WGS 84), which has the same semi-major axis length, but a slightly different flattening factor (1/298.257223563). For most practical purposes, the differences between GRS 80 and WGS 84 are minimal and the two ellipsoids can be considered equivalent.

Most elevations used for topographic purposes are *orthometric heights*, or heights above the irregular surface of the geoid. Since the height above the geoid closely approximates the height above mean sea level, it is extensively used in surveying and mapping, and is widely displayed on most charts and maps.

Elevations produced by the global positioning system (GPS) are referenced to the ellipsoid since it is a predictable, mathematically perfect shape. As a result, there may be significant differences between elevations referenced to the ellipsoid (ellipsoid height, h) as determined by GPS, and elevations referenced to the geoid (orthometric height, H). It is necessary to know the geoidal separation, N, or the distance between the ellipsoid and the geoid, to convert a GPS elevation to an orthometric elevation at any given point. Those differences vary globally by well over a hundred meters. Models of the geoid surface relative to the ellipsoid may be derived from gravity observations or by observation of satellite orbits. Figure 23.2 gives an example of the relationship between the ellipsoid and the geoid. Worldwide models, such as the *Earth Geopotential Model 1996* (EGM 96), provide global coverage, although they are limited to accuracies of about one meter. More precise models have been developed in countries with dense gravity observation networks to provide a difference between the geoid and ellipsoid within a few centimeters. The current objective of the National Oceanic and Atmospheric Administration's (NOAA) U.S. National Geodetic Survey (NGS) is to develop a model for North America with a precision of one centimeter.

Knowing the geoidal separation, an orthometric height can be calculated from a GPS-derived ellipsoid height using Eq. 23.2.

$$H = h - N \qquad 23.2$$

Figure 23.2 *Typical Relationship of the Ellipsoid and Geoid in the United States*

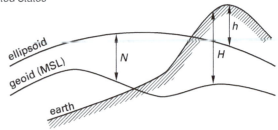

note: geoidal separation is negative

3. GEOGRAPHIC COORDINATE SYSTEMS

Most models of surveyed features use either a graphic or digital format. Survey measurements are typically related to a spatial coordinate system, even if the system is relative and not related to a geodetic datum. However, with modern survey practice, surveys using relative coordinate systems are becoming increasingly rare, as most current survey data must be geo-referenced.

Technological developments such as GPS, geographic information systems (GIS), and the availability of remote sensing data taken from satellites make it increasingly important to not only produce survey data related to a geographic coordinate system but also to be able to convert data among these various systems.

The most familiar geospatial coordinate system is the *plane rectangular Cartesian coordinate system*, which involves linear measurements in two directions from a pair of perpendicular fixed axes. Most plane surveying calculations are based on this system and on the assumption that the earth is a plane. While a plane rectangular system works well with relatively small areas, when large geographical areas are involved, or when a local survey must be related to other areas, allowances must be made for the earth's true shape.

Latitude and Longitude

A *spherical* or *spheroidal grid system* is used when it is necessary to describe a location's features using a global perspective, or when it is necessary to allow for the earth's curvature over a large area. This type of system involves a parallel system of lines of latitude measuring the distance north or south of the equator and a second system of longitude lines measuring the distance east or west of a designated north-south reference line as shown in Fig. 23.3.

Latitude is the angle measured from the center of the earth (or a spheroid representing the earth) northerly or southerly from the equator. Latitude varies from zero at the equator to 90° at the poles and must be designated as either north or south. Latitude grid lines are called *parallels*. Each degree of latitude represents approximately 69 statute miles on the surface of the earth.

approx L of 1° of Longitude = 69.186 miles

Equator D = 7928 mi
Circumf = 24,901 mi

Longitude is the angle measured from the center of the earth (or a spheroid representing the earth) easterly or westerly from a north-south line of reference, known as the prime meridian. Longitude lines are called *meridians*. The *prime meridian* is the meridian passing through the Royal Observatory in Greenwich, England, and is used by international agreement. Along the equator, each degree of longitude represents approximately 69 statute miles on the surface of the earth. However, that distance becomes increasingly smaller toward either pole due to convergence of the meridians. The length of a degree or distance along a meridian at any given latitude can be approximated using Eq. 23.3.

$$d = (69 \text{ mi})\cos\phi \qquad 23.3$$

Although the ellipsoid defines the shape of the earth when used for a geodetic coordinate system, a horizontal datum is needed to identify the ellipsoid's origin and orientation of the coordinate system. While all regions of the earth may use a common geocentric ellipsoid (e.g., GRS 80), typically each country or region utilizes a different datum—one that best fits between the geoid and the ellipsoid in that area. A point of origin is typically chosen where the geoid-ellipsoid separation is zero; therefore, local datums are usually better than datums covering a larger region.

Based on the GRS 80 ellipsoid, the NGS developed the *North American Datum of 1983* (NAD 83). Rather than fitting it to one point, NAD 83 was fit to the ellipsoid using observations taken at over 266,000 control points. The points, located in the United States, Canada, Mexico, Central America, Greenland, Hawaii, and the Caribbean Islands, were fit to the ellipsoid using simultaneous least-squares adjustment, creating the *National Spatial Reference System* (NSRS). Coordinates for the current adjustment of the NSRS are designated as NAD 83 (NSRS 2007). Nations in other regions of the world have developed similarly constructed horizontal datums.

When using spherical coordinates, care must be taken to ensure that all coordinates are based on the same datum. Conversions between datums can be made by using a variety of commercially available software. National or regional vertical datums have also been developed for the standardization of heights. Vertical datums are typically referenced to the geoid and established by reference to one or more tidal gauge stations. Because of the geographic variation in mean sea level, vertical datums vary from country to country. With the advent of GPS, ellipsoidal heights referenced to the ellipsoid are often used in lieu of heights referenced to a local datum. For many applications, ellipsoidal heights are preferable, especially when precise elevations are required, or when used for global or international applications. However, ellipsoidal heights are typically not suitable for topographic surveys, since they are not based on the gravity-based geoid and are therefore not suited for depicting the flow of surface drainage.

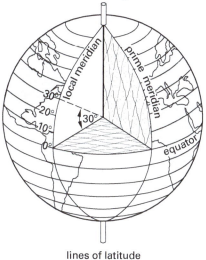

Figure 23.3 *Lines of Latitude and Longitude*

lines of latitude

lines of longitude

4. PROJECTIONS TO PLANE SURFACES

A systematic means is needed to project known ellipsoidal coordinates onto a two-dimensional surface, such as a map or a computer screen. (See Fig. 23.4.) Such a projection is desirable for many surveying and mapping calculations, since calculations with plane coordinates are less complicated and the results more easily used in computer-aided drafting (CAD) and GIS applications. Map projections used for this purpose typically mathematically define the relationship between coordinates on the ellipsoid and a plane.

With map projections, a plane is the ultimate surface where measurements on the surface of the earth are reduced. Although some types of projections involve projecting directly to a plane, most involve projecting to an intermediate mathematical surface, such as a cylinder or cone. That intermediate surface is then cut open and laid flat onto a plane and a coordinate grid overlaid on the plane.

Plane Survey Calculations

Two approaches are used with the projection to the intermediate surface. The first approach is a *tangent projection*, where the intermediate surface is tangent to the spheroid. A single line of contact exists around the spheroid that becomes the standard meridian or parallel for the projection. The second approach is a *secant projection* where the intermediate surface slices through a small portion of the spheroid. Two lines of contact exist with the surface of the spheroid, creating two standard parallels or meridians. The projection process involves calculating x (east) and y (north) Cartesian coordinates for each pair of latitude and longitude coordinates, although there are several alternative approaches used for this process.

Figure 23.4 *Basic Projection Surfaces*

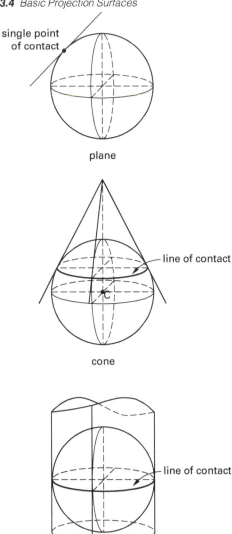

plane

cone

cylinder

Lambert Conformal Conic Projection

Worldwide, the two most widely used projections are the *Lambert conformal conic projection* and the *transverse Mercator projection*. The U.S. State Plane Coordinate System (see Sec. 23.5) uses either Lambert or Mercator projections, depending on the configuration of the area. The worldwide Universal Transverse Mercator System (UTM) uses a Mercator projection.

The Lambert conformal conic projection, as shown in Fig. 23.5, was developed by Johann Heinrich Lambert in 1772. It has two standard parallels, and meridians are converging straight lines that meet at a point outside of the map limits, while parallels are arcs of concentric circles. The meridians and parallels meet at right angles. Scale is true only along the two standard parallels and is compressed between those lines and expanded outside of them. As a result, this projection is best for regions of greater east-west extent.

Figure 23.5 *Lambert Conformal Conic Projection*

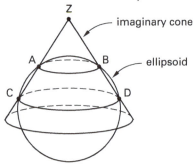

Transverse Mercator Projection

The basic Mercator projection was created by Gerardus Mercator in 1569. Then, in 1772, the same year he developed the conformal conic project, Lambert developed the transverse Mercator projection, as shown in Fig. 23.6. The transverse Mercator projection uses a cylinder with an east-west orientation. As used in the State Plane Coordinate System, the cylinder cuts the sphere along two meridians, parallel to the central meridian. The scale is true along the two standard meridians and is compressed between those lines and expanded outside of them. As a result, this projection is best for regions of greater north-south extent.

The projection process is often described as being similar to taking the peel from an orange and spreading it on a flat surface. Just as it is impossible to flatten the orange peel without tearing it, it is impossible to project a section of a spheroid surface without some distortion. Nevertheless, the distortion can be minimized by the design of the projection system and the restriction of the geographic area covered.

Figure 23.6 *Transverse Mercator Projection*

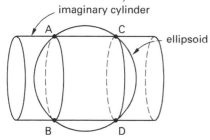

5. STATE PLANE COORDINATE SYSTEM

The *State Plane Coordinate System* (SPCS) is widely used in the United States, as well as in Puerto Rico and the Virgin Islands. It comprises a series of multiple zones that minimize distortion because of its projection process. Each state is given one or more zones, depending on its size and configuration. NGS provides programs on its website (www.ngs.noaa.gov) to obtain the correct zone for a given location, as well as programs that perform conversions to and from the coordinate systems. Software can be downloaded, or that agency's website's "geodetic tool kit" can be used interactively online. Figure 23.7 provides a table of the state plane coordinate zones.

With the SPCS, the Lambert conformal conic projection is used in states or portions of states that are longer in the east-west dimension. This is because it provides the best fit for projecting a rectangular zone greatest in east-west extent from a spherical surface to a plane. The transverse Mercator projection is used for areas that are longer in the north-south dimension, since it provides the best results in such areas. One zone, 0101 in Alaska, uses an oblique Mercator projection where the cylinder is rotated to a predetermined azimuth for better alignment with the shape of the coverage area.

The current SPCS is based on the NAD 83 datum and the GRS 80 ellipsoid. Coordinate units are in meters. This system supersedes the 1927 State Plane Coordinate System, which was based on the NAD 27 datum and the Clarke 1866 spheroid. The 1927 system's coordinates were in U.S. survey foot units, and calculated by multiplying the distance in meters by 3937/1200. There is no direct mathematical relationship between NAD 83 and NAD 27. Conversions between the two systems must be accomplished by comparison of a database of known points. Public domain software known as CORPSCON, available from the U.S. Army Corps of Engineers, can be used for such conversions.

Mapping Angles and Scale Factors

The mapping angle and the scale factor are two parameters in addition to coordinates that are essential for correctly using the State Plane Coordinate System. In any plane projection, the grid north and geodetic north will not coincide except along the central meridian. This is because the meridians converge toward the poles, while north-south grid lines are parallel to the central meridian.

The *mapping angle* (also known as the *convergence angle*, *grid declination*, or *variation*) is the angular difference between grid north and geodetic north. (See Fig. 23.8.) Mapping angles are either positive or negative, depending on whether the location is east or west of the central meridian, regardless of the projection used.

The grid *scale factor* is a measure of the lineal distortion associated with the projection of ellipsoidal distances onto a plane surface. The scale factor is equal to 1.0 along the standard lines of the projection, but is either less or greater than 1.0 in other areas. For Lambert projections, the standard lines are the standard parallels that represent the lines of contact between the conic surface and the surface of the ellipsoid. The scale is true along the two standard parallels and is compressed between those lines and expanded outside of them. Therefore, the scale factor varies with latitude with Lambert projections. For transverse Mercator projections, the standard lines are the standard meridians that represent the lines of contact between the cylinder and the surface of the ellipsoid. The scale is true along the two standard meridians and is compressed between those lines and expanded outside of them. Therefore, the scale factor varies with longitude with transverse Mercator projections. Figure 23.9 shows scale factors in Lambert and transverse Mercator projections.

Reduction of Distances from Ground to Grid

Since measurements on the surface of the earth are arc distances and can be taken at elevations not on the reference spheroid, reductions must be made to use them in plane coordinate systems to ensure a precise survey. This reduction involves correction for elevation above the ellipsoid, as well as using the scale factor to reduce the distance from an ellipsoid arc distance to a plane distance on the grid. The relationship between the grid, geodetic, and ground distances is shown in Fig. 23.10. Since elevations are typically expressed as orthometric elevations (in reference to the geoid), it is necessary to use the location's geoidal separation value for the reduction. The geoidal separation value is readily obtained from benchmark data sheets, or from NGS's online geoid calculators. The formula for reducing observed distances for elevation above the ellipsoid can be expressed as

$$S = D\left(\frac{R}{R + N + H}\right) \qquad 23.4$$

In Eq. 23.4, S is the distance on the ellipsoid, D is the observed distance, R is the mean radius of the ellipsoid (a value of 20,906,000 ft is sufficiently accurate), N is the geoidal separation, and H is the elevation above the sea level (or geoid).

Plane Survey Calculations

Figure 23.7 *State Zones for Plane Coordinates (NAD 83)*

state	zone	projection	central meridian (deg)	(min of arc)
Alabama	E	TM	85	50
	W	TM	87	30
Alaska	1	TM	Oblique	
	2	TM	142	00
	3	TM	146	00
	4	TM	150	00
	5	TM	154	00
	6	TM	158	00
	7	TM	162	00
	8	TM	166	00
	9	L	170	00
	10	TM	176	00
Arizona	E	TM	110	10
	C	TM	111	55
	W	TM	113	45
Arkansas	N	L	92	00
	S	L	92	00
California	1	L	122	00
	2	L	122	00
	3	L	120	30
	4	L	119	00
	5	L	118	00
	6	L	116	15
Colorado	N	L	105	30
	C	L	105	30
	S	L	105	30
Connecticut		L	72	45
Delaware		TM	75	25
Florida	E	TM	81	00
	W	TM	82	00
	N	L	84	30
Georgia	E	TM	82	10
	W	TM	84	10
Hawaii	1	TM	155	30
	2	TM	156	40
	3	TM	158	00
	4	TM	159	30
	5	TM	160	10
Idaho	E	TM	112	10
	C	TM	114	00
	W	TM	115	45
Illinois	E	TM	88	20
	W	TM	90	10
Indiana	E	TM	85	40
	W	TM	87	05
Iowa	N	L	93	30
	S	L	93	30
Kansas	N	L	98	00
	S	L	98	30
Kentucky	N	L	84	15
	S	L	85	45
Louisiana	N	L	92	30
	S	L	91	20
	OF	L	91	20
Maine	E	TM	68	30
	W	TM	70	10
Maryland		L	77	00
Massachusetts	M	L	71	30
	IS	L	70	30
Michigan	N	L	87	00
	C	L	84	22
	S	L	84	22
Minnesota	N	L	93	06
	C	L	94	15
	S	L	94	00
Mississippi	E	TM	88	50
	W	TM	90	20
Missouri	E	TM	90	30
	C	TM	92	30
	W	TM	94	30
Montana		L	109	30
Nebraska		L	100	00
Nevada	E	TM	115	35
	C	TM	116	40
	W	TM	118	35
New Hampshire		TM	71	40
New Jersey		TM	74	30
New Mexico	E	TM	104	20
	C	TM	106	15
	W	TM	107	50
New York	E	TM	74	30
	C	TM	76	35
	W	TM	78	35
	LI	L	74	00
North Carolina		L	79	00
North Dakota	N	L	100	30
	S	L	100	30
Ohio	N	L	82	30
	S	L	82	30
Oklahoma	N	L	98	00
	S	L	98	00
Oregon	N	L	120	30
	S	L	120	30
Pennsylvania	N	L	77	45
	S	L	77	45
Rhode Island		TM	71	30
South Carolina		L	81	00
South Dakota	N	L	100	20
	S	L	100	20
Tennessee		L	85	30
Texas	N	L	101	30
	NC	L	98	30
	C	L	100	00
	SC	L	99	00
	S	L	98	30
Utah	N	L	111	30
	C	L	111	30
	S	L	111	30
Vermont		TM	72	30
Virginia	N	L	78	30
	S	L	78	30
Washington	N	L	120	50
	S	L	120	30
West Virginia	N	L	79	30
	S	L	81	00
Wisconsin	N	L	90	00
	C	L	90	00
	S	L	90	00
Wyoming	E	TM	105	10
	EC	TM	107	20
	WC	TM	108	45
	W	TM	110	05
Puerto Rico		L	66	26

Abbreviations

TM = transverse Mercator
L = Lambert
N = north   S = south
NC = north central   SC = south central
E = east   W = west
EC = east central   WC = west central
C = central   LI = Long Island
M = mainland   IS = island
OF = offshore

Plane Survey Calculations

Figure 23.8 *Angle (Convergence) in a State Plane Coordinate System*

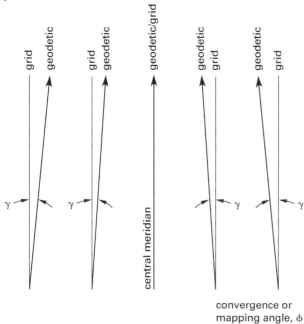

Figure 23.9 *Scale Factors in Lambert and Transverse Mercator Projections*

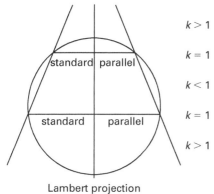

Lambert projection

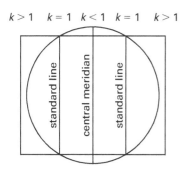

transverse Mercator projection

In addition to the elevation correction, observed distances should be further reduced to grid distances using a location's *average scale factor.* The scale factor can be obtained from state plane coordinate programs available in the "Geodetic Tool Kit" on the NGS website.

Figure 23.10 *Relationship Between Grid, Geodetic (ellipsoidal), and Ground Distances*

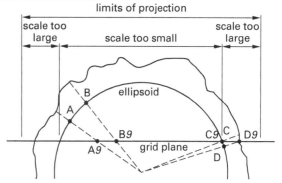

The distance A*9* to B*9* is smaller than geodetic distance A to B.

The distance C*9* to D*9* is larger than geodetic distance C to D.

The grid distance can be calculated using Eq. 23.5.

$$d_{\text{grid}} = kS \qquad \text{23.5}$$

The scale factor is often combined with the elevation factor to form a single multiplier for reducing observed distances to plane coordinate grid distances. The two factors can be combined by taking the product of the two factors. That product, when multiplied by the observed horizontal distance, will provide the grid distance. The area for a parcel of land at ground elevation can be calculated by dividing the plane coordinate area by the square of the combined factor used to reduce measured distances.

Conversion from Geodetic to Grid Azimuth

When calculating state plane coordinates using geodetic azimuths, the geodetic azimuth must be converted to grid azimuth using Eq. 23.6.

$$A_{\text{grid}} = A_{\text{geodetic}} - \theta_{\text{mapping}} + C_{\text{arc-to-chord}} \qquad \text{23.6}$$

6. THE UTM COORDINATE SYSTEM

The *Universal Transverse Mercator* (UTM) system was designed as a worldwide plane coordinate system. The UTM system uses the GRS 80 ellipsoid in North America, but uses various ellipsoids elsewhere. The system has 60 zones. Each zone's longitude has a width of six degrees, starting with zone 1 for the zone from 180° west to 174° west, and increasing eastward to zone 60 for the zone from 174° east to 180° east. Thus, the boundaries of each zone are on those meridians that are integral multiples of six degrees. Table 23.1 gives a list of the UTM zones.

The grid origin for each UTM zone is located at the intersection of the equator and the central meridian for that zone. The scale factor at the central meridian for each zone is 0.9996 and increases with longitude to

Table 23.1 *Universal Transverse Mercator*

UTM zone numbers and corresponding longitude of central meridians

western hemisphere				eastern hemisphere			
zone	central meridian	zone	central meridian	zone	central meridian	zone	central meridian
1	177	16	87	31	−03	46	−93
2	171	17	81	32	−09	47	−99
3	165	18	75	33	−15	48	−105
4	159	19	69	34	−21	49	−111
5	153	20	63	35	−27	50	−117
6	147	21	57	36	−33	51	−123
7	141	22	51	37	−39	52	−129
8	135	23	45	38	−45	53	−135
9	129	24	39	39	−51	54	−141
10	123	25	33	40	−57	55	−147
11	117	26	27	41	−63	56	−153
12	111	27	21	42	−69	57	−159
13	105	28	15	43	−75	58	−165
14	99	29	09	44	−81	59	−171
15	93	30	03	45	−87	60	−177

Note: Each central meridian is in the center of a 6° wide zone.
Scale factor is 0.9996 at each central meridian.

either side of the central meridian. As with SPCS, the mapping angle is zero at the central meridian, negative to the west of the central meridian, and positive to the east of that meridian. The UTM projection is not used for latitudes above 84° north or below 80° south. The *Universal Polar Stereographic* projection is typically used for polar regions.

The UTM scale factor corrections (between ground and grid distances), as well as the mapping angle corrections (between geodetic and grid azimuths), are handled in the same manner as with SPCS corrections. Software to calculate corrections, as well as to convert between geodetic and UTM coordinates, is available on the NGS website and from various commercial sources.

7. PRACTICE PROBLEMS

1. Parallels, or lines of latitude, measure the arc distance

(A) east and west of the prime meridian

(B) westward from the prime meridian

(C) northerly and southerly from the equator to the poles

(D) northerly and southerly from the poles to the equator

2. Meridians, or lines of longitude, measure the arc distance

(A) east and west of the prime meridian

(B) westward from the prime meridian

(C) northerly and southerly from the equator to the poles

(D) northerly and southerly from the poles to the equator

3. Which of the following is used in a Mercator projection?

(A) plane

(B) cone

(C) cylinder

(D) ellipse

4. Which of the following is used in a Lambert projection?

(A) plane

(B) cone

(C) cylinder

(D) ellipse

5. Which of the following is used in a Lambert conformal conic projection?

(A) parallel lines of longitudes

(B) two cones

(C) parallel lines of latitude

(D) one cone that slices the earth at two parallels

6. The mapping angle in a Lambert projection varies with which of the following?

(A) longitude

(B) latitude

(C) latitude and longitude

(D) none of the above

7. The mapping angle in a transverse Mercator projection varies with which of the following?

(A) longitude

(B) latitude

(C) latitude and longitude

(D) none of the above

8. The scale factor in a Lambert projection varies with which of the following?

(A) longitude

(B) latitude

(C) latitude and longitude

(D) none of the above

9. The scale factor in a transverse Mercator projection varies with which of the following?

(A) longitude

(B) latitude

(C) latitude and longitude

(D) none of the above

10. An observed horizontal distance of 2640.00 ft is measured at an average orthometric elevation of 1400 ft. The geoidal separation is -131.23 ft and the scale factor is 0.9994. Most nearly, what grid distance should be used for calculating state plane coordinates with this measurement? (Use 20,906,000 ft for the radius of the earth.)

(A) 2638 ft

(B) 2639 ft

(C) 2640 ft

(D) 2642 ft

11. Most nearly, what grid azimuth is needed to calculate state plane coordinates for a line that has a geodetic azimuth of $118°22'15''$? The mapping angle is $+4'32.1''$, and the arc-to-chord correction is negligible.

(A) $118°16'$

(B) $118°17'$

(C) $118°22'$

(D) $118°26'$

12. One point has state plane coordinates of $N = 202,400$ m and $E = 305,600$ m. A second point has coordinates of $N = 203,300$ m and $E = 306,500$ m. The scale factor is 0.99992, the orthometric elevation is 900 m, and the geoidal separation is -36.2 m. Using a value of 6372 km for the radius of the earth, what is most nearly the expected ground distance between the two points?

(A) 1257 m

(B) 1273 m

(C) 1562 m

(D) 1773 m

Plane Survey Calculations

SOLUTIONS

1. Parallels, or lines of latitude, measure the arc distance as northerly and southerly from the equator to the poles.

The answer is (C).

2. Meridians, or lines of longitude, measure the arc distance east and west of the prime meridian.

The answer is (A).

3. A Mercator projection uses a cylinder as a projection surface.

The answer is (C).

4. A Lambert projection uses a cone as a projection surface.

The answer is (B).

5. A Lambert conformal conic projection uses one cone that slices the earth at two parallels.

The answer is (D).

6. The mapping angle in a Lambert projection varies with longitude.

The answer is (A).

7. The mapping angle in a transverse Mercator projection varies with longitude.

The answer is (A).

8. The scale factor in a Lambert projection varies with latitude.

The answer is (B).

9. The scale factor in a transverse Mercator projection varies with longitude.

The answer is (A).

10. From the problem statement,

$$D = 2640.00 \text{ ft}$$
$$H = 1400 \text{ ft}$$
$$N = -131.23 \text{ ft}$$
$$k = 0.9994$$
$$R = 20,906,000$$

Using Eq. 23.4,

$$S = D\left(\frac{R}{R+N+H}\right)$$
$$= (2640.00 \text{ ft})\left(\frac{20,906,000 \text{ ft}}{20,906,000 \text{ ft} + (-131.23 \text{ ft}) + 1400 \text{ ft}}\right)$$
$$= 2639.84 \text{ ft}$$
$$d_{\text{grid}} = kS$$
$$= (0.9994)(2639.84 \text{ ft})$$
$$= 2638.26 \text{ ft} \quad (2638 \text{ ft})$$

The answer is (A).

11. Using Eq. 23.6,

$$A_{\text{grid}} = A_{\text{geodetic}} - \theta_{\text{mapping}} + C_{\text{arc-to-chord}}$$
$$= 118°22'15'' - 4'32.1''$$
$$= 118°17'42.9'' \quad (118°17')$$

The answer is (B).

12. The grid distance, using the Pythagorean theorem, is

$$d_{\text{grid}} = \sqrt{(N_2 - N_1)^2 + (E_2 - E_1)^2}$$
$$= \sqrt{\begin{array}{c}(203\,300 \text{ m} - 202\,400 \text{ m})^2 \\ + (306\,500 \text{ m} - 305\,600 \text{ m})^2\end{array}}$$
$$= 1272.792 \text{ m}$$

Rearranging from Eq. 23.5,

$$d_{\text{grid}} = kS$$
$$S = \frac{d_{\text{grid}}}{k} = \frac{1272.792 \text{ m}}{0.99992}$$
$$= 1272.894 \text{ m}$$

Using Eq. 23.4,

$$S = D\left(\frac{R}{R+N+H}\right)$$
$$= (1272.89 \text{ m})\left(\frac{(6372 \text{ km})\left(1000 \frac{\text{m}}{\text{km}}\right)}{\left(6372 \text{ km}\right)\left(1000 \frac{\text{m}}{\text{km}}\right) + (-36.2 \text{ m}) + 900 \text{ m}}\right)$$
$$= 1272.72 \text{ m} \quad (1273 \text{ m})$$

The answer is (B).

Topic IV: Land Planning and Development

Land Planning/
Development

24 Land Descriptions

1. LAND DESCRIPTIONS

Throughout history, humankind has created representative systems, such as written music notation and currency, that have allowed great advances in civilization. One such system, the representation of land with a title document, provides security of land ownership and allows land to be considered as a commodity that can be bought and sold or used as collateral for credit (which is one of the most important sources of funds for new businesses in the United States). Therefore, one of the surveyor's most important responsibilities is that of preparing land descriptions. Because of the important role that land descriptions play in society, it is incumbent upon the surveyor to use utmost care in the crafting of these descriptions. Descriptions must allow the precise and unambiguous location of the boundaries of the tracts of land for generations after they are written.

2. TYPES OF LAND DESCRIPTIONS

Public Land Survey System Descriptions

Public Land Survey System descriptions are based on the *Public Land Survey System* (also known as the *U.S. System of Rectangular Surveys*), which today is administered by the U.S. Bureau of Land Management. That system is based on a plan for subdividing the public domain of the United States that began in 1785 when Congress enacted a law that provided for the subdivision of the public lands into townships 6 miles square with townships subdivided into 36 sections, most of which are 1 mile on a side. Thirty states of the United States were subdivided into tracts by this system, known as the Public Land Survey System or the U.S. System of Rectangular Surveys.

The original public lands act that created the Public Land Survey System specified that public lands were to be sold by the section and descriptions were to reference land surveys. Each section was a square mile, with a standard size of 640 ac. Under the 1785 law, townships

were divided into sections numbered from 1 to 36 beginning in the northeast corner and ending in the southeast corner, as shown in Fig. 24.1.

Figure 24.1 *Township Subdivided into Sections*

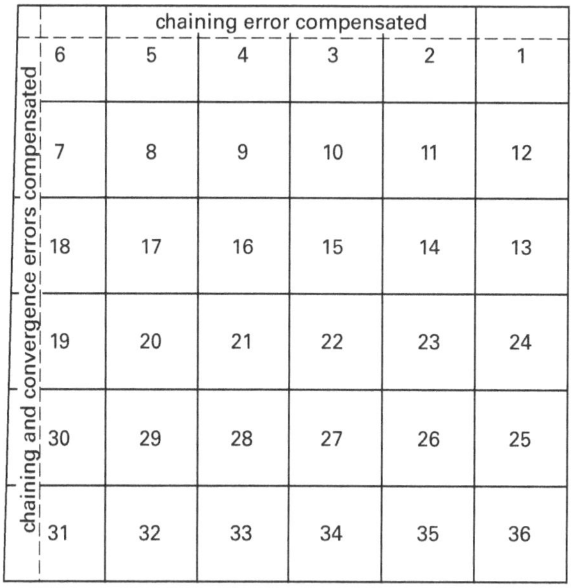

Since each section had its own identity, conveyance of any of these lands involved a relatively simple description providing the pertinent section, township, and range. In addition, the description contained references to the base meridian from which the section had been surveyed and to the official plat that depicted the section.

The following is an example of a Public Land Survey System description.

> Section 6, township 1 north, range 2 east, Tallahassee Baseline as depicted on the official General Land Office plat approved June 1, 1842.

Due to a demand for smaller parcels of public land, successive legislation reduced the minimum size available for purchase to standardized fractions of a section called *aliquot parts*. In 1800, the half section, with a standard size of 320 ac, became the minimum size. Four years later, the sale of land by the quarter section, with a standard size of 160 ac, was authorized. In 1820, sale of land by the half quarter section, with a standard size of 80 ac, was approved. Then in 1832, the minimum area for public land sale became the quarter of a quarter section, with a standard size of 40 ac. That tract became the standard unit of settlement for the United States.

The public land surveys only traversed the perimeter of sections. The boundaries of aliquot parts of sections were not surveyed. Nevertheless, the description of such tracts is a simple process. Figure 24.2 illustrates the convention used. Note that when attempting to visualize the location of aliquot parts of sections, it is best to start at the end of the description and work backward.

Figure 24.2 *Aliquot Parts of Public Land Sections*

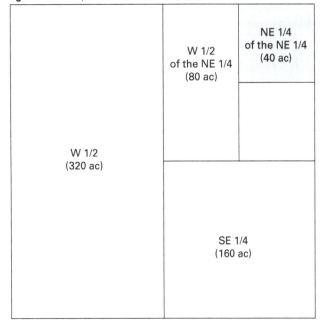

In the platting process for public lands surveys, it was the practice to divide fractional sections along navigable water bodies into small lots. Typically, the boundaries of such lots were drawn along the lines of the fractional parts of the section. This was a cartographic process, and the boundaries of such government lots were not surveyed except for those boundaries coincident with section lines and those coincident with meander lines along the water body. As with descriptions for Public Land Survey System sections and aliquot parts of sections, government lots may be described simply by reference to the lot number, section, township, range, base meridian, and pertinent official plat. The following is an example of a government lot description.

> Government lot 3, section 12, township 1 north, range 2 east, Tallahassee Meridian as depicted on the official General Land Office plat approved June 1, 1842.

Descriptions Referencing Other Subdivisions

A second type of land description references other subdivisions of land. These descriptions are similar to those involving sections and aliquot parts of sections of the Public Land Survey System. Both types of descriptions

involve a reference to a subdivision of land where a plat of survey has been filed in the public records. However, this second type of description may reference subdivisions performed by government agencies outside of the Public Land Survey System, or it may reference subdivisions created by private landowners and recorded in the public records. Such recorded subdivision plats are typically filed in each county courthouse. Most states have statutes that provide standards for the survey and platting of such subdivisions. In addition, many counties have subdivision ordinances with their own requirements.

As a result, modern-day subdivision plats are relatively standardized. Since such subdivision plats are available in the public records, a description on such a plat requires only that the lot or lots be identified, along with the subdivision name and the location of the plat record, as shown in the following example.

> Lots 100 and 101 of Dreblow & Company's Silver Lake Subdivision as recorded at Plat Book A, p. 3 of the public records of Jefferson County, Florida.

Metes and Bounds Descriptions

In the 20 states not subdivided under the Public Land Survey System, most of the surveys for the original private conveyances were made by the metes and bounds method. *Metes and bounds descriptions* are also used to describe irregular areas that are severed from lands originally subdivided under the Public Land Survey System. A bounds description defines the boundaries of a tract of land by identifying adjoiners or monuments but does not typically provide a direction, as shown in the following example.

> All of that land lying north of State Road 99; bounded on the north by land of Albert Bowie, bounded on the east by lands of Betty Anderson; and bounded on the west by Trout Creek.

A metes description identifies a beginning point and then describes each course in sequence around the perimeter of the tract until the point of beginning is reached again to complete the description of the perimeter. Such a description includes a distance and direction for each course, as shown in the following example.

> A tract of land in section 12, township 3 north, range 6 east in Jefferson County, Florida, more particularly described as follows: For a point of beginning, commence at an old axle marking the southeast corner of said section 12; then go N00°01'E for 200 ft; then go N89°59'W for 400 ft; then go S00°01'W for 200 ft; then go S89°59'E for 400 ft to the point of beginning.

As may be seen from the examples, metes and bounds descriptions are somewhat more complex than descriptions that merely reference a plat of survey in the public records, such as those descriptions from the Public Land Survey System or those that describe a lot in a private subdivision. Since the survey that forms the foundation of a metes or bounds description is not typically part of official records, the description must, in effect, communicate the results of the survey.

Other Types of Descriptions

There are a number of other types of descriptions used for conveying land. An example of these is a *strip description*. This type is frequently used to describe a road easement or right of way. Other common descriptions are those where a well-described tract of land is divided based on a man-made or natural *monument*, such as a road or stream; based on a certain distance or width; based on a certain area; or based on a certain fraction of the total area of the tract. Another type of description that is increasingly used today relies on *geographic coordinates*. Many other descriptions are a combination of these types.

Descriptions may also include *qualifying clauses* or *augmentation clauses* that take away or add something to a tract, respectively. Examples of these other types of descriptions follow.

- *strip description:* a right of way for ingress and egress purposes across a strip of land lying 30 ft on each side of the described center line

- *division by monument*: all of section 12 lying northerly of U.S. Highway 90

- *division by distance:* the westerly 50 ft of lot 2

- *division by area*: the southern 10 ac of government lot 2

- *division by fraction:* the western one-half of lot 4

- *geographic coordinates:* For a point of beginning, commence at an old concrete monument marking the south quarter corner of said section 24 and having a north coordinate of 1,972,048.50 and an east coordinate of 563,589.10; then go N00°01′E for 200 ft to a point having a north coordinate of 1,972,248.62 and an east coordinate of 563,589.16; then go N89°59′W for 400 ft to a point having a north coordinate of 1,972,248.50 and an east coordinate of 563,189.16; then go S00°01′W for 200 ft to a point having a north coordinate of 1,972,048.50 and an east coordinate of 563,189.10; then go S89°59′E for 400 ft to the point of beginning. Coordinates are in feet and are based on the Florida State Plane Coordinate System, North Zone, NAD88.

3. ESSENTIALS OF LAND DESCRIPTION WRITING

Some essential considerations in preparing land descriptions include

- *format:* A land description should include both a caption that provides the general location, city,

county, and state, and a body that provides a detailed description of the area. In addition, the land description may include qualifying clauses that take away from the area outlined in the body or augmenting clauses that add something to the area outlined in the body.

- *monuments:* Any monuments called, especially monuments called as the reference points for metes and/or bounds descriptions, should be permanent in character, visible, and stable.

- *directions:* The basis for any directions called should be stated. For example, bearings should be identified as based on astronomic coordinates, state plane coordinate grids, magnetic north, or a specifically defined line.

- *coordinates:* If coordinates of corners are called, the datum for the coordinates should be stated.

- *curves:* For any curves called, at least two elements of the curve should be stated. The most frequently used elements are the radius, arc length, central angle, and tangent length. In addition, the relationship of the curve to the previous line, the direction of the curve (e.g., concave to the south), and the direction of travel along the curve (e.g., easterly) should be stated, as described in the example.

  Then go 1000 ft along a tangent curve to the left, said curve being concave to the east and having an interior angle of 45° and a radius of 2000 ft.

4. INTERPRETING LAND DESCRIPTIONS

Most surveyors have encountered problems in boundary determination due to a land description with ambiguous or conflicting calls within it, or due to conflicts between that description and another. Fortunately, case law over the years has resulted in a body of guidelines for the best action in such situations. The most salient of these guidelines follow.

Priority of Calls

In the absence of a clear intention to the contrary, the order of priority for conflicting calls is as follows.

1. calls for natural objects such as rivers, creeks, and mountains

2. calls for artificial objects such as concrete monuments and adjoining lands

3. calls for course

4. calls for distance

5. calls for area

Land Planning/Development

Senior and Junior Rights

Many original metes and bounds surveys resulted in overlaps or gaps between surveys. It became the rule of law in the early history of the United States that where an overlap occurred, the holder of the first grant or patent, the *senior awardee*, retained ownership of the overlap area, and the holder of the latter grant or patent, the *junior awardee*, lost the area. These rights are known as *senior rights* and *junior rights*. A common expression summarizes the priority of these rights: "The first deed is the best deed." Where two parties have title to the same land, the party holding the senior conveyance has the right of possession.

As an example, Smith owns 200 ac of land, or thinks that he owns that amount. He sells one half of his land to Jones by a metes and bounds description that calls for 100 ac. Later, Smith moves to sell the other half of his land to Brown with a description that calls for one-half of the original tract. At that time, a survey reveals he had only 195 ac originally. He cannot sell more than 95 ac to Brown. Jones has the senior deed and is entitled to the full 100 ac that Smith sold to him. Smith cannot recover any of Jones's land. This principle in law assures the first buyer that his land cannot be taken from him as it was conveyed. It is the basis of senior and junior rights.

Distances

Unless otherwise stated, distances are horizontal and measured in a straight line along the shortest distance.

Excess and Deficiency in Subdivision

Distances and directions between found original measurements take priority over plat distances and directions. Excess or deficiency of such lines between original monuments should be distributed proportionally throughout the line within each block.

Calls That Use the Word "To"

When calls use the word "to"—such as "to a concrete monument" or "to a river" or "to an adjoining property boundary"—it should be interpreted that the actual direction and distance to the object called takes priority over any direction and distance cited in the description. When the call is to an adjoiner, this means that the adjoining boundary must be located before the description can be correctly interpreted.

Indeterminate Fractional or Area Calls

Indeterminate fractional calls or area calls are those where the direction of the dividing line is not given or implied. Principles have been developed for such situations.

A called area of land on the side of a tract, such as "the northern 80 ac," should be interpreted as including the called area in the form of a parallelogram with the dividing line parallel to the side of the tract on which the area is located.

A called area of land in a specified corner of a tract should be interpreted as being a corner quadrangle with equal sides.

- If the easterly and westerly sides of a lot are nearly parallel and either the easterly or westerly half is described, the dividing line should be considered to be on a mean bearing between the two sides.

- If the easterly and westerly sides of a lot are not parallel and either the easterly or westerly half is described, the dividing line should be considered to be north-south.

- If the eastern boundary of a lot is nearly north-south, and if the easterly half of the lot is conveyed and a description calls the lot except for the eastern half, the dividing line should be parallel to the eastern side.

5. PRACTICE PROBLEMS

1. What is the area of the southeast quarter of the northeast quarter of the southwest quarter of a typical public land survey section?

(A) 5 ac

(B) 10 ac

(C) 20 ac

(D) 40 ac

2. A tract of land is described as the southeast quarter of the northeast quarter of the southwest quarter of a typical public land survey section. In which direction would this tract of land lie in relation to the southwest quarter of the northwest quarter of the southwest quarter of the same section?

(A) north

(B) east

(C) south

(D) west

3. Which is the controlling call in the following description?

> N54°E for a distance of 298 ft to the shore of Wolf Creek

(A) N54°E

(B) 298 ft

(C) the thread of Wolf Creek

(D) the shore of Wolf Creek

4. The recorded plat of the Silver Lake Subdivision shows lot 2 to be a 100 ft by 200 ft rectangular lot. The owner conveyed the northern one-half of the lot and later conveyed the southern 100 ft of the lot. A survey conducted for the second conveyance revealed that the western and eastern lines of the lot are 198.5 ft long. Where should the northeast corner of the later conveyance be placed?

(A) 89.25 ft from the southeast corner of lot 2

(B) 98.50 ft from the southeast corner of lot 2

(C) 99.25 ft from the northeast corner of lot 2

(D) 100.00 ft from the southeast corner of lot 2

5. In the event of conflict among calls in a description, which of the following calls has the lowest priority?

(A) call for an artificial monument

(B) call for distance

(C) call for a natural monument

(D) call for a course

6. In the event of conflict among calls in a description, which of the following calls has the second highest priority?

(A) call for an artificial monument

(B) call for distance

(C) call for a natural monument

(D) call for a course

7. In the following description, what would be considered an example of a bounds?

> A parcel of land situated in Jefferson County, Florida being part of Section 12, Township 3 North, Range 5 East and being more particularly described as follows: Begin at an old axle marking the southwest corner of said Section 12; then go N80°42′E for 542.0 ft to the westerly right of way of County Road 246; then go N2°35′W for 120.0 ft along said right of way; then go N50°15′W for 502.5 ft to an old iron pipe; then go S15°09′W for 547.8 ft to the point of beginning.

(A) N80°42′E

(B) 542.0 ft

(C) westerly right of way of County Road 246

(D) There are no bounds in this description.

8. Referring to the illustrated recorded plat for Wild Turkey Subdivision, a survey has found no interior lot corners fronting on Wing Road, but finds the original front block corners 500.80 ft apart. At what distance from the block corners should the adjacent lot corners be set?

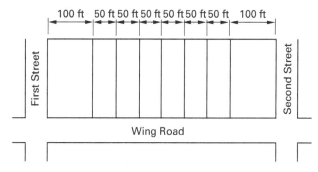

(A) 100.00 ft

(B) 100.10 ft

(C) 100.16 ft

(D) 100.80 ft

9. The answer for Prob. 8 is correct because the

(A) excess should be placed in the corner remnant lots

(B) excess is distributed equally in each lot

(C) excess is distributed by proportional measurement

(D) plat dimensions control remonumentation

SOLUTIONS

1. The total section area is 640 ac.

$$\text{quarter section} = \frac{640 \text{ ac}}{4} = 160 \text{ ac}$$

$$\begin{array}{l}\text{quarter of a} \\ \text{quarter section}\end{array} = \frac{160 \text{ ac}}{4} = 40 \text{ ac}$$

$$\begin{array}{l}\text{quarter of a} \\ \text{quarter of a} \\ \text{quarter section}\end{array} = \frac{40 \text{ ac}}{4} = 10 \text{ ac}$$

The answer is (B).

2. For the illustration shown, the described tract (southeast quarter of the northeast quarter of the southwest quarter) lies easterly of the southwest quarter of the northwest quarter of the southwest quarter.

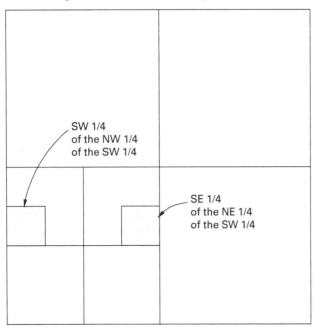

SW 1/4
of the NW 1/4
of the SW 1/4

SE 1/4
of the NE 1/4
of the SW 1/4

The answer is (B).

3. The controlling call is to "the shore of Wolf Creek." The creek is a natural monument that has precedence over the calls for distance and bearing.

The answer is (D).

4. The senior deed called for one-half of the lot. The east and west lines would be divided in half to set the monument.

$$\frac{198.5 \text{ ft}}{2} = 99.25 \text{ ft}$$

The answer is (C).

5. In general, calls for distance should be given the lowest priority of the options given.

The answer is (B).

6. Calls for artificial monuments should be given the second highest priority of the options given.

The answer is (A).

7. In the description provided, the westerly right of way of County Road 246 is an example of a bounds, which is a monument or adjoiner called in the deed.

The answer is (C).

8. Excess and deficiency should be proportioned within the block. The total frontage excess within the block is

$$\begin{array}{l}\text{measured distance} \\ -\text{recorded distance}\end{array} = 500.80 \text{ ft} - 500.00 \text{ ft}$$

$$= 0.80 \text{ ft}$$

For a 100 ft lot, the excess is

$$\left(\frac{100 \text{ ft}}{500 \text{ ft}}\right)(0.80 \text{ ft}) = 0.16 \text{ ft}$$

The required lot corner should be set 100.16 ft from the block corner.

The answer is (C).

9. Excess between found original monuments in a subdivision is distributed by proportional measurement.

The answer is (C).

25 Subdivisions

1. DEFINITION

The act of subdivision is the division of any tract or parcel of land into two or more parts for the purpose of sale or building development.

2. REGULATION OF SUBDIVISIONS

Unlike retracement surveys, surveys made for the purpose of the subdivision of land are creative in nature. Therefore, they are the original surveys for the subdivided land and will have a key role in the future ownership of the land being subdivided. Further, the care and imagination used in the planning of the subdivision process affect the entire community for many years to come. Creation of a subdivision involves more than furnishing a location for a home for a new member of a community and a profit for the developer. It requires planning for traffic, transportation, the location of schools, churches, and shopping centers, and the health and happiness of the citizens of the community.

There are many instances in the United States in the past where developers have created subdivisions that are a credit to their foresight and integrity without any regulatory laws. Yet there are also some poorly planned subdivisions that have caused communities to spend excessive amounts on street widening and resurfacing, reconstruction of sewer lines, establishment of additional drainage facilities, and increase in size of water mains. Such poor planning in the past has caused an acute awareness among state, county, and municipal officials of the need for the regulation of subdivision development. That awareness, together with increasing population and decreasing availability of land for development, have made it necessary for the states and the federal government to adopt laws regulating land divisions. Yet most subdivision regulation today is at the county level, and in some cases, the city level.

3. SUBDIVISION AND PLATTING LAWS

Subdivision law includes regulations for land use, types of streets and their dimensions, arrangement of lots and their sizes, land drainage, sewage disposal, protection of nature, and many other details. *Platting laws* include regulations for the recording of the subdivision plat, monumenting the parcels, establishing the accuracy of the survey, and means of identifying the parcels and their dimensions.

Most counties, and major cities, deal with developers of subdivisions through a planning commission. Therefore, the first step for a developer is to contact the planning commission of the designated approving agency for consultation. Many planning commissions require, at the outset, a data sheet that indicates the general features of the developer's plan and a location map that locates the proposed subdivision in relation to zoning regulation and existing community facilities.

4. SURVEYING FOR SUBDIVISIONS

The first step in the subdivision surveying process is a boundary survey. Unlike the interior monuments of a subdivision which, after platting, are correct and have no error of position, the exterior boundary monuments of a subdivision can be located in error. Establishing a subdivision does not take away the rights of the adjoiner. It is the surveyor's role to eliminate future boundary difficulties with a sound and defensible boundary survey before any detailed planning is made. Investigation as to any conflict with senior title holders should be made, and after certainty of location is established, the corners should be monumented.

The next step in the subdivision process is to establish a system of benchmarks for vertical control and prepare a topographic map. The map is essential in planning the subdivision, especially in regard to drainage and sanitary sewer plans. The topo survey provides the base for the preparation of a draft of the planned subdivision showing contours, street locations, and lots for preliminary approval by the planning commission.

The next step is preparation of a formal *preliminary plat*. That is actually the detailed plan for the subdivision and includes the names of the subdivider, engineer, or surveyor; a legal description of the tract; location and dimensions of all streets, lots, drainage structures, parks, and public areas; easement, lot, and block numbers; contour lines; scale of map; north arrow; and date of preparation. After approval of the preliminary plat, the interior boundaries of the subdivision may be monumented and the developer can proceed with stake-out and construction operations.

5. FINAL PLAT

The *final plat* conforms to the preliminary plat, except for any changes imposed by the planning commission. It must be prepared and filed for record in accordance with the platting laws of the state. It establishes a legal description of the streets, residential lots, and other sites in the subdivision. This plat will serve as the basis for the description of the lots for formal conveyance.

Land Planning/
Development

Plan and Profile Sheets

Nomenclature

E	elevation	ft	m
g	gradient	%	%
L	length	sta	sta
r	radius	ft	M

1. PLAN AND PROFILE SHEETS

On linear construction projects, such as highways and railroads, as well as utility corridors, both the surveys and the construction plans for the project are typically based on a centerline along the proposed route. Distances along the centerline are designated by stationing, with a station representing 100 ft in the English system. To provide data for location studies as well as for earthwork calculation, a profile is usually run along the centerline and cross profiles taken at each station at right angles to the center line. Other significant topographic features are also located based on the stationing along the centerline.

Based on the surveys, plans for such projects are usually depicted on *plan-profile sheets* (Fig. 26.1). Such plans provide a plan view of the centerline with surrounding topography on the top half of the sheet and elevation profile on the bottom half.

Plan and profile sheets often depict both the existing topographic elevations, based on the route survey, and the desired finished elevations, to provide guidance for necessary earthwork. For highway plans where vertical curves are involved, the points of intersection (PIs) as well as the planned finish elevations are typically shown on plan and profile sheets, as depicted in Fig. 26.1. The finish elevations along the vertical curves are calculated, as described in Chap. 22, by the basic equations for a parabola (Eq. 26.1 and Eq. 26.2) where g_1 and g_2 are the gradients for the incoming and outgoing slopes, L is the length of the curve, and x is the distance in stations from the beginning of the vertical curve (BVC).

$$y = \left(\frac{r}{2}\right)x^2 + xg_1 + E_{BVC} \qquad \text{26.1}$$

$$r = \frac{g_2 - g_1}{L} \qquad \text{26.2}$$

$$E_{BVC} = E_{PI} - g_1\frac{L}{2} \qquad \text{26.3}$$

Since vertical curves are symmetrical, the PC and PT may be located by measuring one-half the length of the vertical curve in each direction from the PI. The midpoint of the vertical curve can be found by drawing a straight line from the PC to the PT and measuring one-half the distance from the PI to this line. In determining whether y-distances should be added to or subtracted from tangent elevations, look at the plotted profile to see whether the curve is higher or lower than the tangent at a particular station.

Example 26.1

Using the information given, calculate tangent and finish elevations for each full station, plot the finish elevation profile, and show pertinent information needed for construction.

- gradient, sta 32+00 to 34+00 = −2.000%
- gradient, sta 34+00 to 39+50 = +2.182%
- gradient, sta 39+50 to 43+00 = −2.125%
- curve 1: PI, sta 34+00, elev 470.00, 300 ft length
- curve 2: PI, sta 39+50, elev 482.00, 500 ft length

Solution

Curve 1:

$$BVC = PI_2 - \frac{L}{2} = 34 + 00 - 2 + 50 = 32{+}50$$

From Eq. 26.2,

$$r = \frac{g_2 - g_1}{L}$$
$$= \frac{2.182\% + 2.000\%}{3 \text{ sta}}$$
$$= 1.394\%$$

From Eq. 26.3,

$$E_{BVC} = E_{PI} - g_1\frac{L}{2}$$
$$= 470.00\,\text{ft} - (-2.000\%)\left(\frac{3\,\text{ft}}{2}\right)$$
$$= 473.00\text{ ft}$$

Figure 26.1 *Plan-Profile Sheet*

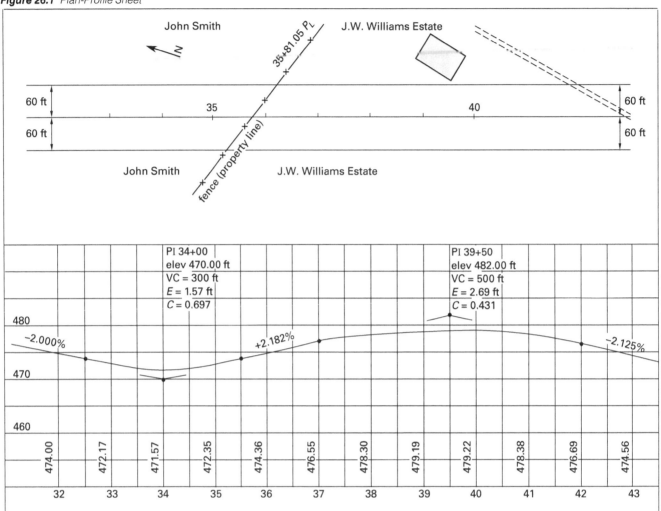

Curve 2:

$$BVC = PI_2 - \frac{L}{2} = 39 + 50 - 2 + 50 = 37 + 00$$

$$r = \frac{g_2 - g_1}{L}$$

$$= \frac{-2.125\% - 2.182\%}{5 \text{ sta}}$$

$$= -0.861\%$$

$$E_{BVC} = E_{PI} - g_1\frac{L}{2}$$

$$= 482.00\,\text{ft} - (2.182\%)\left(\frac{5\,\text{ft}}{2}\right)$$

$$= 476.55\,\text{ft}$$

Using Eq. 26.1, Eq. 26.2, and Eq. 26.3, both the tangent and finish elevations may be calculated as follows:

		elevation (ft)		
point	station	dist from BVC	tangent	finish
	32+00		474.00	474.00
BVC	32+50		473.00	473.00
	33+00	0.5	472.00	472.17
PI	34+00	1.5	470.00	471.57
	35+00	2.5	472.18	472.36
EVC	35+50	3.0	473.27	473.27
	36+00	3.5	474.36	472.36
BVC	37+00		476.55	473.27
	38+00	1.0	478.73	478.30
	39+00	2.0	480.91	479.19
PI	39+50	2.5	482.00	479.31
	40+00	3.0	480.94	479.22
	41+00	4.0	478.81	478.38
PC	42+00	5.0	476.69	476.69
	43+00	6.0	474.56	474.56

The results may be plotted on a plan and profile sheet, as illustrated in Fig. 26.1.

Plan-Profile Sheet for Problem 1

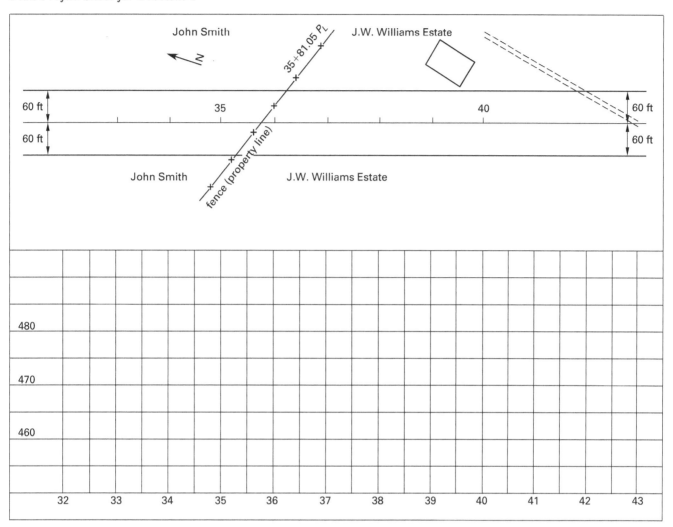

2. PRACTICE PROBLEMS

1. From the information given, complete the profile half of the plan-profile sheet using the *Plan-Profile Sheet for Problem 1* shown.

station	finish elevation (ft)
32+00	476.00
33+00	474.00
34+00	472.18
35+00	471.65
36+00	472.58
37+00	474.80
38+00	476.82
39+00	478.09
40+00	478.61
41+00	478.37
42+00	477.37
43+00	475.63

Solution for Problem 1

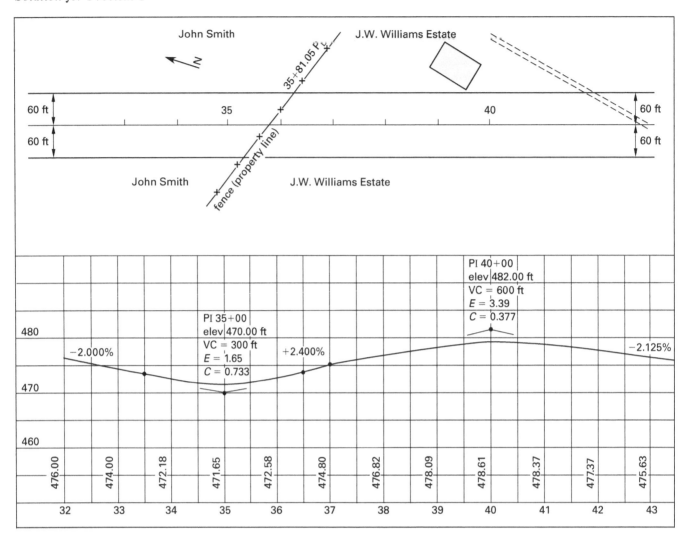

SOLUTIONS

1. See the completed plan-profile sheet in *Solution for Problem 1.*

27 Geographic Information Systems

1. INTRODUCTION

In the revolution that computers have brought to surveying, geographic information systems (GIS) are one of the most rapidly evolving and most significant developments. The capacity of GIS to collect, store, analyze, and display a great diversity of information about the earth's surface and its occupants allows the systems to be sophisticated models of the real world. With such models, the interrelationships of various layers of information can be examined, and trends observed and quantified. GIS has the potential to provide significant increases in our understanding of human beings and the environment.

Although the roots of GIS lie in the 1960s, it was not until 1988 that the first true GIS was developed by the government of Canada to support the mapping and assessment of its land resources. Today, GIS technology has developed into a large industry and impacts nearly all aspects of our lives. Among its many applications, GIS has become an essential component of property tax assessment. It is widely used in agriculture to allow precise application of fertilizers at the right location and at the right time for maximum efficiency and maximum yields. It is used by almost all utilities to manage the distribution of their products, and it is used by most transportation industries for route planning and fleet management.

GIS is especially important to the surveying profession, since surveyors work almost exclusively with geographic data. Many surveyors today are actively involved in the production of these systems. Even those surveyors who are not should realize that almost all of the survey data they produce will end up in some type of GIS. In the near future, it is likely that all survey data will be presented in a GIS environment. For these reasons, it is important that surveyors have a thorough understanding of this technology.

2. WHAT IS GIS?

Many definitions of GIS have been suggested over the years, although none seem totally adequate to describe the multiple facets of GIS. A definition used by the U.S. Geological Survey states that a GIS is "a computer system capable of capturing, storing, analyzing, and displaying geographically referenced information." Another leading definition is "a database in which every object has a precise geographical location, together with software to perform functions of input, management, analysis, and output." Despite the lack of consensus on an all-encompassing definition, there is agreement on several key requirements of a GIS:

- It must have the capability to integrate data from different sources and at different scales.

- It must allow the graphic display of multiple layers of geographic information.

- It must link to databases of various attribute information tied to geographic positions.

- It must be able to analyze the spatial relationship between various features and respond to queries regarding those relationships.

3. TYPES OF GIS DATA MODELS

The heart of a GIS is the data model that represents objects and processes. Most systems use either a raster or vector data model, or combinations and variations of those models. A *raster data model* consists of arrays of grid cells, or pixels, usually referenced by row and column numbers, along with an identifier representing the attribute being mapped. With such a model, points are represented by a single cell; lines are represented by a string of neighboring cells, and areas are represented by collections of neighboring cells. A digitized aerial photograph is an example of a raster data model.

A *vector data model* assumes a continuous coordinate space, not quantified as with a raster model, and therefore allows positions, lines, and area boundaries to be more precisely defined. With such a model, points on a plane are defined by x- and y-coordinate pairs, as opposed to grid cells in a raster model. Line segments on a plane are defined by a pair of x- and y-coordinates for the beginning point and end point, allowing a precise calculation of their length and bearing. Areas are similarly defined by a series of x- and y-coordinate pairs, one pair for each inflection point on the perimeter. A computer-aided design and drafting (CADD) file is an example of a vector data model.

Both types of data models are valid methods for representing spatial data, and there are advantages and disadvantages associated with each. Traditionally, the raster model has been considered more useful for spatial analysis, due to the ease of operations such as polygon intersection of features. Disadvantages of the raster model include the relatively large computer memory requirements for obtaining an acceptable level of spatial resolution. The vector model poses technical difficulties in operations such as polygon intersections, but it does allow greater precision. In addition, some types of spatial analysis processes, such as transport network analysis, are only possible with vector models. With the recent development of more efficient computer hardware, improved computational methods, and improved routines for data transformation between model types, distinctions between the two types have become less significant, and some systems utilize both types of data models.

There are other, less commonly used data models, and one of special note is the triangulated irregular network (TIN) data model. It is frequently used for three-dimensional data. A TIN is composed of a series of cells as with the raster data model, but unlike raster data, cells in a TIN are irregular in shape. A TIN model can be best visualized as triangles connecting a series of sample points with measured elevations that are irregularly spaced across a given area. The points are typically measured at inflection points in the topography. The structure of a TIN makes it easy to calculate topographic slope and aspect. This strength means TIN models are widely used in applications such as volume calculation for earthwork and drainage studies.

4. DATA CAPTURE FOR GIS

A diversity of geographic data from different sources can be integrated into a GIS. As a result, there are a variety of methods used to capture data for these systems.

One frequently used method of data capture for raster models is remote sensing. This process measures properties of objects without direct physical contact. Remote sensing uses passive sensors that rely on reflected solar radiation or emitted terrestrial radiation and active sensors, such as radar, that generate their own sources of electromagnetic radiation. Remote sensing is generally associated with measurements from earth-orbiting satellites, but also includes aircraft-based surveys. Though often considered separately from remote sensing, aerial photography is an example of remote sensing using a passive sensor. Light detection and ranging technology (LiDAR) is an example of an active sensor used in remote sensing. All remote sensing processes measure a

horizontal position together with one or more attributes associated with that position. The attributes derived by remote sensing may be a spectral signature or, in the case of a system using LiDAR, may be an elevation.

Land surveying is an example of data capture for a vector data model. Data capture occurs in surveys using the Global Positioning System (GPS), conventional leveling, and total station surveys. GPS is used most frequently for data capture for GIS, since this technology can directly produce geographic coordinates.

In addition to data capture for GIS from primary sources, such as remote sensing and land surveying, geographic data may be captured from secondary sources. Examples of secondary data capture for raster models are scans of existing maps or photographs. Examples of secondary data capture for vector models are tablet digitizing, stereo photogrammetry, and coordinate geometry (COGO) data entry.

5. SPATIAL ANALYSIS

Most GIS software packages provide capability for various spatial analysis processes. In many ways, that capability represents the greatest value of a GIS. It adds value to the raw geographic data by revealing patterns and anomalies not otherwise visible.

The most basic type of spatial analysis is the query, in which the GIS finds answers to simple questions. Typical questions might be "How many people live within 5 mi of a potential hazard?" or "What areas in a county have elevations greater than 200 ft?" Other types of simple spatial analysis include measurements of properties, such as length, area, shape, and distance or direction between objects, and transformations that use simple geometric rules to generate new data sets based on existing data.

An important type of transformation is spatial interpolation, such as that used to develop contour lines from a series of irregularly spaced ground elevation points. One process used for spatial interpolation is inverse distance weighting (IDW). This process is used to estimate elevations at regularly spaced points on a fine grid covering an area and then systematically thread elevation contours through the grid points. The estimated elevation at each grid point is determined by finding the sum of the elevations at all random data points and multiplying by a parameter based on the inverse of the distance from the grid point to the random point.

Another process that is used for spatial interpolation is Kriging. This is more theoretically sound than IDW, but considerably more complex. Kriging interpolates the attribute value at unobserved points across a grid by modeling the attenuating effect of distance and fitting the points to smooth, continuous curves or surfaces described by mathematical functions.

Another example of spatial interpolation related to topographic elevations is the calculation of slope and aspect. This process can be used to predict runoff volume and direction. The process calculates the elevation difference between each cell and the eight surrounding cells. From this information, both the slope and the direction that surface water would flow may be determined.

6. ERRORS IN GIS

Two types of error occur in GIS: mistakes in positional accuracy and mistakes in attribute accuracy. *Positional accuracy* is the accuracy of absolute or relative mapping coordinates for a particular feature. *Attribute accuracy* is the accuracy of the qualitative or quantitative values attached to the feature. Note that elevation, while it could also be thought of as positional data, is normally considered to be an attribute of a horizontal position, and therefore an error in elevation values would be classified in the attribute accuracy category.

The most common sources of inaccurate information in GIS fall into four general categories: errors in original field measurements, data input errors, processing errors, and other errors.

- *Original field measurement errors:* These errors, which may affect both positional and attribute accuracy, are typically caused by weak measurement techniques or judgment, or by faulty equipment.

- *Data input errors:* These errors in the digital representation of graphic data are often encountered in the digitizing or geocoding processes, conversions between raster and vector data types, and feature coding and topological matching.

- *Processing errors:* These mistakes are often rounding errors, errors caused by limitations of the computer hardware or software, errors in interpolation due to assumption of linearity for non-linear applications, and errors due to overlay and boundary intersections.

- *Other errors:* Mistakes in this category may be due to changes in the data since they were measured, use of data at scales significantly different from the compilation scale, and insufficient density of data.

7. TYPICAL APPLICATIONS

A study of a simple GIS model will make it easier to understand the processes and capabilities of the technology. This example will demonstrate the basic concepts and some potential applications using data familiar to the surveyor.

The basic information for the GIS used for the model was established by a county property appraisal office in Jefferson County, Florida. The base map layer, or shape file, illustrated in Fig. 27.1, is a land parcel map of the county. The base map was created in two steps: a survey location of many of the controlling public land survey section corners, followed by the use of COGO and CADD routines to create a map from recorded subdivision plats and deeds of record. Aerial photography for the county is available as another layer of information, shown in Fig. 27.2.

Figure 27.1 *GIS Base Map[1]*

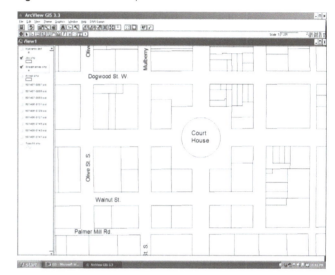

Source: Jefferson County (Florida) Property Appraiser

Figure 27.2 *GIS Aerial Photo*

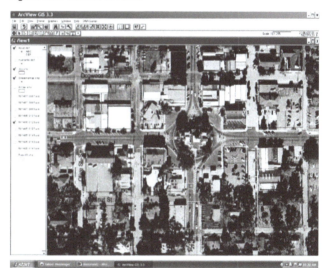

Source: Jefferson County (Florida) Appraiser

Land Planning/Development

[1]The illustrated base map and database was developed by the Jefferson County (Florida) Property Appraiser's Office under the direction of County Property Appraiser David Ward.

Each parcel on the base map is linked by parcel number to a database of information. Information for each parcel includes the name and mailing address of each owner of record, area of the parcel, appraised value, tax district, the location of the deed in the public records, and a number of additional items associated with the appraisal process. A small portion of the database is shown in Table 27.1.

The basic information in this GIS allows a complete overview of land ownership and land value in the county. Such a GIS is obviously of value to land surveyors, in addition to tax officials and real estate professionals. All parcels can be readily viewed on the parcel map as well as in the aerial photo, and information for each parcel is available with a click of a mouse, as shown in Fig. 27.3.

Figure 27.3 *Individual Parcel Information*

Source: Jefferson County (Florida) Property Appraiser

In addition, the GIS allows queries to be made regarding information in the system. In response to a query, the relevant parcel is highlighted on the map as well as in the database. Some examples of possible queries follow.

- Where is a tract of land with a given parcel number?
- What parcels are owned by a given individual?
- Which parcels are appraised at a value greater than a given figure?

The system can serve many other functions in a typical community. The system can serve as a base map for land regulation and zoning. With the addition of a street address field to the parcel database, the system can serve as a base map for emergency response. An emergency dispatcher or emergency responder can immediately locate an emergency with a query for a street address. On the photo layer, the dispatcher can view structures, terrain, and access points for the property and determine at a glance if there are nearby locations for landing emergency helicopters. The addition to the GIS database of a file of phone numbers tied

to street addresses allows the system to be used with 911 emergency response systems that capture the phone numbers of emergency calls.

With the addition of a simple database of surveyed coordinates for fire hydrants, such as those shown in Table 27.2, the location of the closest fire hydrant can be viewed, as well as information regarding that hydrant. This would provide guidance for dispatching fire emergency units. A typical view of how such features may be graphically displayed is provided in Fig. 27.4, with the hydrant locations represented by targets and identification codes. Note that, as depicted in Table 27.2, a database can be as basic as a file of x- and y-coordinates along with a feature identifier. Such files may be easily prepared in commonly used spreadsheet programs. Conversely, a database may be much more complex and include many additional attributes associated with the position.

Figure 27.4 *Map of Fire Hydrants*

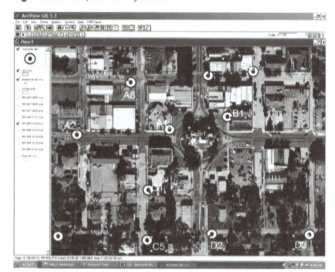

Source: Jefferson County (Florida) Appraiser

Table 27.2 *Fire Hydrant Coordinates*

northing	easting	hydrant no.
562,516	2,166,635	A1
563,077	2,166,516	A10
563,068	2,165,398	A11
563,504	2,166,524	A13
563,761	2,166,207	A14
563,751	2,165,500	A15
564,265	2,165,692	A17

Table 27.1 Portion of GIS Database

area (ft^2)	parcel identification	just value	assessed value
20,828,481.31320	01-3N-3E-0000-0010-0000	$1,581,000	$167,183
19,104,179.05649	01-3N-3E-0000-0010-0000	$1,581,000	$167,183
329,532.29695	01-3N-3E-0000-0020-0000	$38,961	$38,961
898,197.80266	01-3N-3E-0000-0030-0000	$18,000	$1,404
3,700,148.56154	01-3N-4E-0000-0060-0000	$175,862	$39,111
29,880,471.89741	01-3N-4E-0000-0010-0000	$1,532,444	$278,222
367,263.93574	01-3N-4E-0000-0040-0000	$10,600	$10,600
502,491.70116	01-3N-4E-0000-0190-0000	$22,922	$22,922
204,572.19697	01-3N-4E-0000-0030-0000	$27,220	$2922

Law enforcement agencies can also use the system as a graphic database of crime. This can be accomplished by adding a database containing geographic coordinates for crime scenes measured with an inexpensive handheld GPS receiver. Such a system provides a view of geographic trends in crime, as well as a simple means of retrieving information regarding certain crimes.

Local road departments can use the GIS for creating a graphic inventory of signs or drainage structures. Those features could be located using a hand-held GPS receiver. The addition of public domain shape files and associated databases of soil types, topographic elevations (see Fig. 27.5), and land use could be added to the system to make it invaluable to agricultural, engineering, construction, and land development interests. Examples of typical queries that could be addressed with these additions follow.

- Where are tracts of land of a given size with soil suitable for a certain crop?

- Where are wetland soils located?

- Where are tracts of land with high elevations suitable for cell tower location?

- Where are vacant tracts of a given size with soil suitable for a certain type of construction?

Figure 27.5 Topographic Map

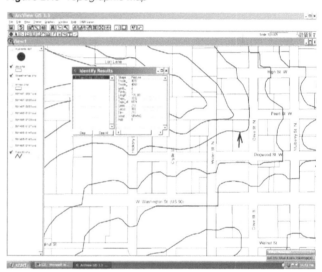

Map source: Jefferson County (Florida) Property Appraiser
Contour source: U.S. Geological Survey

These example queries represent relatively simple applications. With the addition of census tract data, environmental data, transportation statistics, and similar information, the potential applications for such a community-based GIS are almost endless. GIS offers a means of using survey technology to create models of the earth that can be used in almost all aspects of society.

Land Planning/ Development

8. PRACTICE PROBLEMS

1. All the assessor plats for a municipality are scanned into a file for a GIS. Which is the most probable structure for the resulting file?

    (A)   vector data structure

    (B)   metadata structure

    (C)   raster data structure

    (D)   none of the above

2. A CADD file is an example of which of the following?

    (A)   vector data structure

    (B)   metadata structure

    (C)   raster data structure

    (D)   none of the above

3. Inverse distance weighing and Kriging are examples of which of the following?

    (A)   geographic data models

    (B)   database linkage methods

    (C)   processes for visualization of geographic data

    (D)   spatial interpolation

4. A TIN is often used for which of the following?

    (A)   photographic data

    (B)   census data

    (C)   topographic data

    (D)   land use data

5. An error in elevation is an example of which of the following?

    (A)   positional error

    (B)   attribute error

    (C)   qualitative value error

    (D)   none of the above

SOLUTIONS

1. Scanning typically creates a grid or raster structure file.

The answer is (C).

2. A CADD file is an example of a vector data structure.

The answer is (A).

3. Inverse distance weighing and Kriging are examples of spatial interpolation.

The answer is (D).

4. A TIN, or triangulated irregular network, is often used for topographic data.

The answer is (C).

5. An error in elevation is an example of an attribute error. While elevation is also representative of position, it is more typically considered an attribute of a horizontal position.

The answer is (B).

28 Construction Staking

Nomenclature

C	chord length	ft	m
HI	height of instrument	ft	m
I	deflection angle	deg	deg
L	length of curve	ft	m
LC	length of long chord	ft	m
R	radius	ft	m
T	tangent distance	ft	m

1. DEFINITION

Construction surveying involves locating and marking locations of structures that are to be built. It is often referred to as *giving line and grade*. A transit or theodolite is used in establishing line (horizontal alignment), and a level is used in establishing grade (elevation).[1]

2. CONSTRUCTION STAKES

Construction stakes are a critical component of construction surveying. They are used to define the location and nature of earthwork as well as for construction.

Thus, they are the surveyor's means for communication of instructions for construction, and, as a result, stake locations and markings are critical. Standards for the use and marking of stakes vary somewhat from agency to agency.

Common types of construction stakes are listed. In addition to the listed types, stakes are also used to mark the limits of foundations, structures, curbing, and numerous other planned features.

Hub stakes are typically 2 in by 2 in diameter square stakes, and usually have a tack in the top to mark the precise position. Hub stakes are reference stakes and are driven flush with the ground. They contain no markings.

Witness stakes, or *guard stakes*, are used to locate and identify hub stakes. They serve to call attention to the nearby hub or other type of construction stake but do not themselves identify a specific point. The front and back sides of witness stakes are used to provide location and position information regarding the nearby hub. When elevations for the hub are to be provided, they are written on the narrow edge of the witness stake.

Alignment stakes are used to indicate stationing and centerline alignment along proposed corridors. They are usually placed at every station along tangent sections with uniform grade but at more frequent intervals along horizontal and vertical curves.

Offset stakes are used to mark the horizontal limits of earthwork or construction. They are typically offset from the actual edge to protect them during the construction process. The offset distance is circled on the stake and separated from the other locational data by a double line.

Grade stakes are used to control the vertical limits of earthwork at particular locations. After determining the elevation of the top of the grade stake relative to the desired construction grade, the required cut (C) or fill (F) is marked on the witness stake for the grade stake.

Slope stakes are used to indicate the intersection of the natural ground and the proposed cut or fill (referred to as the catch point) for excavation and embankment operations. The front sides of slope stakes are marked with a "C" for cut or "F" for fill to indicate the nature of the planned

---

[1] The word *grade* is not used consistently, sometimes meaning slope and sometimes meaning elevation above a datum. In this text, *gradient* will be used for rate of slope, and *finish elevation* will be used for the elevation above a datum to which a part of the structure is to be built.

Land Planning/Development

earthwork, the elevation difference between the grade point and the finished grade, and the distance offset from the centerline, with an "L" or "R" indicating the direction of the stake point from the centerline when looking ahead on stationing. In addition, they are marked with the planned slope. Stationing is indicated on the back sides of slope stakes. Figure 28.1 provides an illustration of a typical slope stake.

Figure 28.1 *Typical Slope Staking*

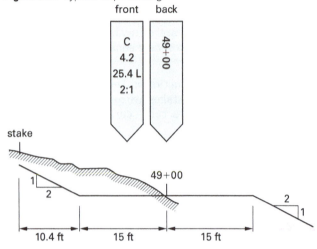

3. CONVERSION BETWEEN INCHES AND DECIMALS OF A FOOT

Construction stakes are usually set to the nearest hundredth of a foot for concrete, asphalt, pipelines, and so on. For earthwork, stakes are set to the nearest tenth of a foot. Constructors use foot and inch rules. So, in deference to them, surveyors set the stakes for their convenience.

In converting measurements in feet, inches, and fractions to feet, it may be easier to convert the inches and fractions of an inch separately, and then to add the parts. See Sec. 3.4 for more on conversion between inches and decimals of a foot.

Example 28.1

Convert the measurements given to feet and decimals of a foot.

(a) 1 ft 4 in
(b) 11 ft 9⅛ in
(c) 7 ft 5⅛ in
(d) 2 ft 8⅞ in
(e) 5 ft 11½ in

Solution

(a) 1.33 ft
(b) 11.76 ft
(c) 7.43 ft
(d) 2.74 ft
(e) 5.96 ft

Example 28.2

Convert the given measurements to feet and inches.

(a) 3.79 ft
(b) 6.34 ft
(c) 5.65 ft
(d) 3.72 ft

Solution

(a) 3 ft 9½ in
(b) 6 ft 4⅛ in
(c) 5 ft 7¾ in
(d) 3 ft 8⅝ in

4. STAKING OFFSET LINES FOR CIRCULAR CURVES

Stakes set for the construction of pavement or curbs must be set on an offset line so that they will not be destroyed by construction equipment. However, they must be close enough for short measurements to the actual line. The offset line may be 3 ft or 5 ft, or any convenient distance from the edge of pavement or back of curb. Stakes are set at 25 ft or 50 ft intervals, and tacks are set in the stakes to designate the offset line.

In setting stakes on a parallel circular arc, the central angle is the same for parallel arcs. The radius to the centerline of the street or road is usually the design radius. The PC of the design curve and the PC of a parallel offset curve, whether right or left, will fall on the same radial line. Likewise, the PTs of the parallel arcs will fall on the same radial line. In computing the stations for PCs and PTs, the design curve data (design radius) should be used. Then, the PC and PT stations for an offset line will be the same as for the design curve (centerline of road or street), even though the lengths of offset curves will not be the same as the length of the centerline curve.

The design curve data will be used to compute deflection angles. These angles will be the same for offset lines, since the central angle between any two radii is the same for the parallel arcs.

Because chord lengths are a function of the radius of an arc ($C = 2R\sin(I/2)$), the chord length between two stations on the design curve and two corresponding stations on the offset line will not be the same.

By using design curve data in computing PC and PT stations, deflection angles for curves can be recorded in the field book. Such angles will be the same whether a right offset line, a left offset line, or both right and left offset lines are used.

In performing field work, centerline PIs, PCs, and PTs are located on the ground before construction. Offset PCs and PTs are located at right angles to the centerline PCs and PTs. Offset PCs and PTs should be carefully referenced.

Example 28.3

Stakes are to be set on 4 ft offsets for each edge of a pavement that is 36 ft wide. The curve has a deflection angle of 60° to the right and a centerline radius of 300 ft. PI is at station 12+44.32, and stakes are to be set for each full station and half-station, and at the PC and PT. Calculate PC and PT stations, deflection angles, and chord lengths, and tabulate field notes for the curve.

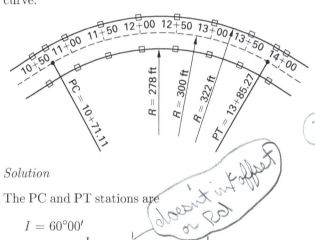

Solution

The PC and PT stations are

$$I = 60°00'$$

$$T = R\tan\frac{I}{2} = (300\text{ ft})(\tan 30°) = 173.21\text{ ft}$$

doesn't in offset or Rd

$$L = \left(\frac{60°}{360°}\right)2\pi R = \left(\frac{60°}{360°}\right)2\pi(300\text{ ft}) = 314.16\text{ ft}$$

(1)

$$PI = 12+44.32$$
$$T = -1+73.21$$
$$PC = 10+71.11$$
$$L = 3+14.16$$
$$PT = 13+85.27$$

The deflection angles are *(2)* $\dfrac{\text{Total Dist}}{L}\left(\dfrac{\theta}{2}\right)$

sta 10+71.11: 0°00' *28.89*

sta 11+00: $\left(\dfrac{11+00 - 10+71.11}{314.16\text{ ft}}\right)\left(\dfrac{60°}{2}\right)$

$$= \left(\dfrac{28.89\text{ ft}}{314.16\text{ ft}}\right)(30°)$$

$$= 2.7588°\quad(2°45')$$

sta 11+50: $\left(\dfrac{28.89+50\text{ ft}}{314.16\text{ ft}}\right)\left(\dfrac{60°}{2}\right)$

$$= \left(\dfrac{78.89\text{ ft}}{314.16\text{ ft}}\right)(30°)$$

$$= 7.5334°\quad(7°32')$$

sta 12+00: $\left(\dfrac{128.89\text{ ft}}{314.16\text{ ft}}\right)(30°) = 12.3081°\quad(12°18')$

sta 12+50: $\left(\dfrac{178.89\text{ ft}}{314.16\text{ ft}}\right)(30°) = 17.0827°\quad(17°05')$

sta 13+00: $\left(\dfrac{228.89\text{ ft}}{314.16\text{ ft}}\right)(30°) = 21.8573°\quad(21°51')$

sta 13+50: $\left(\dfrac{278.89\text{ ft}}{314.16\text{ ft}}\right)(30°) = 26.6320°\quad(26°38')$

sta 13+85.27: $\left(\dfrac{314.16\text{ ft}}{314.16\text{ ft}}\right)(30°) = 30.0000°\quad(30°00')$

(3) The outside chord lengths are $C = 2R\sin(\text{def }\theta)$

$R_o = R_{C\ell} + \left(\dfrac{Rd W}{2}\right) + \text{offset} = 300 + \dfrac{36}{2} + 4 = 322'$

station to station	computation	length
10+71.11 to 11+00:	(2)(322 ft) × (sin 2.7588°)	= 31.00 ft
11+00 to 11+50:	(2)(322 ft) × (sin 4.7746°)	= 53.60 ft
13+50 to 13+85.27:	(2)(322 ft) × (sin 3.3680°)	= 37.83 ft

← 7.5334 − 2.7588

30° − 26.6320°

(4) The inside chord lengths are $C = 2R\sin(\text{def }\theta)$

$R_{in} = R_{C\ell} - \dfrac{Rd W}{2} - 4 = 300 - \dfrac{36}{2} - 4 = 278$

station to station	computation	length
10+71.11 to 11+00:	(2)(278 ft) × (sin 2.7588°)	= 26.76 ft
11+00 to 11+50:	(2)(278 ft) × (sin 4.7746°)	= 46.28 ft
13+50 to 13+85.27:	(2)(278 ft) × (sin 3.3680°)	= 32.66 ft

Field Notes

point	station	angle	C_{out} (ft)	C_{in} (ft)	curve data
PT	13+85.27	30°00′			
			37.83	32.66	
	+50.00	26°38′			
			53.60	46.28	
	13+00.00	21°51′			$I = 60°00′$
			53.60	46.28	
+50.00		17°05′			$R = 300$ ft
			53.60	46.28	
12+00.00		12°18′			$T =$ 173.21 ft
			53.60	46.28	
+50.00		7°32′			$L =$ 314.16 ft
			53.60	46.28	
	11+00.00	2°45′			
			31.00	26.76	
PC	10+71.11	0°00′			

5. CURB RETURNS AT STREET INTERSECTIONS

Curb returns are the arcs made by the curbs at street intersections. The radius of the arc is selected by the designer with consideration given to the speed and volume of traffic. A radius of 30 ft to the back of curb is common. Streets that intersect at a right angle have curb returns of one-quarter circle. The arcs can be swung from a radius point (center of circle). The radius point can be located by finding the intersection of two lines, each of which is parallel to one of the centerlines of the streets and at a distance from the centerline equal to half the street width plus the radius, as shown in Fig. 28.2. One stake at the radius point is sufficient for a curb return.

Figure 28.2 *Curb Returns*

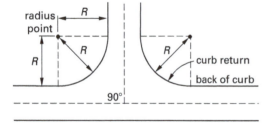

6. STAKING OFFSET LINES AT STREET INTERSECTIONS

In setting stakes for curbs and gutters for street construction on an offset line, stations for the PC and PT of a curb return at a street intersection are computed along the centerline of the street. In Ex. 28.4, stakes are to be set on a 5 ft offset line from the back of the left curb along Elm Street. In setting stakes at the intersection of 24th Street, the PCs and PTs of the curb return are computed from the centerline stations. However, it must be remembered that in calculating the long chord on the offset line, the radius is not the design radius, but the design radius minus the offset distance. Stakes are set at the PC, the PT, and the radius point of each curb return arc.

At street intersections not at 90°, there will be two deflection angles, one being the supplement of the other.

Example 28.4

Elm Street and 24th Street intersect as shown. Both streets are 26 ft wide, back to back of curb.

Compute the PC and PT stations, deflection angles from PC to PT, and long chord measured from the offset line for each return.

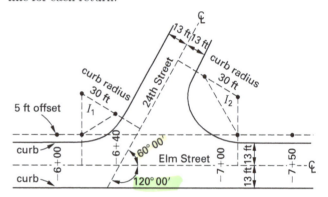

Solution

PC Stations Along Elm Street

$$I_1 = 60°00′$$

$$T_1 = R_1 \tan \frac{I_1}{2}$$

$$= (43 \text{ ft}) \tan \left(\frac{60°}{2} \right)$$

$$= 24.83 \text{ ft}$$

$$PI = 6+40.00$$

$$T_1 = -24.83$$

$$PC = \overline{6+15.17}$$

R = Curb R + R
30 + 13
does not
include
offset

Land Planning/ Development

②

$$I_2 = 120°00'$$

$$T_2 = R_2 \tan\frac{I_2}{2}$$

$$= (43 \text{ ft})\tan\left(\frac{120°}{2}\right)$$

$$= 74.48 \text{ ft}$$

$$\text{PI} = 6+40.00$$

$$T_2 = +74.48$$

$$\cancel{\text{PC}} = \overline{7+14.48}$$
PT

handwritten: R = Curb R + Rdw, R = 30 + 13 = 43

PT Stations Along 24th Street

③

$$\begin{array}{ll}
\text{PI} & = 0+00.00 \\
T_1 & = +24.83 \\
\text{PT} & = 0+24.83 \\
\text{PI} & = 0+00.00 \\
T_2 & = +74.48 \\
\text{PT} & = 0+74.48
\end{array}$$

The deflection angles and long chords are

④

$$\text{LC} = 2R\sin\frac{I_1}{2}$$

$$= (2)(25 \text{ ft})\sin\frac{60°}{2}$$

$$= 25.00 \text{ ft}$$

$$I = \frac{60°}{2} = 30°$$

$$\text{LC} = 2R\sin\frac{I_2}{2}$$

$$= (2)(25 \text{ ft})\sin\left(\frac{120°}{2}\right)$$

$$= 43.30 \text{ ft}$$

$$I = \frac{120°}{2} = 60°$$

handwritten: R = Curb R − offset, 30 − 5 = 25

7. ESTABLISHING FINISH ELEVATIONS OR "GRADE"

Establishing the elevation above a datum to which a structure, or part of a structure, is to be built is usually accomplished by the following steps.

step 1: Set the top of a grade stake to the exact elevation (nearest one hundredth). Mark the top of the stake with blue keel.

step 2: Set the top of a grade stake at an exact distance above or below finish elevation. Mark the top of it with blue keel, and mark this exact distance above or below (called *cut* or *fill*) on another

stake known as a *guard stake*, usually driven at an angle, beside the grade stake.

step 3: Use the line stake as a grade stake driven to a random elevation. Compute the difference in elevation between that elevation and the finish elevation, and mark this difference as cut or fill on a guard stake.

Marks on a wall, such as the wall of forms for a concrete structure, may be used instead of the tops of stakes. This is illustrated in Ex. 28.5.

8. GRADE ROD

A *grade rod* is the rod reading determined by finding the difference in elevation between the height of instrument (height of the level), HI, and the finish elevation. In Fig. 28.3, the finish elevation is 441.23 ft, and the HI of the level is 445.55 ft. The grade rod is the difference in these two numbers, 4.32 ft. A stake is driven so that when the level rod is placed on the top of it, the rod reading is 4.32 ft.

Figure 28.3 *Use of Grade Rod*

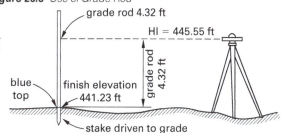

The procedure to set the stake is to place the rod on the ground where the stake is to be driven and determine the distance (in tenths of a foot) that the top of the stake should be above the ground. Then, drive the stake until the top of the stake is at the finish elevation, stopping to check the rod reading so that the top of the stake is not too low. When the grade rod reading is reached, the top of the stake is marked with blue keel, and thus the name *blue top* is given to this type of stake. A witness stake is driven beside the blue top in a slanting position. The witness stake is marked "G" to indicate the stake is driven to grade (finish elevation).

If the finish elevation is just below ground level, the blue top can be left above ground. The cut from the top of the stake to the finish elevation should be marked on a witness stake.

Where line stakes are also used as grade stakes driven to random elevations, a grade rod is not used. The elevation of the top of the stake is determined by leveling. The difference in elevation between the top of the stake and the finish elevation is determined and marked on the witness stake.

(vertical margin text: Land Planning/Development)

Example 28.5

The finish elevation is to be marked on the inside wall of the form for the concrete cap of the bridge shown. The finish elevation of the cap is 466.97 ft, and the HI is 468.72 ft.

What rod reading should be used to mark the line of finish elevation?

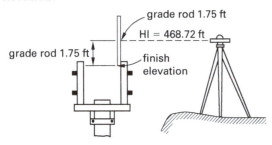

Solution

HI	= 468.72 ft
finish elevation	= 466.97 ft
grade rod	= 1.75 ft

The rod is held against the side of the form and raised or lowered until the rod reading is 1.75 ft. A nail is driven at the bottom of the rod, and the rod is placed on the nail so that the rod reading can be checked to see that the nail is correctly placed. Another finish elevation nail is driven at the other end of the form, and a string line is drawn between the two nails, then chalked and snapped to mark the grade line on the form. A chamfer strip is nailed on the form along this line, and the strip is used to finish the concrete to grade (finish elevation).

9. SETTING STAKES FOR CURB AND GUTTER

Separate stakes are often set for line and grade. In Fig. 28.4, a line stake is driven so that a tack is exactly 3 ft from the back of a curb. These line stakes are set on any convenient offset to avoid disturbance by construction equipment. A separate grade stake is driven so that the top of the stake is either at finish elevation or at an elevation that makes it an exact distance above or below finish elevation.

A guard stake is driven near the grade stake and marked to show this exact distance as cut or fill, and the top of the grade stake is marked with blue keel. Grade stakes can be driven so that the cut or fill is in multiples of half a foot. If the grade stake is driven to finish elevation, the guard stake is marked "G" for grade. The cut or fill can be determined by considering the finish elevation and a ground rod reading at each station.

Figure 28.4 *Line and Grade Stakes for Curb and Gutter*

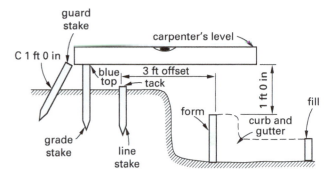

The top of the grade stake should be above ground. Then, the builder can lay a carpenter's level on top of the stake and measure from the established level line to establish the tops of curb forms. In Ex. 28.6, the guard stake is marked for a cut as "C 1 ft, 0 in" so that the builder will measure 1 ft 0 in down from the level line to the top of the forms. Horizontal alignment will be maintained by measuring 3 ft from each tack point to the back of curb line.

Example 28.6

Grade stakes for curb and gutter have been driven to grade or to a multiple of 6 in above or below grade. Part of the level notes recorded in setting the stakes is shown. Also shown are finish elevations that have been taken from construction plans. Computations for grade rod and for the cut or fill marks on witness stakes are to be made and recorded. Rod readings on grade stakes as driven are to be recorded in the column marked "rod." (Note: A more detailed explanation of this procedure can be found in Sec. 28.14.)

station	+	−	finish elevation	ground elevation	rod
BM no. 1	3.42		452.36		
0 + 00				454.10	0.4
+ 50				454.58	1.6
1 + 00				455.06	3.2
+ 50				455.53	3.3
2 + 00				456.01	2.5
+ 50				456.49	1.8
3 + 00				456.97	0.6
TP	8.21	3.75			
+ 50				457.44	3.6
4 + 00				457.83	1.7

Solution

Finished field notes are shown in Table 28.1. A graphical solution is illustrated in *Finished Grade Line for Example 28.6*. Solutions for sta 0+00, 0+50, 1+00, and 2+50 are shown in *Sample Staking for Example 28.6*.

Land Planning/ Development

Table 28.1 *Field Notes for Curb and Gutter Grades (all measurements in ft)*

station	+	HI	−	rod	elevation	finish elevation	grade rod	ground	mark stake
BM no. 1	3.42	455.78			452.36	r.r. spike in 12″ oak-100′ lt. sta 0 ǀ 00			
0+00				0.18	455.60	454.10	1.68	0.4	C 1 ft 6 in
0+50				1.20	454.58	454.58	1.20	1.6	F 1 ft 6 in
1+00				2.72	453.06	455.06	0.72	3.2	F 2 ft 0 in
1+50				2.75	453.03	455.53	0.25	3.3	F 2 ft 6 in
2+00				2.27	453.51	456.01	−0.23	2.5	F 2 ft 6 in
2+50				1.79	453.99	456.49	−0.71	1.8	F 2 ft 6 in
3+00				0.31	455.47	456.97	−1.19	0.6	F 1 ft 6 in
T.P.	8.21	460.24	3.75		452.03				
3+50				3.30	456.94	457.44	2.80	3.6	F 0 ft 6 in
4+00				1.41	458.83	457.83	2.41	1.7	C 1 ft 0 in
4+50				0.70	459.54	458.04	2.20	1.0	C 1 ft 6 in
5+00				1.16	459.08	458.08	2.16	1.5	C 1 ft 0 in
5+50				0.31	459.93	457.93	2.31	0.7	C 2 ft 0 in
6+00				1.04	459.20	457.70	2.54	1.4	C 1 ft 6 in
BM no. 2			2.06		458.18				
	11.63		5.81						
	11.63		458.18						
	−5.81		−452.36						
	5.82		5.82						

Finished Grade Line for Example 28.6

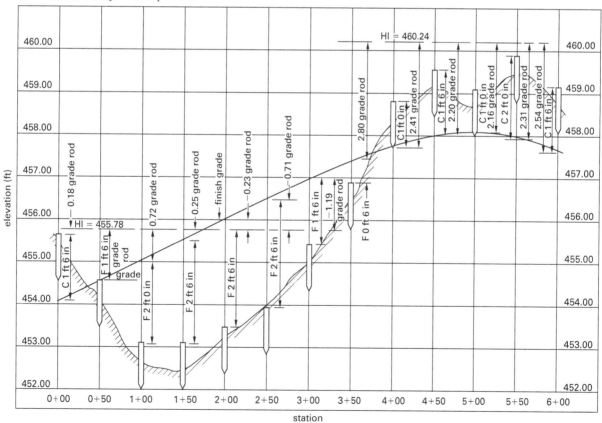

Sample Staking for Example 28.6

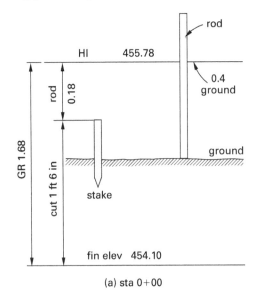

(a) sta 0+00

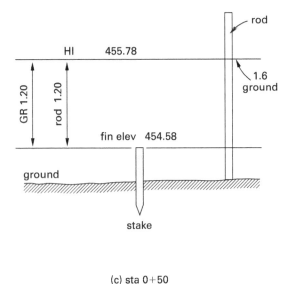

(c) sta 0+50

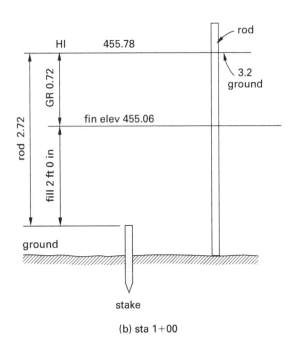

(b) sta 1+00

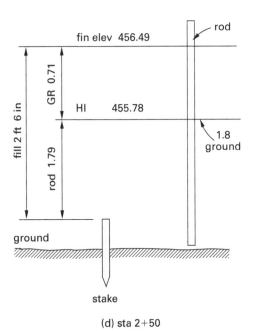

(d) sta 2+50

(not to scale)

10. STAKING CONCRETE BOX CULVERTS ON HIGHWAYS

Tack points for concrete box culverts can be set on offsets from the outside corners of the culvert headwalls. For normal culverts (centerline of culvert at right angle to centerline of roadway), the distance from the centerline of the roadway to the outside of the headwall is equal to one-half the clear roadway width plus the width of the headwall. Tacks should also be set on the centerline of the roadway, offset from the outside of the culvert wall. The offset distance from the outside walls depends on the depth of cut.

Stakes for wingwalls and aprons (see Fig. 28.5(a)) are not necessary, although stakes to establish the centerline of the culvert can be set if desired. Cuts to the flowline of the culvert can be marked on guard stakes at the tack points.

In staking skewed culverts, the distance from the centerline of the roadway to the outside of the headwall (along the centerline of the culvert) is equal to one-half the clear roadway plus the headwall thickness divided by the cosine of the skew angle (see Fig. 28.5(b)). The *skew angle* is the angle between the normal and the centerline of the culvert.

Figure 28.5 *Staking Box Culverts (plan view)*

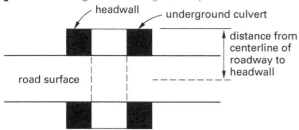

(a) normal culvert (headwall perpendicular to culvert)

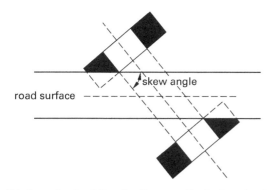

(b) skewed culvert (headwall perpendicular to culvert)

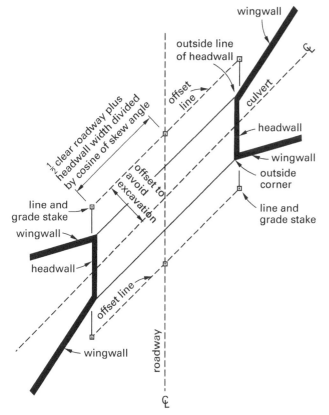

(c) skewed culvert (headwalls parallel to roadway)

11. SETTING SLOPE STAKES

Before earthwork construction is started, the extremities of a cut or fill must be located at numerous places for the benefit of machine operators engaged in the earthwork.

With the centerline as a reference, the edge (*toe*) of a fill must be established on the natural ground. This point is known as the *toe of slope*. Likewise, the top edge of a cut must be established on the natural ground. (See Fig. 28.6.)

Figure 28.6 *Slope Staking*

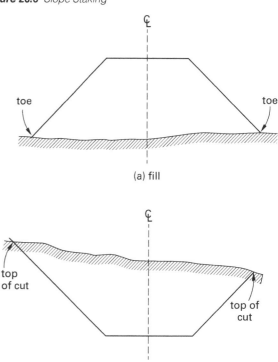

(a) fill

(b) cut

Because the natural ground may slope from left to right or from right to left, the distance from the centerline to the left toe of slope of a fill is usually different from the distance from the centerline to the right toe of slope at any particular station. The same is true of the top of a cut. This fact, plus the fact that the height of fill or depth of cut varies along the centerline, makes toe and top lines irregular when seen in plan view, as shown in Fig. 28.7.

The toe of a fill or the top of a cut is found by a measure-and-try method. The horizontal distance from centerline to toe or top is determined by horizontal tape measurements combined with vertical distance measurements derived by use of level and rod.

Dimensions of the top of a fill or bottom of a cut and the slope of the sides of the fill or cut must be known. They are shown on construction plans, as in Fig. 28.8.

Figure 28.7 *Fill Views*

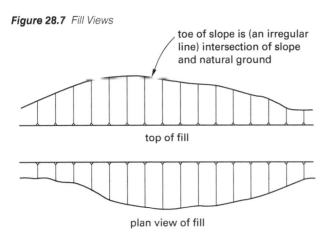

Figure 28.8 *Fill and Cut Dimensioning*

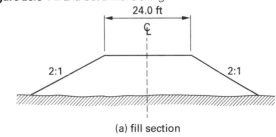

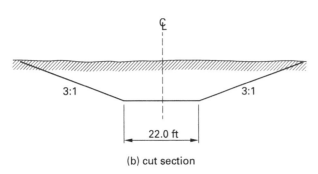

The side slopes of a fill and the back slopes of a cut are expressed as a ratio of horizontal to vertical distance. Thus, a 4:1 slope means a rise or fall of 1 ft for each 4 ft of horizontal distance. Slopes of 1:1, 2:1, and 3:1 are illustrated in Fig. 28.9.

With the centerline finish elevation, width of top of fill or bottom of cut, and side slopes all known, the intersection of the side slopes and the natural ground is located at each station or intermediate point.

When the intersection is found, it is marked by a *slope stake*. The stake is driven so that it slopes away from the fill or cut and is marked with its horizontal distance (left or right) from the centerline and the vertical distance from the ground at the stake to the finish elevation. A stake marked "C 3.2-48.2" means that the stake is 48.2 ft from the centerline, and the ground at the stake is 3.2 ft above the finish elevation. The station number is shown on the side of the stake facing the ground.

Figure 28.9 *Calculation of Slopes*

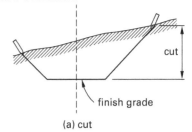

12. GRADE ROD FOR SLOPE STAKES

In setting slope stakes, as in setting finish elevation for pavement, sewer lines, and so on, the grade rod is used to determine the difference in elevation between the HI and the finish elevation. To determine the cut at a particular point, the rod is read on the ground, and the ground rod is subtracted from the grade rod at that point. To determine the fill at a particular point, the grade rod is subtracted from the ground rod if the HI is above the finish elevation. The grade rod is added to the ground rod if the HI is below the finish elevation. (See Fig. 28.10, Fig. 28.11, Ex. 28.8, and Ex. 28.9.)

Figure 28.10 *Stake Orientations*

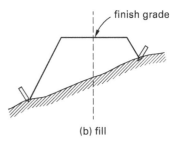

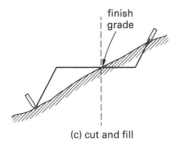

Figure 28.11 *Use of Grade Rod to Determine Cut and Fill*

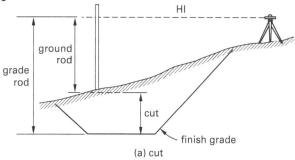

(a) cut

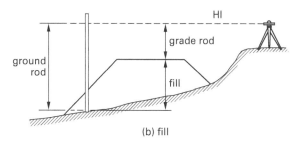

(b) fill

13. SETTING SLOPE STAKES AT CUT SECTIONS

An explanation of setting slope stakes without the benefit of a demonstration in the field is difficult. In Ex. 28.7, a scale illustration is used at a cut section at which the HI and finish elevation are known and plotted on the illustration. The width of the ditch bottom and the side slopes (also referred to as *back slopes*) are also known. In this example, the level and rod are replaced by the plotted HI and an engineer's scale. The scale is used to measure vertical distance from HI to ground, just as the level and rod are used.

Example 28.7

In the cut shown, slope stakes are to be set at sta 3+00. The bottom of the cut is to be at elev 462.00 ft and is 10 ft wide. The side slopes are 2:1; all measurements are in feet.

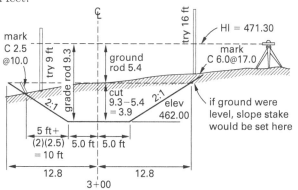

The calculated distance from the centerline to set the slope stake is

$$5 \text{ ft} + (2)(2.5 \text{ ft}) = 10.0 \text{ ft}$$

Solution

step 1: Establish the level near sta 3+00 and determine the HI (471.30 ft).

step 2: Calculate the grade rod by subtracting the elevation at the bottom of the cut from the HI.

$$471.30 \text{ ft} - 462.00 \text{ ft} = 9.30 \text{ ft}$$

step 3: Determine the ground rod by placing the rod on the ground at the centerline. Read 5.4 ft.

step 4: Calculate the cut at the centerline by subtracting the ground rod from the grade rod.

$$9.3 \text{ ft} - 5.4 \text{ ft} = 3.9 \text{ ft}$$

step 5: Calculate the distance to the left slope stake from the centerline as if the ground were level at this station.

$$5 \text{ ft} + (2)(3.9 \text{ ft}) = 12.8 \text{ ft}$$

step 6: Note that the ground on the left slopes down and the side of the cut slopes up, indicating that the distance to the stake will be less than that for level ground.

step 7: Try a distance less than 12.8 ft, say 9.0 ft, and read the rod at this distance. The rod reading is 6.6 ft.

$$\text{grade rod} - \text{ground rod} = 9.3 \text{ ft} - 6.6 \text{ ft}$$
$$= 2.7 \text{ ft}$$

The distance calculated from this rod reading is

$$5 \text{ ft} + (2)(2.7 \text{ ft}) = 10.4 \text{ ft}$$

Move toward 10.4 ft; try 10.0 ft. (Move less because slopes are opposite.)

step 8: The ground rod at 10.0 ft is 6.8 ft.

$$9.3 \text{ ft} - 6.8 \text{ ft} = 2.5 \text{ ft}$$

The calculated distance is

$$5 \text{ ft} + (2)(2.5 \text{ ft}) = 10.0 \text{ ft}$$

The calculated distance agrees with the measured distance.

step 9: Set the stake at 10.0 ft left of centerline and mark "C 2.5 @ 10.0" on the top face of the stake and "3+00" on the bottom.

step 10: Move to the right side. Try a distance greater than that for level ground because the ground and sides both slope up.

step 11: Try 16.0; the ground rod is 3.4 ft.

$$9.3 \text{ ft} - 3.4 \text{ ft} = 5.9 \text{ ft}$$
$$5 \text{ ft} + (2)(5.9 \text{ ft}) = 16.8 \text{ ft}$$

Land Planning/Development

step 12: Try 17.0 ft. Move beyond 16.8 ft because the slopes are in the same direction. The ground rod is 3.3 ft.

$$9.3 \text{ ft} - 3.3 \text{ ft} = 6.0 \text{ ft}$$
$$5 \text{ ft} + (2)(6.0 \text{ ft}) = 17.0 \text{ ft}$$

step 13: Set the stake at 17.0 ft. Mark "C 6.0 @ 17.0."

14. SETTING SLOPE STAKES AT FILL SECTIONS

In setting slope stakes for fills, two situations may arise: (a) the HI may be below the finish elevation, as shown in Ex. 28.8 and Ex. 28.9, or (b) the HI may be above the finish elevation, as shown in Fig. 28.11. If the HI is below the finish elevation, the fill is the sum of the grade rod and the ground rod. If the HI is above the finish elevation, the fill is the difference between the ground rod and the grade rod.

Example 28.8

The ground cross section at a station at which slope stakes are to be set for a fill is shown.

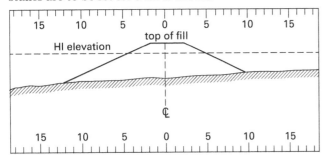

The HI has been established, and the finish elevation, width of top of fill, and side slopes have been obtained from the construction plans. The centerline of fill has also been established. Known information is tabulated as follows.

finish elevation of top of fill = 452.36 ft

top of fill width = 4 ft

side slopes = 2:1

HI = 450.54 ft

Determine the correct cut and distance.

Solution

The solution is shown. The correct cut and distance can be found on the marked stake.

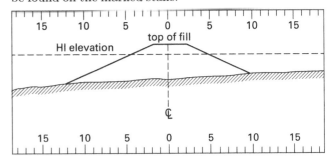

	F 5.2 @ 12.4					F 4.1 @ 10.2	
try 12.5	try 13.0	try 11.0	₵	try 9.0	try 11.0	try 10.4
1.82	1.82	1.82	1.82	1.82	1.82	1.82
+3.4	+3.5	+3.3	+2.7	+2.3	+2.2	+2.3
5.2	5.3	5.1	4.5	4.1	4.0	4.1
×2	×2	×2	×2	×2	×2	×2
10.4	10.6	10.2	9.0	8.2	8.0	8.2
+2.0	+2.0	+2.0	+2.0	+2.0	+2.0	+2.0
12.4	12.6	12.2	11.0	10.2	10.0	10.2

Example 28.9

In the illustration shown, slope stakes are to be set at sta 10+00. The top of fill is to be at elevation 468.00 ft and is 10 ft wide. Side slopes are 1½:1 (h:v). What is the distance from the centerline and rod reading for the top of fill?

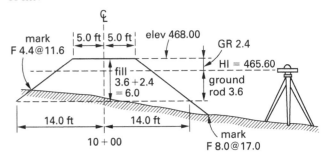

Solution

step 1: Establish the level near sta 10+00 and determine the HI (465.60 ft in this example).

step 2: Calculate the grade rod by subtracting the HI from the elevation of the top of the fill.

$$468.00 \text{ ft} - 465.60 \text{ ft} = 2.4 \text{ ft}$$

step 3: Determine the ground rod by placing the rod on the ground at the centerline. Read 3.6 ft.

step 4: Compute the fill at the centerline by adding the grade rod and the ground rod.

$$3.6 \text{ ft} + 2.4 \text{ ft} = 6.0 \text{ ft}$$

step 5: Compute the distance to the left slope stake from the centerline as if the ground were level at this station.

$$5 \text{ ft} + (1.5)(6.0 \text{ ft}) = 14.0 \text{ ft}$$

step 6: Note that slopes are opposite, indicating that the distance will be less than that for level ground.

step 7: Try 11.0 ft. The rod reads 2.2 ft.

$$2.2 \text{ ft} + 2.4 \text{ ft} = 4.6 \text{ ft}$$
$$5 \text{ ft} + (1.5)(4.6 \text{ ft}) = 11.9 \text{ ft}$$

Move toward 11.9 ft, but less because slopes are opposite.

step 8: Try 11.5 ft. The ground rod is 2.0 ft.

$$2.0 \text{ ft} + 2.4 \text{ ft} = 4.4 \text{ ft}$$
$$5 \text{ ft} + (1.5)(4.4 \text{ ft}) = 11.6 \text{ ft}$$

This is close enough.

step 9: Set the stake at 11.6 ft left of centerline and mark "F 4.4 @ 11.6." Mark "10+00" on the bottom face of the stake.

step 10: Move to the right side. Try a distance greater than that for level ground because ground and slope are in the same direction.

step 11: Try 15.0 ft. The ground rod is 5.3 ft.

$$5.3 \text{ ft} + 2.4 \text{ ft} = 7.7 \text{ ft}$$
$$5 \text{ ft} + (1.5)(7.7 \text{ ft}) = 16.6 \text{ ft}$$

Move toward 16.6 ft and beyond because the slopes are both down.

step 12: Try 17.0 ft. The ground rod is 5.6 ft.

$$5.6 \text{ ft} + 2.4 \text{ ft} = 8.0 \text{ ft}$$
$$5 \text{ ft} + (1.5)(8.0 \text{ ft}) = 17.0 \text{ ft}$$

step 13: Set the stake at 17.0 ft and mark "F 8.0 @ 17.0."

15. SETTING STAKES FOR UNDERGROUND PIPE

Stakes for line and grade for underground pipe, like stakes for roads and streets, are set on an offset line. One hub stake with tack can be used at each station for both line and grade, or separate stakes can be set for line and grade. If only one stake is to be used, the elevation of the top of that stake is determined. Then, the cut from the top of the stake to the flowline (*invert*) is computed and marked on a guard stake. This method is faster.

It is often desirable to set a grade stake close to the tacked line stake. This may be set so that the cut from the top of the stake to the flowline is at some multiple of a half-foot.

In setting a cut stake for underground pipe, the surveyor first sets up the level and determines its HI. Using the flowline of the pipe at a particular station, the grade rod at that station is computed and recorded in the field book. A rod reading on the ground is taken at the point where the stake is to be driven. This is the ground rod. Using the grade rod and the ground rod, the rod reading on top of the stake that will give a half-foot cut from the top of the stake to the flowline is computed. A stake is driven to the rod reading that gives this cut. The stake is blued, and the cut is marked on the guard stake.

Example 28.10

A grade stake is to be set to show the cut to the flowline of a pipe. The HI is 472.36 ft, the flowline is 462.91 ft, and the ground rod at the point of stake is 5.1 ft. Determine the grade rod and rod reading that will give a half-foot cut to the flowline.

Solution

$$HI = 472.36 \text{ ft}$$
$$\text{flowline} = 462.91 \text{ ft}$$
$$\text{grade rod} = 9.45 \text{ ft}$$

The rod reading on the grade stake to give a half-foot cut from the top of the stake to the flowline could be 8.95 ft, 8.45 ft, ... , 5.45 ft, 4.95 ft, and so on, and the corresponding cuts would be 0 ft 6 in, 1 ft 0 in, 4 ft 0 in, 4 ft 6 in, and so on. The ground rod is 5.1 ft; therefore, the cut is approximately 9.5 ft − 5.1 ft = 4.4 ft. Therefore, the rod reading for this cut will be either 5.45 ft (which will give a cut of 4 ft 0 in) or 4.95 ft (which will give a cut of 4 ft 6 in).

A rod reading of 5.45 ft cannot be used because the top of the stake would be 0.3 ft below the surface of the ground. A rod reading of 4.95 ft would place the top of the stake about 0.1 ft above ground, which is satisfactory. The stake is driven so that the rod reading is 4.95 ft. The top is marked with blue keel, and the guard stake is marked "C 4 ft, 6 in" since 9.45 ft − 4.95 ft = 4.50 ft (4 ft 6 in). In the illustration shown, it can be seen that the rod reading on the stake must be less than the ground rod in order that the top of the stake be above ground.

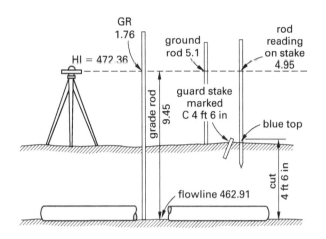

16. FLOWLINE AND INVERT

The bottom inside of a drainage pipe is known as the *flowline*.[2] It is also referred to as the *invert*. Invert is more commonly used to describe the bottom of the flow channel within a manhole.

Vertical control is of prime importance in laying pipe for gravity flow, especially sanitary sewer pipe. In order to facilitate vertical alignment, excavation of the trench often extends a few inches below the bottom of the pipe so that a bedding material, such as sand, can be placed in the trench for the pipe to lie on. Because of various methods of using bedding material in laying pipe, stakes are always set for the flowline, or invert, of the pipes. Excavation depth allows for the amount of bedding specified.

17. MANHOLES

Sanitary sewers are not laid along horizontal or vertical curves. Horizontal and vertical alignments are straight lines. Where a change in horizontal alignment or a change in slope is necessary, a *manhole* is required at the point of change. Therefore, a vertical drop within the manhole is needed. In staking, two cuts are often recorded on witness stakes: one for the incoming sewer and one for the outgoing sewer.

Gravity lines, such as sanitary sewers, flow only partially full. The slope of the sewer determines the flow velocity, and the velocity and size of the pipe determine the quantity of flow. Manholes are used to provide a point of change in conditions. Sewers must be deep enough below the surface of the ground to prevent freezing of their contents and damage to the pipe by construction equipment.

18. PRACTICE PROBLEMS

1. Stakes are to be set on 4 ft offsets for each edge of pavement (which is 28 ft wide) for a curve that has a deflection angle of 55°00′ and a centerline radius of 250 ft. The PI is at station 8+56.45. Stakes are to be set on full-stations and half-stations, and at the PC and PT. (a) Calculate T and L. (b) Determine the deflection angles used to stake the curve. (c) Calculate the outside and inside chord lengths.

2. Prepare a set of field notes to be used in staking a street curve on the quarter-stations from 3 ft offset lines on both sides of the street.

$$\text{PI} = 4{+}55.00$$
$$\text{I} = 60°00′ \text{ (angle to the left)}$$
$$\text{R} = 100 \text{ ft (centerline)}$$
$$\text{pavement width} = 28 \text{ ft}$$

3. The intersection of Ash Lane and 32nd Street is to be staked for paving from an offset line 4 ft left of the left edge of the pavement. The pavement is 28 ft wide, and the radius to the edge of the pavement is 30 ft. From this information and information in the sketch shown, calculate PC and PT stations and deflection angles, along with chord lengths from PC to PT.

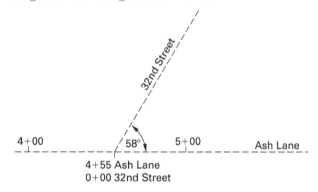

---

[2] *Flowlines* are the lines used as finish elevation for pipes.

SOLUTIONS

1.

(a)

$$T = R \tan \frac{I}{2}$$

$$= (250 \text{ ft}) \tan\left(\frac{55°}{2}\right) = 130.14 \text{ ft}$$

$$L = \frac{\pi R I}{180°}$$

$$= \frac{(55°) 2\pi (250 \text{ ft})}{360°} = 239.98 \text{ ft}$$

$$\text{PI} = 8+56.45$$
$$T = 1+30.14$$
$$\text{PC} = 7+26.31$$
$$L = 2+39.98$$
$$\text{PT} = 9+66.29$$

(b)

point	station		deflection angle
PC	7+26.31		0° (0°00')
	7+50	$\left(\dfrac{23.69 \text{ ft}}{239.98 \text{ ft}}\right) \times (27°30')$	2.7147° (2°43')
	8+00	$\left(\dfrac{73.69 \text{ ft}}{239.98 \text{ ft}}\right) \times (27°30')$	8.4443° (8°27')
	8+50	$\left(\dfrac{123.69 \text{ ft}}{239.98 \text{ ft}}\right) \times (27°30')$	14.1740° (14°10')
	9+00	$\left(\dfrac{173.69 \text{ ft}}{239.98 \text{ ft}}\right) \times (27°30')$	19.9036° (19°54')
	9+50	$\left(\dfrac{223.69 \text{ ft}}{239.98 \text{ ft}}\right) \times (27°30')$	25.6333° (25°38')
PT	9+66.29	$\left(\dfrac{239.98 \text{ ft}}{239.98 \text{ ft}}\right) \times (27°30')$	27.5000° (27°30')

(c) The outside chord lengths are

> 7+26.31 to 7+50:
> $(2)(268 \text{ ft})(\sin 2.7147°) = 25.39 \text{ ft}$
> 7+50 to 8+00:
> $(2)(268 \text{ ft})(\sin 5.7296°) = 53.51 \text{ ft}$
> 9+50 to 9+66.29:
> $(2)(268 \text{ ft})(\sin 1.8667°) = 17.46 \text{ ft}$

The inside chord lengths are

> 7+26.31 to 7+50:
> $(2)(232 \text{ ft})(\sin 2.7147°) = 21.98 \text{ ft}$
> 7+50 to 8+00:
> $(2)(232 \text{ ft})(\sin 5.7296°) = 46.32 \text{ ft}$
> 9+50 to 9+66.29:
> $(2)(232 \text{ ft})(\sin 1.8667°) = 15.11 \text{ ft}$

2.

			Elm Street		
point	station	deflection angle	C-in (ft)	C-out (ft)	curve data
PT	5+01.98	30°00'			
			1.65	2.33	
	5+00	29°26'			$I = 60°$ to the left
			20.70	29.17	$R = 100 \text{ ft}$
	4+75	22°16'			$T = 57.72 \text{ ft}$
			20.70	29.17	$L = 104.72 \text{ ft}$
	4+50	15°07'			
			20.70	29.17	
	4+25	7°57'			
			20.70	29.17	
	4+00	0°47'			
			2.27	3.19	
PC	3+97.26	0°00'			

Land Planning/ Development

3.

$$I_1 = 58°00' \text{ left}$$

$$T_1 = (44 \text{ ft})\tan\left(\frac{58°}{2}\right) = 24.39 \text{ ft}$$

$$PI = 4{+}55.00$$
$$T_2 = 24.39 \text{ ft}$$
$$PC = 4{+}30.61$$
$$PI = 0{+}00.00$$
$$T_1 = 24.39 \text{ ft}$$
$$PT = 0{+}24.39$$

$$LC_1 = (2)(26 \text{ ft})\sin\left(\frac{58°}{2}\right) = 25.21 \text{ ft}$$

$$I = 29°00'$$

$$I_2 = 122°00'$$

$$T_2 = (44 \text{ ft})\tan\left(\frac{122°}{2}\right) = 79.38 \text{ ft}$$

$$PI = 4{+}55.00$$
$$T_2 = 79.38 \text{ ft}$$
$$PC = 3{+}75.62$$
$$PI = 0{+}00.00$$
$$T_2 = 79.38 \text{ ft}$$
$$PT = 0{+}79.38$$

$$LC_2 = (2)(26 \text{ ft})\sin\left(\frac{122°}{2}\right) = 45.48 \text{ ft}$$

$$I = 61°00'$$

29 Earthwork

Nomenclature

A	area	ft^2	m^2
L	length	ft	m
V	volume	ft^3	m^3

Subscripts

m mean

1. DEFINITION

Earthwork is the excavation, hauling, and placing of soil, rock, gravel, or other material found below the surface of the earth. This also includes the measurement of such material in the field, the calculation of the volume of such material, and the determination of the most economical method of performing such work.

2. UNIT OF MEASURE

The *cubic yard* (i.e., the "yard") is the unit of measure for earthwork. However, the volume and density of earth changes under natural conditions and during the operations of excavation, hauling, and placing.

3. SWELL AND SHRINKAGE

A cubic yard of earth measured in its natural position will be more than a cubic yard after it is excavated. If the earth is compacted after it is placed, the volume may be less than a cubic yard. The volume of the earth in its natural state is known as *bank-measure*. The volume in the vehicle is known as *loose-measure*. The volume after compaction is known as *compacted-measure*.

The change in volume of earth from its natural to loose state is known as *swell*. Swell is expressed as a percent of the natural volume.

The change in volume of earth from its natural state to its compacted state is known as *shrinkage*. Shrinkage also is expressed as a percent decrease from the natural state.

As an example, 1 yd^3 in the ground may become 1.2 yd^3 loose-measure and 0.85 yd^3 after compaction. The swell would be 20%, and the shrinkage would be 15%. Swell and shrinkage vary with soil types.

4. CLASSIFICATION OF MATERIALS

Excavated material is usually classified as *common excavation* or *rock excavation*. Common excavation is soil.

In highway construction, common road excavation is soil found in the roadway. *Common borrow* is soil found outside the roadway and brought in to the roadway. Borrow is necessary where there is not enough material in the roadway excavation to provide for the embankment.

5. CUT AND FILL

Earthwork that is excavated, or is to be excavated, is known as *cut*. Excavation that is placed in embankment is known as *fill*.

Payment for earthwork is normally either for cut and not for fill, or for fill and not for cut. In highway work, payment is usually for cut; in dam work, payment is usually for fill. To pay for both would require measuring two different volumes and paying for moving the same earth twice.

6. FIELD MEASUREMENT

Cut and fill volumes can be calculated from slope-stake notes, plan cross sections, or digital models produced by laser scanning or by aerial surveying methods such as photogrammetry or airborne light detection and ranging (LiDAR).

Cross sections are profiles of the earth taken at right angles to the centerline of an engineering project (such as a highway, canal, dam, or railroad). A cross section for a highway is shown in Fig. 29.1.

Figure 29.1 *Typical Highway Cross Section*

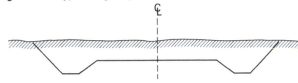

To obtain volume measurement, cross sections are taken before construction begins and after it is completed. By plotting the cross section at a particular station both before and after construction, a sectional view of the change in the profile of the earth along a certain line is obtained. The change along this line appears on the plan as an area. By using these areas at various intervals along the centerline, and by using distance between the areas, volume can be calculated.

7. ESTIMATING EARTHWORK

Earthwork quantities for a highway, canal, or other project can be estimated by superimposing a template on the original plotted cross section, which is drawn to represent the final cross section. The template is obtained from the *typical section sheet* of the construction plans. Figure 29.2 shows a typical completed section.

8. TYPICAL SECTIONS

Typical sections show the cross section view of the project as it will look on completion, including all dimensions. Highway projects usually show several typical sections, including cut sections, fill sections, and sections showing both cut and fill. Interstate highway plans also show access-road sections and sections at ramps.

9. DISTANCE BETWEEN CROSS SECTIONS

Cross sections are usually taken at each full station and at breaks in the ground along the centerline. In taking cross sections, it must be assumed that the change in the earth's surface from one cross section to the next is uniform, and that a section halfway between the cross sections is an average of the two. If the ground breaks appreciably between any two full stations, one or more cross sections between full stations must be taken. This is referred to as *taking sections at pluses.* Figure 29.3 shows the stations at which cross sections should be taken.

In rock excavation, or any other expensive operation, cross sections should be taken at intervals of 50 ft or less. Cross sections should always be taken at the PC and PT of a curve. Plans should also show a section on each end of a project (where no construction is to take place) so that changes caused by construction will not be abrupt.

Where a cut section of a highway is to change to a fill section, several additional cross sections are needed. Such sections are shown in Fig. 29.4.

10. GRADE POINT

The point where a fill section meets the natural ground (where a cut section begins) is known as a *grade point*.

11. METHODS FOR CALCULATING VOLUME

The most common method for calculating volume of earthwork is the *average end area method*. Another approach, the *prismoidal formula*, is more complex but furnishes more accurate results, especially when used with irregular ground surfaces.

The *average end area method* is based on the assumption that the volume of earthwork between two vertical cross sections A_1 and A_2 is equal to the average of the two end areas multiplied by the horizontal distance L between them, as illustrated in Eq. 29.1. If the areas used are in square feet, the results of Eq. 29.1 should be divided by 27 ft^3/yd^3 to convert to cubic yards.

$$V = \frac{L(A_1 + A_2)}{2} \qquad 29.1$$

The *prismoidal formula method* uses the mean area, A_m, midway between the two end sections. Equation 29.2 may be used for the prismoidal method. If field measurements for mean area are not available, the dimensions may be estimated by averaging the similar dimensions of the two end areas (not the average area of the two end sections), as illustrated in Fig. 29.5. As with the

Figure 29.2 *Typical Completed Section*

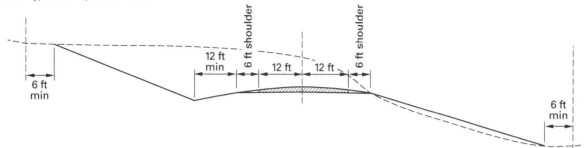

Figure 29.3 *Cross-Section Distances*

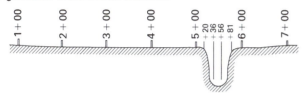

Figure 29.4 *Cut Changing to Fill*

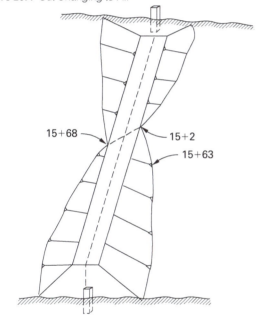

Figure 29.5 *Example of Soil Prismoid Calculation*

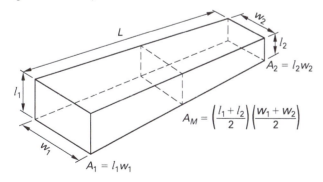

average end area method, if the areas used are in square feet, the results should be divided by 27 ft³/yd³ to convert to cubic yards.

$$V = \left(\frac{L}{6}\right)(A_1 + 4A_m + A_2) \qquad 29.2$$

12. FIELD NOTES

Figure 29.6 shows a sample sheet from cross-section field notes. The left half of the page is the same as for any set of level notes. The right half shows rod readings over horizontal distance measured from the centerline for each point on the ground that requires a reading. These readings should always include shots on the centerline, at each break in the ground, and at the right-of-way on each side.

13. PLOTTING CROSS SECTIONS

When drawn by hand, cross sections are plotted on specially printed cross-section paper. A scale of 1 in = 5 ft is usually used for both the horizontal and vertical. For wide sections, a scale of 1 in = 10 ft or 1 in = 20 ft can be used. The vertical scale can also be exaggerated if necessary.

A vertical line in the center of the sheet is drawn to represent the centerline of the project. Shots taken in the field are plotted to the proper elevation and distance from the centerline.

Each cross section is plotted as a separate section. Sufficient space is allowed between cross sections so that they do not overlap. The station number for each cross section is recorded just under the centerline shot, and the elevation at the centerline is recorded in a vertical direction just above the centerline shot.

The heavy lines on the paper are used to represent an elevation ending in 0 or 5 ft. With the elevation of the centerline recorded, these heavy lines can be identified as the elevation they represent.

The notes can be plotted by first reducing the level shots to elevation and then plotting by elevation. Alternatively, the rod shots can be plotted directly from a

Figure 29.6 Typical Field Notes for Cross-Section Work

station	+	HI	−	rod elevation	℄
BM no. 12	3.30	468.21		464.91	r.r. spike in 12 in elm 125 rt sta 16+75
11+00					$\frac{4.4}{50}$ $\frac{7.1}{15}$ $\frac{4.9}{12}$ $\frac{7.3}{5}$ 7.9 $\frac{9.1}{20}$ $\frac{12.0}{25}$ $\frac{9.7}{30}$ $\frac{11.2}{50}$
TP$_4$	5.27	462.97	10.51	457.70	
12+00					$\frac{2.0}{50}$ $\frac{4.0}{20}$ $\frac{1.3}{15}$ $\frac{4.5}{10}$ 5.0 $\frac{6.0}{15}$ $\frac{10.1}{20}$ $\frac{5.7}{27}$ $\frac{8.0}{50}$
13+00					$\frac{4.9}{50}$ $\frac{6.0}{25}$ $\frac{4.2}{20}$ $\frac{7.7}{15}$ 9.9 $\frac{9.8}{8}$ $\frac{12.6}{15}$ $\frac{11.0}{24}$ $\frac{11.2}{50}$
TP$_5$	1.76	458.22	6.51	456.46	
13+50					$\frac{2.3}{50}$ $\frac{4.1}{30}$ $\frac{1.2}{25}$ $\frac{5.0}{20}$ 6.7 $\frac{7.1}{3}$ $\frac{12.2}{10}$ $\frac{8.3}{19}$ $\frac{11.1}{50}$
14+00					$\frac{5.2}{50}$ $\frac{6.0}{42}$ $\frac{10.2}{35}$ $\frac{7.9}{20}$ 10.1 $\frac{11.0}{30}$ $\frac{8.1}{50}$
15+00					$\frac{5.0}{50}$ $\frac{5.8}{48}$ $\frac{9.6}{40}$ $\frac{7.3}{20}$ 9.2 $\frac{10.1}{22}$ $\frac{7.5}{50}$
BM no. 13			5.15	453.07	

line on the paper representing the HI. As an example, if the HI is 447.6 ft (rounded off) and the rod shot is 5.4 ft, subtracting 5.4 ft from 447.6 ft gives an elevation of 442.2 ft.

Figure 29.7 illustrates the process of plotting cross sections.

14. DETERMINING END AREAS

End areas are commonly determined by dividing the area into triangles and trapezoids for ease of calculation.

Cut areas and fill areas must be kept separate. After the areas have been determined, the sum of each two adjacent areas is placed in a column. The distance between two sections is recorded, and the volume for each sum is calculated from Eq. 29.1.

After the volume has been calculated, shrinkage must be added to fill quantities to balance with cut quantities. Shrinkage will vary from 30% for light cuts and fills to 10% for heavy cuts and fills.

Excavated rock will occupy larger volume when placed in a fill, and this swell will be subtracted from the fill quantity.

15. VOLUMES FROM PROFILES

For preliminary estimates of earthwork, volumes can be calculated from the centerline profiles. After the ground profile and finish grade profile are plotted, the area of cut can be measured on the profiles and the average determined by dividing by the length of cut. Using the average cut, a template can be drawn, and the end area can also be measured. This area times the length of the cut will give the volume.

16. BORROW PIT

As mentioned previously, it is often necessary to borrow earth from an adjacent area to construct embankments. Normally, the *borrow pit* area is laid out in a rectangular grid with 10 ft, 50 ft, or even 100 ft squares. Elevations are determined at the corners of each square by leveling before and after excavation so that the cut at each corner can be calculated. Points outside the cut area are established on the grid lines so that the lines can be reestablished after excavation is completed.

As an example, volumes for two of the prisms shown in Fig. 29.8 are calculated by multiplying the average cut by the area of the figure. The volume of the prism A0-B0-B1-A1 is

$$V = \left(\frac{(50 \text{ ft})(50 \text{ ft})}{27 \frac{\text{ft}^3}{\text{yd}^3}} \right) \times \left(\frac{3.2 \text{ ft} + 3.4 \text{ ft} + 3.0 \text{ ft} + 2.6 \text{ ft}}{4} \right)$$
$$= 282 \text{ yd}^3$$

The volume of the triangle E2-F2-E3 is

$$V = \left(\frac{(50 \text{ ft})(15 \text{ ft})}{(2)\left(27 \frac{\text{ft}^3}{\text{yd}^3}\right)} \right) \left(\frac{2.3 \text{ ft} + 2.4 \text{ ft} + 2.4 \text{ ft}}{3} \right)$$
$$= 33 \text{ yd}^3$$

Figure 29.7 Plotting Cross Sections

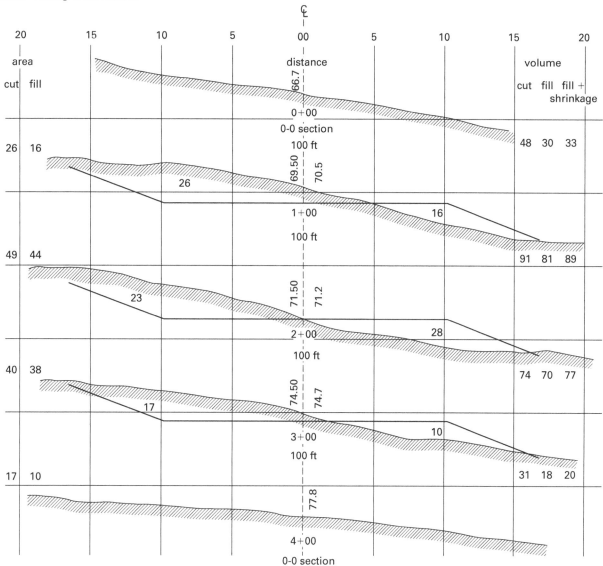

Figure 29.8 Borrow Pit Areas

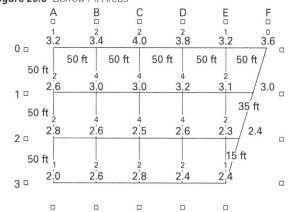

Instead of calculating volumes of prisms represented by squares separately, all square-based prisms can be calculated collectively by multiplying the area of one square by the sum of the cut at each corner by the number of times that cut appears in any square, divided by 4. For instance, on the second line from the top in Fig. 29.8, which is line 1, 2.6 appears in two squares, 3.0 appears in four squares, 3.2 appears in four squares, and 3.1 appears in two squares. In the figure, the small number above the cut indicates the number of times the cut is used in averaging the cuts for the prisms.

Example 29.1

Calculate the volume of earth excavated from the borrow pit shown in Fig. 29.8.

Solution

The volume of the squares is

$$V = \left(\frac{50 \text{ ft}}{\left(27 \frac{\text{ft}^3}{\text{yd}^3}\right)(4)}\right)(50 \text{ ft})$$

$$\times \begin{pmatrix} 3.2 \text{ ft} + (2)(3.4 \text{ ft}) + (2)(4.0 \text{ ft}) \\ + (2)(3.8 \text{ ft}) + 3.2 \text{ ft} + (2)(2.6 \text{ ft}) \\ + (4)(3.0 \text{ ft}) + (4)(3.0 \text{ ft}) + (4)(3.2 \text{ ft}) \\ + (2)(3.1 \text{ ft}) + (2)(2.8 \text{ ft}) + (4)(2.6 \text{ ft}) \\ + (4)(2.5 \text{ ft}) + (4)(2.6 \text{ ft}) + (2)(2.3 \text{ ft}) \\ + 2.0 + (2)(2.6 \text{ ft}) + (2)(2.8 \text{ ft}) \\ + (2)(2.4 \text{ ft}) + 2.4 \text{ ft} \end{pmatrix}$$

$$= 3194 \text{ yd}^3$$

The volume of the trapezoids is

$$V = \left(\frac{50 \text{ ft} + 35 \text{ ft}}{(2)\left(27 \frac{\text{ft}^3}{\text{yd}^3}\right)}\right)(50 \text{ ft})$$

$$\times \left(\frac{3.2 \text{ ft} + 3.6 \text{ ft} + 3.1 \text{ ft} + 3.0 \text{ ft}}{4}\right)$$

$$= 254 \text{ yd}^3$$

$$V = \left(\frac{35 \text{ ft} + 15 \text{ ft}}{(2)\left(27 \frac{\text{ft}^3}{\text{yd}^3}\right)}\right)(50 \text{ ft})$$

$$\times \left(\frac{3.1 \text{ ft} + 3.0 \text{ ft} + 2.3 \text{ ft} + 2.4 \text{ ft}}{4}\right)$$

$$= 125 \text{ yd}^3$$

The volume of the triangle is

$$V = \left(\frac{(15 \text{ ft})}{(2)\left(27 \frac{\text{ft}^3}{\text{yd}^3}\right)}\right)(50 \text{ ft})$$

$$\times \left(\frac{2.3 \text{ ft} + 2.4 \text{ ft} + 2.4 \text{ ft}}{3}\right)$$

$$= 33 \text{ yd}^3$$

The total volume of earth excavated is

$$3194 \text{ yd}^3 + 254 \text{ yd}^3 + 125 \text{ yd}^3 + 33 \text{ yd}^3 = 3606 \text{ yd}^3$$

17. HAUL

In some contracts for highways and railroads, the contractor is paid per cubic yard for excavation (which includes the cost of excavation, hauling, placing in embankment, and compaction of embankment). However, the cost of hauling one cubic yard of earth over a long distance can easily become greater than the cost of excavation, so it is often practical to pay a contractor for excavating and hauling earth.

18. FREE HAUL

It is common not to pay for hauling if the material is hauled less than a certain distance, usually 500 ft to 1000 ft. An additional price is paid for hauling the earth beyond the prescribed limit. The haul distance for which no pay is received is known as *free haul*.

19. OVERHAUL

The hauling of material beyond the free haul limit is known as *overhaul*. The unit of overhaul measure is yard-stations or yard-quarters. A *yard-quarter* is the hauling of one cubic yard of earth a distance of 1/4 mi. For example, if six yards of earth were hauled one mile, the overhaul would be twenty-four yard-quarters. Thus, the word "haul" may have two meanings. It may mean linear distance or volume times distance.

It should be mentioned that the distance is measured along the centerline. Distance from the extremity of the right-of-way to the centerline is not considered.

20. BALANCE POINTS

It is important in planning and construction to know the points along the centerline of a particular section of cut that will balance a particular section of fill. For example, assume that a cut section extends from sta 12+25 to sta 18+65, and a fill section extends from sta 18+65 to sta 26+80. Also, assume that the excavated material will exactly provide the material needed to make the embankment. Then, cut balances fill, and sta 12+25 and sta 26+80 are *balance points*.

21. MASS DIAGRAMS

A method of determining economical handling of material, quantities of overhaul, and location of balance points is the mass-diagram method.

The *mass diagram* is a graph that has distance in stations as the abscissa and the cumulative earthwork (i.e., the algebraic sums of cut and fill) as the ordinate. The *x*-axis parallels the centerline, and the cut and fill (plus shrinkage) quantities are taken from the cross-section sheets. Often, the mass diagram is plotted below the centerline profile so that like stations are vertically in line.

To add cut and fill algebraically, cut is given a plus sign, and fill is given a minus sign.

22. PLOTTING THE MASS DIAGRAM

After volumes of cut and fill between stations have been calculated, they are tabulated as shown in Table 29.1. The cuts and fills are then added, and the cumulative yardage at each station is recorded in the table. It is this cumulative yardage that is plotted as an ordinate. In Fig. 29.9, the baseline serves as the *x*-axis. Cumulative yardage that has a plus sign is plotted above the baseline, and cumulative yardage that has a minus sign is plotted below the baseline.

The scale is not important. In Fig. 29.9, the horizontal scale is 1 in = 5 sta, and the vertical scale is 1 in = 5000 yd^3. A larger scale would be more practical in actual computations. The mass diagram is plotted on the lower half of the sheet, and the centerline profile of the project is plotted on the upper half.

23. BALANCE LINE

Any horizontal line cutting off a loop of the mass curve intersects the curve at two points, between which the cut is equal to the fill. Figure 29.10 shows a portion of the mass diagram of Fig. 29.9 with an enlarged scale.

Figure 29.9 *Baseline and Centerline Profile Mass Diagram*

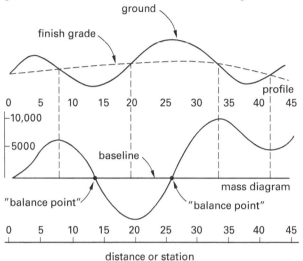

24. SUB-BASES

Sub-bases are horizontal balance lines that divide an area of the mass diagram between two balance points into trapezoids for the purpose of more accurately computing overhaul. In Fig. 29.10, the top sub-base is less than 600 ft in length. Therefore, the volume of earth represented by the area above this line will be hauled a distance less than the free-haul distance, and no payment will be made for overhaul. All the volume represented by the area below this top sub-base will receive payment for overhaul.

The area between two sub-bases is very nearly trapezoidal. The average length of the bases of a trapezoid can be measured in feet, and the altitude of the trapezoid can be measured in cubic yards of earth. The product of these two quantities can be expressed in yard-quarters. If free haul is subtracted from the length, the quantity can be expressed as overhaul.

Sub-bases are drawn at distinct breaks in the mass curve. Distinct breaks in Fig. 29.10 can be seen at sta 1+00(184), 2+00(806), and 5+00(4340). After subbases are drawn, a horizontal line is drawn midway between the sub-bases. This line represents the average haul for the volume of earth between the two sub-bases. If a horizontal scale of 1 in = 100 ft is used, the length of the average haul can be determined by scaling. This line, shown as a dashed line in Fig. 29.10, scales 875 ft for the area between the top sub-bases. The free haul is subtracted from this in Table 29.1.

The volume of earth between the two sub-bases is found by subtracting the ordinate of the lower sub-base from the ordinate of the upper sub-base. These ordinates are found in Table 29.1.

Multiplying average haul minus free haul in feet by volume of earth in cubic yards gives overhaul in yard-feet. Dividing by 1320 ft gives yard-quarters.

Table 29.1 *Typical Cut and Fill Calculations (all volumes are in cubic yards)*

sta	cut +	fill −	cum sum	sta	cut +	fill −	cum sum
0			0	23			−4710
	184				1377		
1			+184	24			−3034
	622				1676		
2			+806	25			−1358
	1035				1860		
3			+1841	26			+502
	1268				1917		
4			+3109	27			+2419
	1231				1839		
5			+4340	28			+4258
	919				1611		
6			+5259	29			+5869
	503				1338		
7			+5762	30			+7207
	164	21			1029		
8			+5905	31			+8236
	12	190			652		
9			+5727	32			+8888
		616			357		
10			+5111	33			+9245
		942			150	39	
11			+4169	34			+9356
		1150			52	236	
12			+3019	35			+9172
		1500				465	
13			+1519	36			+8707
		1773				712	
14			−254	37			+7995
		1755				904	
15			−2009	38			+7091
		1540				904	
16			−3549	39			+6187
		1262				757	
17			−4811	40			+5430
		932				516	
18			−5743	41			+4914
		546				280	
19			−6289	42			+4634
		203				127	
20			−6461	43			+4507
		101				98	
21			−6283	44			+4409
		18				20	
22			−5715	45			+4389

25. LOCATING BALANCE POINTS

A balance point occurs where the mass curve crosses the baseline. In Fig. 29.10, it can be seen that a balance point falls between sta 13 and 14. The ordinate of 13 is +1519; the ordinate of 14 is −254. Therefore, the curve fell $1519 + 254 = 1773$ yd^3 in 100 ft or 17.73 yd^3/ft. The curve crosses the baseline at a distance of $1519/17.73 = 86$ ft from sta 13 (13+86).

26. CHARACTERISTICS OF THE MASS DIAGRAM

Several important characteristics that should be considered in using the mass diagram in planning the economical hauling of earth include:

- A horizontal line connecting two points on the mass curve cuts off a loop in which the cut equals the fill.

- A loop that rises and then falls from left to right indicates that the haul from cut to fill will be from left to right.

- A loop that falls and then rises from left to right indicates the haul will be from right to left.

- A high point on a mass curve indicates a change from cut to fill.

- A low point on a mass curve indicates a change from fill to cut.

- High and low points on the mass curve occur at or near grade points on the profile.

Figure 29.10 *Mass Diagram Showing Sub-bases*

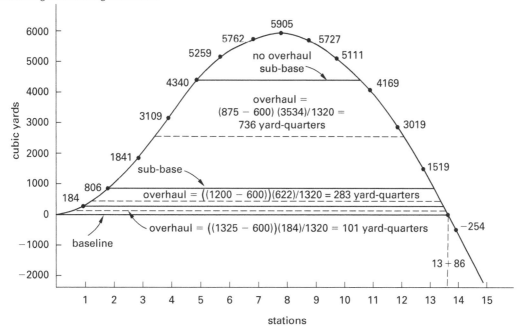

27. PRACTICE PROBLEMS

Figure for Practice Problems 1 and 2

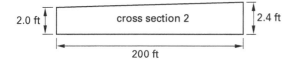

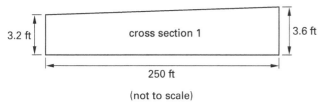

(not to scale)

1. The distance between the two cross sections shown is 150 ft. Most nearly, what is the volume of earthwork between the two cross sections based on the average end area method?

(A) 3530 yd³

(B) 3580 yd³

(C) 3600 yd³

(D) 3790 yd³

2. If the distance between the two cross sections is 150 ft, what is most nearly the volume of earthwork between the two figures based on the prismoidal formula method?

(A) 3530 yd³

(B) 3580 yd³

(C) 3600 yd³

(D) 3790 yd³

3. Earth removed from a 10,000 yd³ borrow pit has a swell factor of 20% and a compaction shrinkage factor of 15%. Most nearly, what is the compacted yield of earth removed from the borrow pit?

(A) 8500 yd³

(B) 10,000 yd³

(C) 10,200 yd³

(D) 12,000 yd³

SOLUTIONS

1. The area of cross section 1 is

$$A_1 = (3.2 \text{ ft})(250 \text{ ft}) + \left(\frac{1}{2}\right)(0.4)(250 \text{ ft}) = 850 \text{ ft}^2$$

The area for cross section 2 is

$$A_2 = (2.0 \text{ ft})(200 \text{ ft}) + \left(\frac{1}{2}\right)(0.4)(200 \text{ ft}) = 440 \text{ ft}^2$$

From Eq. 29.1,

$$V = \frac{L(A_1 + A_2)}{2}$$

$$= \frac{\dfrac{(150 \text{ ft})(850 \text{ ft}^2 + 440 \text{ ft}^2)}{2}}{27 \dfrac{\text{ft}^3}{\text{yd}^3}}$$

$$= 3583 \text{ yd}^3 \quad (3580 \text{ yd}^3)$$

The answer is (B).

2. The area of cross section 1 is

$$A_1 = (3.2 \text{ ft})(250 \text{ ft}) + \left(\frac{1}{2}\right)(0.4)(250 \text{ ft}) = 850 \text{ ft}^2$$

The area for cross section 2 is

$$A_2 = (2.0 \text{ ft})(200 \text{ ft}) + \left(\frac{1}{2}\right)(0.4)(200 \text{ ft}) = 440 \text{ ft}^2$$

Averaging the similar dimensions of the two end areas,

$$\frac{3.2 \text{ ft} + 2.0 \text{ ft}}{2} = 2.6 \text{ ft}$$

$$\frac{200 \text{ ft} + 250 \text{ ft}}{2} = 225 \text{ ft}$$

$$\frac{3.6 \text{ ft} + 2.4 \text{ ft}}{2} = 3.0 \text{ ft}$$

The area for the middle cross section is

$$A_m = (2.6 \text{ ft})(225 \text{ ft}) + \left(\frac{1}{2}\right)(0.4)(225 \text{ ft}) = 630 \text{ ft}^2$$

From Eq. 29.2,

$$V = \left(\frac{L}{6}\right)(A_1 + 4A_m + A_2)$$

$$= \frac{\left(\dfrac{150 \text{ ft}}{6}\right)(850 \text{ ft}^2 + (4)(630 \text{ ft}^2) + 440 \text{ ft}^2)}{27 \dfrac{\text{ft}^3}{\text{yd}^3}}$$

$$= 3528 \text{ yd}^3 \quad (3530 \text{ yd}^3)$$

The answer is (A).

3. With the swell factor of 20%, the 10,000 yd^3 would yield $(1.20)(10,000 \text{ yd}^3) = 12,000 \text{ yd}^3$ after removal from the borrow pit. But after compaction with a shrinkage factor of 15%, the 10,000 yd^3 (in its natural state) would be reduced to

$$(0.85)(10,000 \text{ yd}^3) = 8500 \text{ yd}^3$$

The answer is (A).

Land Planning/
Development

30 Engineering Surveying Law

1. INTRODUCTION TO ENGINEERING SURVEYING LAW

In the state of California, laws are in place to protect the welfare and best interest of the general public in regard to the practice of professional engineering, including engineering surveying, and the division and development of land. The laws regulating the practice of engineering are in the California Business and Professions (B&P) Code, Sections 6700 through 6799, also known as the *Professional Engineers Act* (*PE Act*). The laws regulating the division and development of real property are in the California Government Code, Sections 66410 through 66499.58, also known as the *Subdivision Map Act* (*SMA*).

2. THE SUBDIVISION MAP ACT

The Subdivision Map Act, often referred to simply as the Map Act, is the portion of the California Government Code that requires local agencies (cities or counties) to regulate the design and improvement of land subdivisions within their jurisdictional boundaries.

> The purpose of the Map Act is to regulate and control the design and improvement of subdivisions with proper consideration for their relation to adjoining areas; require subdividers to install streets and other improvements; prevent fraud and exploitation; and protect both the public and purchasers of subdivided lands. (Pratt v. Adams (1964) 229 Cal.App.2d 602.)

The primary objectives of the SMA include

- Ensure that real property is subdivided legally and without fraud.

- Promote the orderly development, design, and improvement of any real property being subdivided, with respect to the local agency's general plan, surrounding land use, and future or existing adjacent development.

- Require the subdivider to place improvements within those portions of the subdivision boundary that are dedicated for public use, such as roadways, parkways, and common use areas.

Every local agency has ordinances that empower them to regulate subdivisions in accordance with the SMA requirements. A local agency may impose additional requirements of subdivisions, as long as those requirements do not conflict with any SMA requirements.

A *subdivision* is the division of a property into two or more separate parcels. Per the SMA, for any subdivision of land, the subdivider submits a tentative map and a final map, or a parcel map, for review and approval.

The SMA distinguishes between subdivisions creating five or more parcels or lots and those creating four or fewer parcels or lots. A tentative map and a final map are required for subdivisions creating five or more parcels. A tentative map is always required when a final map (subdivision map or tract map) is required. A parcel map is required for subdivisions creating four or fewer parcels.

A *tentative map* (*tentative tract map*) is submitted to the local agency prior to approval of the map and project. A tentative map shows the proposed design and improvements of the proposed subdivision, existing topographic conditions in and around the map boundaries, the proposed street alignment, proposed roadway right-of-way, proposed grades and widths, alignment and widths of easements, rights-of-way for drainage and easements, and lot layout. The tentative map typically also contains enough information for the agency to make its review of the proposed subdivision, such as

- a legal description defining the boundaries of the proposed tract

- existing and proposed land use zoning per the agency's specific or general plan

- the proposed use of the property

- the locations, names, and existing widths of all adjoining highways, streets, and road rights-of-way

- the locations, widths, and proposed grades of all streets within the proposed subdivision

- the width and approximate location of all existing and proposed utilities and easements for roads, drainage, sewers, and other public utility purposes

- the tentative lot layout, number of lots, and dimensions and area of each lot

Land Planning/ Development

- names and contact information of the owner/developer, engineer, and all utility purveyors
- the approximate locations of all areas subject to storm inundations and the locations of flow of all watercourses within or adjacent to the property
- any proposed public areas

The tentative map is often limited to only the property layout (map boundary, property lines, and right-of-way), and the grading, improvements, and building layout are shown on separately submitted plans, such as a preliminary grading plan or a site plan (plot plan). A tentative map does not need to be based on an accurate and final survey of the property.

After the local agency approves the tentative map, the subdivider submits a final map for approval and recordation. The final map should substantially match the approved tentative map in layout design and proposed improvements. A final map may only be prepared by a licensed civil engineer or a licensed land surveyor and must be based on an accurate land survey. The SMA requires that a final map include

- signatures of all parties with any record interest in the property being subdivided, and other certificates or statements for signing by the approving agency
- the map number (e.g., "Tract 34289")
- the legal description of the property being divided
- the map boundary
- all proposed and existing right-of-way and property lines, lot numbers (or letters), and street names
- all survey, mathematical information, and data necessary to locate and retrace all right-of-way and property lines shown on the map including bearing, distance, radii, and chord lengths

The map must be printed with black ink on long-lasting media, such as polyester film (Mylar), on C-Size sheets (18 in × 26 in) with border, correct scale, page numbers, and match lines.

Unlike a tentative map, a final map does not typically show topography or proposed improvements, but primarily the centerline, right-of-way, property line, easements, and boundary line work. The proposed improvements are usually shown on separate plans such as street improvement plans or grading plans.

When the local agency approves tentative and final maps, the subdivision is approved with conditions. These *conditions of approval* are requirements made by the local agency that require the subdivider to, for example, dedicate right-of-way and construct the roadway improvements (traffic signals, street lights, pavement, etc.) for all roadways within or fronting the project, install all necessary drainage improvements (channels, inlets, storm drains, etc.), place overhead

utility lines underground, follow grading ordinances, submit a drainage and geotechnical study, ensure the minimum number of ingress and egress points for fire and other emergency vehicles, install landscaping, keep structures to a maximum height, and so on. These improvements must be completed before any buildings are occupied.

When a subdivider must construct improvements as part of the project's conditions of approval, the subdivider (or developer) submits a security to the local agency to guarantee the completion of the improvements.

A *parcel map* is different from a tract map in several ways. A parcel map does not necessarily require the submittal of a tentative map; some agencies do require a tentative parcel map, but the SMA does not specifically require agencies to mandate one. Conditions of approval for a parcel map are less stringent than those of a final map: a parcel map's conditions of approval may not include the installation of onsite and offsite improvements. Requirements for the construction of such improvements are instead imposed when the parcels are developed and grading and building permits are sought by the developer. The form and content of a parcel map is otherwise very similar to that of a final map.

The SMA allows some exceptions from the more stringent requirements and conditions of approval typically required of subdivisions. Some of these exemptions include

- *lot line adjustment:* the reconfiguration of common lot lines between four or fewer parcels where, by relocating the property lines between parcels, area is taken from one parcel and added to another. For a lot line adjustment to be exempt from SMA requirements, no new lots or parcels may be created, the lots must be adjoining, and the new lots must be developable and conform to the local agency's general plan.
- *parcel merger:* the combining of multiple parcels into fewer parcels without enforcing the conditions required of maps. To qualify for a parcel merger, the parcels must be owned by the same owner and must be contiguous, and at least one of the parcels must be of substandard size per local standards.
- *second units:* the construction of second units, sometimes called "granny units," on a single family property is exempt from the SMA requirements as long as the structure is not sold, houses up to two adults 62 years or older, and the square footage does not exceed the lesser of either 30 percent of the square footage of the primary living structure or 1200 ft².
- *land conveyances to or from public entities:* if land or a portion of land is being transferred to or from a government agency, public entity, or public utility, the requirements of the SMA may be waived.

- *remainder parcels:* sometimes the subdivider may opt not to include the entire property within the proposed map boundary. The leftover area is called a *remainder parcel.* Remainder parcels are not subject to conditions of approval; a remainder parcel receives conditions at a later date when development is proposed. A remainder parcel is not included in parcel count; for instance, a map proposing the division of four parcels and a remainder parcel would be considered a parcel map (four parcels, not five).

Other, less common, exceptions include land use for cemetery purposes, apartment to condominium conversions (under certain cases), certain agricultural uses, small temporary buildings, and others.

See Chap. 25 for more about subdivisions.

3. LEGAL DESCRIPTIONS

When a property or portions of a property are conveyed via documents (instruments) like those described in the SMA (e.g., grant deed, lot line adjustment, easement), it is necessary that the instrument clearly and specifically describe the location and limits of the property being conveyed. Legal descriptions are written descriptions that accurately and uniquely describe parcels of land. Legal descriptions are written by a professional land surveyor.

It is important that the legal description defining a property be written so that the limits of the property being described can be located, even many years later. For instance, legal descriptions should refer to permanent features such as recorded maps or set monuments, not trees, rocks, or fences. Land descriptions are explained further in Chap. 24.

4. THE PROFESSIONAL ENGINEERS (PE) ACT

The Professional Engineers Act (PE Act) contains the laws, rules, and regulations pertaining to the practice of professional engineering in California. These laws ensure that only competent and qualified licensed individuals practice civil engineering, safeguarding the life, health, property, and welfare of the public. California's Professional Engineers Act was adopted in 1929, prompted by the St. Francis Dam collapse in 1928.

Per the PE Act, civil engineering includes:

...studies or activities in connection with fixed works for irrigation, drainage, waterpower, water supply, flood control, inland waterways, harbors, municipal improvements, railroads, highways, tunnels, airports and airways, purification of water, sewerage, refuse disposal, foundations, grading, framed and homogeneous structures, buildings, or bridges. (PE Act §6731)

As defined in the PE Act, civil engineering also includes "engineering surveying," which consists of

The practice or offer to practice, either in a public or private capacity, all of the following: (a) Locates, relocates, establishes, reestablishes, or retraces the alignment or elevation for any of the fixed works embraced within the practice of civil engineering, as described in Section 6731. (b) Determines the configuration or contour of the earth's surface or the position of fixed objects above, on, or below the surface of earth by applying the principles of trigonometry or photogrammetry. (c) Creates, prepares, or modifies electronic or computerized data in the performance of the activities described in subdivisions (a) and (b). (d) Renders a statement regarding the accuracy of maps or measured survey data pursuant to subdivisions (a), (b), and (c). (PE Act §6731.1)

The PE Act, along with the Professional Land Surveyors Act, lists and distinguishes those duties and functions unique to licensed (professional) engineers and those unique to licensed (professional) land surveyors.

Pursuant to §6731 of the PE Act, licensed civil engineers registered prior to January 1, 1982 (civil engineer registration number 33,965 or lower), are authorized to practice all professional land surveying functions. All licensed civil engineers registered January 1, 1982, or later (registration number 33,966 or higher) may practice only civil engineering as defined in §6731 of the PE Act and engineering surveying as described in §6731.1 of the PE Act.

Civil engineers registered after January 1, 1982, can offer land surveying work related to their civil engineering practice, provided the land surveying work is performed by, or under the direction of, a licensed land surveyor or authorized civil engineer (i.e., one registered before 1982). The engineering surveying that civil engineers are authorized to do, as listed in §6731.1, includes construction staking for the location of fixed engineering works, establishing the alignment of fixed engineering works, and determining contours for topographic maps.

As a general rule, civil engineers licensed after January 1, 1982, may not perform any survey or prepare any document that would locate, relocate, or establish property lines, right-of-way lines, easement lines, or boundary lines.

Professional land surveyors retrace property lines, perform boundary line adjustments, prepare topographic maps, prepare legal descriptions and subdivision maps or plats, perform construction surveys and staking, establish the alignment and elevations of engineering works, set or reset monuments, and perform photogrammetric surveying or aerial mapping, among other duties. It is unlawful for anyone to perform land surveying unless he or she is currently licensed as a land surveyor or an authorized civil engineer. Only licensed land surveyors may use the title of land surveyor.

5. PRACTICE PROBLEMS

1. A licensed civil engineer registered before 1982 is authorized to

   I.   use the title of land surveyor

   II.   perform boundary line surveys

   III.   locate survey monuments

   IV.   perform construction staking

   (A)   I

   (B)   I and II

   (C)   III and IV

   (D)   II, III, and IV

2. A subdivider proposes to subdivide a 50 acre property into four 10 acre parcels with a remainder 10 acre parcel. Per the Subdivision Map Act, the subdivider is required to submit a

   (A)   tentative map

   (B)   parcel map

   (C)   tentative map and a final map

   (D)   tentative parcel map

3. A competent civil engineer who has been licensed since 1985 is authorized to

   I.   stake the limits of earthwork on a roadway project

   II.   determine the position of a fixed engineering work

   III.   establish the alignment of a proposed retaining wall

   IV.   perform aerial topographic mapping

   (A)   I and II

   (B)   I and III

   (C)   II and III

   (D)   II and IV

4. Per the Subdivision Map Act, which of the following statements is always true?

   (A)   A tentative tract map must be approved before a tract map can be recorded.

   (B)   A tentative parcel map must be approved before a parcel map can be recorded.

   (C)   A parcel merger of five or more lots requires a tentative map.

   (D)   A lot line adjustment of five or more lots is not subject to the map act requirements.

Land Planning/Development

SOLUTIONS

1. Assuming competency, a civil engineer registered before 1982 (registration number 33,965 or lower) may practice all forms of surveying. However, he or she may not use the title of land surveyor.

The answer is (D).

2. Per the SMA, the remainder parcel is not included in the lot count. Therefore, the subdivision is creating four legal parcels. Per the SMA, a parcel map must be submitted. Some agencies require a tentative parcel map for this proposal, but it is not a requirement of the SMA.

The answer is (B).

3. Per the PE Act, §6731.1, civil engineers licensed after 1982 may determine the position of fixed engineering works, establish the alignment of fixed engineering works, and determine contours of the earth's surface.

The answer is (C).

4. A tentative map (or tentative tract map) is always required for approval before a final map (or tract map) can be recorded. The SMA does not mandate that a tentative parcel map be submitted (although some agencies require one). There is no limit to the number of parcels in a parcel merger, and if a parcel merger is approved, no tentative map is required. A lot line adjustment must be between four or fewer lots.

The answer is (A).

Topic V: Support Material

Appendices

Support Material

APPENDIX 1.A
Miscellaneous Conversion Factors

multiply	by	to obtain
ac	0.40469	hectare
ac	10.0	chains^2
ac	43,560	ft^2
ac	1/640	mi^2
ac	4046.87	m^2
ac-ft	43,560	ft^3
ac-ft	1233.5	m^3
centiare	1.0	m^2
cm	1×10^{-5}	km
cm	1×10^{-2}	m
cm	10	mm
cm	0.03281	ft
cm	0.39370	in
chain	792.0	in
chain	66.0	ft
chain	22	yd
chain	4.0	rods
cm^3	0.06102	in^3
cubit	18	in
day	86,400	sec
degree (angular)	$2\pi/360$	radians
degree (angular)	17.778	mils
degree/sec	0.1667	rev/min
engineer's link	1	ft
fathom	6.0	ft
ft (international)	30.48	cm
ft (international)	0.3048	m
ft (U.S. Survey)	1200/39.37	cm
ft (U.S. Survey)	12/39.37	m
ft (U.S. Survey)	1.645×10^{-4}	mi (nautical)
ft (U.S. Survey)	1.894×10^{-4}	mi (statute)
ft/min	0.5080	cm/s
ft/sec	0.592	knot
ft/sec	0.6818	mi/hr
ft^3	7.4805	gal
ft^3	0.02832	m^3
furlongs	660.0	ft
furlongs	0.125	mi (statute)
gal	0.13368	ft^3
gal	3.7854	L
grad	0.90	degree (angular)
grad	0.01570797	radian
hectare	2.4711	ac
hectare	10,000	m^2
hr	4.167×10^{-2}	days

multiply	by	to obtain
hr	5.952×10^{-3}	wk
in	2.540	cm
in	25.40	mm
in	1.578×10^{-5}	mi
km	3281.0	ft
km	1000.0	m
km	0.6214	mi
km/h	0.5396	knots
knot	6076.0	ft/hr
knot	1.0	mi/hr (nautical)
knot	1.151	mi/hr (statute)
labor	177.14	ac
league (nautical)	7627.9	ac
link	7.92	in
L	1000.0	cm^3
L	61.024	in^3
L	0.26417	gal (U.S. liquid)
L	1000.0	mL
L	2.113	pints
m	100.0	cm
m	1/0.3048	ft (international)
m	39.37/12	ft (U.S. Survey)
m	1×10^{-3}	km
m	5.396×10^{-4}	mi (nautical)
m	6.214×10^{-4}	mi (statute)
m	1000	mm
micron	1×10^{-6}	m
mi (nautical)	6076	ft
mi (nautical)	1.853	km
mi (nautical)	1.1507	mi (statute)
mi (statute)	80	chains
mi (statute)	5280.0	ft
mi (statute)	1.609	km
mi (statute)	0.8689	mi (nautical)
mi (statute)	320.0	rods
mil (angular)	0.05625	degree (angular)
mil (angular)	3.375	min
radian	$180/\pi$	deg
radian	3437.7	min (angular)
rod	16.50	ft
rod	5.029	m
rod	1	perch
rod	1	pole
yd	0.91440	m
yd	4.934×10^{-4}	mi (nautical)

APPENDIX 1.A *(continued)*
Miscellaneous Conversion Factors

multiply	by	to obtain
yd	5.682×10^{-4}	mi (statute)
vara (California)	33	in
vara (Texas)	$33 \frac{1}{3}$	in

APPENDIX 1.B
Glossary

A

Acre: A measure of land area equal to 160 rod^2, 4480 yd^2, or 43,560 ft^2.

Add tape: A measuring tape with an extra foot beyond the zero mark, graduated in tenths of a foot or tenths and hundredths of a foot.

Alignment stake: A stake used to indicate stationing and centerline alignment along proposed corridors.

Angle: The difference in direction between two convergent lines. It may be classed as horizontal, vertical, oblique, spherical, or spheroidal, according to whether it is measured in a horizontal, vertical, or inclined plane, or in a curved surface.

Angle and distance method: A method of preparing area maps where angles and distance are measured from a base line. Also known as the azimuth-stadia method.

Automatic level: A level that makes use of a compensator to ensure that the line of sight remains horizontal once the operator has roughly leveled the instrument. Also known as a self-leveling level.

Autonomous GPS: The use of a stand-alone GPS receiver for positioning without differential corrections or post-processing. May provide sufficient accuracy for navigation, but usually considered inadequate for surveying applications.

Average end area method: A method for calculating volume of earthwork, based on the assumption that the volume of earthwork between two vertical cross sections is equal to the average of the two end areas multiplied by the horizontal distance between them.

Azimuth: The horizontal angle of a line as measured clockwise from the meridian; usually measured from the north.

B

Backsight (BS): The first rod reading taken by the surveyor after the leveling instrument is set up and leveled; generally taken on a point of known elevation, such as a benchmark. Short for "backsight reading."

Bank-measure: In earthwork, the volume of the earth in its natural state, before excavation.

Barleycorn: An old measure of length, equal to the average length of a grain of barley (approximately a third of an inch).

Base line: A surveyed line established with more than usual care, to which surveys are referred for coordination and correlation.

Bathymetry: The study of underwater depths, typically using the water surface as a reference plane.

Bearing: The horizontal angle between the meridian and the line.

Benchmark: A permanent object, natural or artificial, with a known elevation.

Blunders: Incorrect measurements caused by carelessness or confusion on the part of the observer.

Boundary monument: A material object placed on or near a boundary line to preserve and identify the location of the boundary line on the ground.

C

Cardinal directions: The directions on the surface of the earth: north, south, east, and west.

Cartesian coordinates: A plane coordinate system that allows the specifying of the location of a point on the plane or in three-dimensional space using linear measurements that specify the distances from a designated coordinate axis. Also known as rectangular coordinates.

Chain: A unit of length used in the subdivision of public lands of the United States. The Gunter's chain is 66 ft long and divided into 100 links, each 7.92 in long.

Chaining pin: A steel pin used in taping to mark tape lengths. Also known as a marking pin.

Chord: A straight line connecting two points on the perimeter of a circular curve or circle.

Closed traverse: A traverse that starts and ends at the same point. Because it is a closed polygon, its interior angles can be checked for accuracy.

Collimation error: An error in leveling caused by the difference between the line of sight for a level and a truly horizontal line.

Common borrow: Fill material obtained outside the excavation site and brought into the excavation site.

Common law: Unwritten law, as opposed to statute, or written, law. Developed in England and based upon immemorial usage.

Compacted-measure: In earthwork, the volume of fill after compaction.

Compass rule: A method for adjusting a traverse that distributes the differences in the sums of the latitudes and departures over each course, based on the ratio of the length of that course to the total length of the traverse. Also known as the Bowditch method.

Compound curve: A curve consisting of two or more simple curves of different radii having their curvature in the same direction and joined together at a common tangent point.

Support Material

APPENDIX 1.B (*continued*)
Glossary

Connecting traverse: A traverse that starts on one established control point and ends on another.

Continuously operating reference stations (CORS): GPS stations operating continuously at known points that can be used to correct GPS observation data.

Contour interval: The vertical distance between contour lines on a topographic map.

Controlling points method: A method of locating topographic contours for an area based on measuring the elevations at key topographic points, such as stream junctions, intermediate points in stream beds between junctions, and along ridge lines. Interpolation between such control points is used for drawing contour lines.

Convergence angle: The angular difference between grid north and geodetic north for plane coordinates. Also known as the mapping angle, grid declination, or variation.

Coordinate method of area calculation: A method of calculating the area within a traverse based on coordinates for each point in the traverse. Also known as the "criss-cross" or "shoe lace" method.

Coordinate method of plotting: A method for plotting a traverse with each point being plotted independently based on coordinates for the points.

Crest curve: A vertical curve connecting two tangents with their point of intersection higher than the points of curvature and tangency of the tangents.

Cross sections: Profiles of the earth taken at right angles to the centerline of a linear construction project.

Cross-section method: A method of locating topographic contours for an area based on measuring cross sections at right angles to a centerline or baseline. Interpolation between cross sections is then used for drawing contour lines.

Cubit: A measure of length, originally the length of the forearm from the elbow to the extremity of the middle finger. In English measure, a cubit is 18 in (45.72 cm).

Curvature error: An error in leveling, especially with long sight distances, caused by the curvature of the earth.

Cut: Earthwork that is excavated or to be excavated.

Cut (subtract) tape: A measuring tape with the last foot at each end graduated in tenths of a foot or tenths and hundredths of a foot.

D

Datum: Any numerical or geometrical quantity or set of quantities that serve as a reference or base for other quantities.

Datum of reference correction: A correction to depth measurements to adjust for the water level difference between the water level at the time of measurement and that of datum to be used in the survey. In the United States, the standard datum of reference for hydrographic surveys is mean lower low water (MLLW).

Datum surface: A reference plane with respect to which survey points are determined.

Declination: The horizontal angle between the magnetic meridian and the true (geodetic) meridian.

Deflection angle: The angle between the prolongation of one tangent of a circular curve and that of the following tangent. It is also the central angle of a circular highway curve.

Degree of curve (arc definition): The central angle of a curve that subtends a 100 ft arc (used for highways).

Degree of curve (chord definition): The central angle of a curve that subtends a 100 ft chord (used for railroads).

Differential GPS: The use of a second set of GPS observations, taken simultaneously with the first and at a known position, for corrections.

Differential leveling: The process of measuring the difference in elevation between points.

Digital level: A level utilizing electronic image processing to automatically read a distant bar-coded level rod.

Double meridian distance (DMD) method: A traditional method of calculating the area within a traverse based on latitudes and departures.

Draft correction: A correction applied to fathometer soundings to adjust for the difference between the surface of the water and the level of the transducer.

E

Earthwork: The excavation, hauling, and placing of soil, rock, gravel, or other material found below the surface of the earth.

Electronic distance measuring (EDM) instrument: A device for measuring distances electronically by measuring time required for a transmitted light signal to reach a distant target and return.

Ellipsoid: An elliptical shape that approximates the figure of the earth and which is used as a basis for calculating geodetic horizontal coordinates.

Ellipsoid height: An elevation referenced to the ellipsoid (as opposed to an orthometric height, which is referenced to the geoid). Also known as ellipsoidal elevation.

Expansion error: An error in leveling caused by the expansion and contraction of the metallic strip on the rod due to temperature changes.

APPENDIX 1.B *(continued)*
Glossary

External distance: The vertical distance from the point of intersection of two tangent segments for a circular highway curve to the middle of the arc of the curve.

F

Fathom: A unit of distance equivalent to 6 ft, used primarily in marine depth measurements.

Fathometer: A hydrographic surveying instrument that measures depth by determining the time required for a sound wave to travel from its point of origin and back, and converting the measured time to depth. Also known as an echo sounder.

Fill: Excavated material that is placed into an earthwork project.

Flattening factor: A factor used in describing an ellipsoid based on the ratio of the difference between the lengths of the semimajor and semiminor axis to the length of the semimajor axis.

Flood plain: Land along the course of a stream that is subject to inundation during periods of high water that exceed normal bank-full elevation.

Foresight (FS): The rod reading taken on the point whose elevation is to determined. It is subtracted from the height of instrument to determine the elevation of the point.

Four-pole chain: A chain 66 ft long (the length of four poles). Also known as a Gunter's chain.

Furlong: A measure of length equal to $\frac{1}{4}$ mile or 220 yd.

G

Geodetic Reference System of 1980 (GRS 80): The earth-centered ellipsoid most commonly used for geodetic control in the United States. The GRS 80 has a semimajor axis length of 6,378,137 m and a flattening factor of 1/298.257222101.

Geographic information system (GIS): A computer system capable of capturing, storing, analyzing, and displaying geographically referenced information.

Geoid: A figure of the earth that approximates the level of the undisturbed sea, and where the gravity potential is equal, used as a reference for orthometric heights.

Geoidal separation: The vertical distance between the heights of the ellipsoid and the geoid at a specific location on the earth's surface.

Global Navigation Satellite System (GNSS): Any satellite system that can be used to determine a precise location on the surface of the earth. The U.S. system is known as the *NAVSTAR Global Positioning System* (GPS); the Russian system is known as the *Globalnaya*

Navigatsionnaya Sputnikovaya Sistema, or GLONASS; the European Space Agency system is known as GALILEO.

Grade: In roadway design, the slope of the line that is the profile of the centerline. Also known as the gradient.

Grade point: The point where a fill section meets the natural ground or where a cut section begins.

Grade stake: A stake used to control the vertical limits of earthwork at particular locations.

Gradient: An inclined surface. The change in elevation per unit of horizontal distance.

Grid method: A method of locating topographic contours for an area based on establishing a grid and measuring the elevation at each of the grid line intersections. The location of the point where the contour line crosses each side of each grid cell may then be determined by interpolation between the elevations at the grid intersections.

Gunter's chain: A 66 ft long chain consisting of 100 links, each link being 7.92 in long, used for most of the early U.S. Public Land Surveys.

H

Height of instrument (HI): The elevation of the line of sight.

Hub stake: A reference stake (typically 2 in by 2 in in size with a tack in the top) used to mark a precise position for a control point.

Hydrographic survey: A survey having for its principal purpose the determination of data relating to bodies of water, including depth of water, configuration of bottom, directions and force of current, heights, times and water stages, and location of fixed objects for survey and navigation purposes.

I

Index error: An error in leveling occurring when the bottom of the rod does not precisely correspond with zero on the rod scale.

Inscribed angle: An angle formed by two chords in a circle with a common endpoint, which serves as the vertex for the angle.

Invar tape: A tape made of a steel and nickel alloy to minimize expansion and contraction in length due to changes in temperature. Used for precise measurements.

Inverse distance weighing (IDW): A process used to estimate elevations.

Support Material

APPENDIX 1.B *(continued)*
Glossary

K

Kriging: A process used for spatial interpolation. Kriging interpolates the attribute value at unobserved points across a grid by modeling the attenuating effect of distance and fitting the points to smooth, continuous curves or surfaces described by mathematical functions.

L

Lambert projection: A projection used with plane coordinates that uses a cone with a north-south orientation, two standard parallels, converging straight meridians that meet at a point outside of the map limits, and arcs of concentric circles serving as parallels. Since scale is true only along the two standard parallels and is compressed between those lines and expanded outside of them, this projection is best for regions of greater east-west extent.

Laser level: A leveling instrument that projects a fixed red or green laser beam along a horizontal and/or vertical axis.

Laser scanning: The process of using a laser to measure existing conditions in the built or natural environment. Also known as high-definition surveying and three-dimensional imaging.

Latitude: The angle measured from the center of the earth (or a spheroid representing the earth) northerly or southerly from the equator.

Lead line: A traditional means of measuring water depths using a lead weight attached to a graduated line.

Length error: An error in leveling due to the graduations of the rod not precisely matching national length standards.

Length of curve: The distance from the point of curvature to the point of tangency along the arc of a circular curve.

Length of long chord: The straight line distance from the point of curvature to the point of tangency of a circular curve.

Level rod plumb error: An error caused by the level rod not being plumb at the time of sighting.

Link: A unit of linear measure, one hundredth of a chain long (7.92 in). (*See also* chain.)

Longitude: The angle measured from the center of the earth (or a spheroid representing the earth) easterly or westerly from a north-south line of reference known as the prime meridian.

Loose-measure: In earthwork, the volume of earth once excavated and loaded.

M

Mapping angle: The angular difference between grid north and geodetic north for plane coordinates. Also known as the convergence angle, grid declination, or variation.

Mass diagram: A method of determining economical handling of material, quantities of overhaul, and location of balance points.

Mean: The sum of all of the measurements in a data set divided by the number of measurements. Also known as the average.

Mean high water (MHW): The average of all the high tides occurring each day over a 19-year tidal epoch.

Mean higher high water (MHHW): The average of the higher of the high tides occurring each day over a 19-year tidal epoch.

Mean low water (MLW): The average of all of the low tides occurring over a 19-year tidal epoch.

Mean lower low water (MLLW): The average of the lower of the low tides occurring each day over a 19-year tidal epoch.

Median: The measurement in the middle of a data set that is arranged in increasing (or decreasing) order.

Meridians: A reference line. Used in geometry to determine the direction of a line and in geographic coordinate systems to determine longitude.

Metes and bounds description: A land description that defines the boundaries of a tract of land by identifying a beginning point and then describing the direction and distance for each course around the perimeter of the tract (metes) and by adjoining boundaries or monuments (bounds).

Middle ordinate: The vertical line between the middle of the chord (between the point of curvature to the point of tangency of a circular highway curve) and the middle of the curve itself.

Mode: The value that occurs most frequently in a data set.

Multi-beam fathometer: A fathometer equipped with an array of transducers oriented to varying angles off the vertical axis, which allows for coverage of a wide swath.

Multipath error: An error in GPS measurements caused by deflection of the satellite signal by objects near the receiver.

Support Material

APPENDIX 1.B *(continued)*
Glossary

N

National tidal datum epoch: A specific 19-year period, currently 1983–2001, used to calculate tidal datum values in the United States.

North American Datum of 1983 (NAD 83): A datum for the measurement of horizontal geodetic coordinates based on the GRS 80 ellipsoid.

North American Vertical Datum of 1988 (NAVD 88): The currently used vertical geodetic datum in the United States. It was established using the elevation of mean sea level as fixed at a point at Father Point in Quebec, Canada.

O

Offset stake: A stake used to mark the horizontal limits of earthwork or construction, typically offset from the actual edge to protect the stake during the construction process.

Online positioning user service (OPUS): A service provided by the National Geodetic Survey for online correction of stand-alone GPS observations by use of corrections from nearby continuously operating base stations.

Open traverse: A traverse that begins at one point and ends at a different point. Because an open traverse is not a closed figure, no mathematical check can be made of its angles and distances.

Orthometric height: An elevation referenced to the geoid (as opposed to an ellipsoid height, which is referenced to the ellipsoid). Since the height above the geoid closely approximates the height above mean sea level, it is extensively used in surveying and mapping, and is widely displayed on most charts and maps.

P

Parallax: The apparent change in the position of the crosshair as viewed through a telescope.

Perch: A measure of length, varying locally in different countries, but by statute in Great Britain and the United States equal to 16.5 ft. It was used extensively in the early public land surveys and is equivalent in length to a rod or pole.

Philadelphia rod: A rod used in leveling, usually graduated in feet, tenths of a foot, and hundreds of a foot, with two sliding parts that makes it easy to transport and suitable for a variety of terrains and projects.

Plan-profile sheets: Plans for transportation or utility corridors, which provide a plan view of the centerline with surrounding topography on the top half of the sheet and elevation profile on the bottom half.

Platting laws: Regulations for recording of subdivision plats, which define requirements for the monumentation of parcels, the accuracy of the survey, and the means of identifying the parcels and their dimensions.

Point of common curvature: The junction point of two curves making up a compound curve.

Point of curvature: The intersection of the tangent preceding a circular highway curve and the curve itself.

Point of tangency: The intersection of a circular highway curve and the tangent leading out of the curve.

Position dilution of precision (PDOP): A numerical representation of the predicted accuracy of GPS measurement, based on the strength of figure of the GPS satellites used. The term represents the quality of the satellite geometry with respect to the receiver location; a value of 3 or less is generally considered necessary for the highest survey quality.

Prime meridian: The meridian of longitude passing through the Royal Observatory in Greenwich, England, used by international agreement.

Prismoidal formula: A method for calculating volume of earthwork based on the section length, the cross section at either end, and the cross section at a midpoint.

Protractor method: A method for plotting a traverse by plotting the measured angle and distances.

R

Radian: An angle that, when situated as a central angle of a circle, is subtended by an arc whose length is equal to the radius of the circle.

Radius: A straight line from the center point of a circle to the circumference.

Random errors: Differences between observed values and true values that remain after blunders and systematic errors have been removed from the data.

Range: The difference between the highest value and lowest value in a data set. Also known as dispersion.

Range ratio method: A mathematical process used to adjust tidal observations to the equivalent of a 19-year mean value using simultaneous observations at the tide station at which the observations were made and another tide station where 19-year mean values are known.

Raster data model: A type of data file consisting of arrays of grid cells, or pixels, usually referenced by row and column numbers, along with an identifier representing the attribute being mapped.

Support Material

APPENDIX 1.B *(continued)*
Glossary

Real-time differential GPS: A variation of differential GPS that involves equipping the base station with a transmitter to broadcast the corrections and equipping the subordinate point with a receiver to accept such corrections for real-time correction of GPS observations.

Refraction: An error in leveling caused by the bending of light rays due to differences in air temperature.

Reverse curve: A compound curve with deflections in opposite directions.

Right-of-way survey: A survey performed for the purpose of laying out an acceptable route for an easement or right-of-way for a road, pipeline, utility, or transmission line.

Right-angle offset method: A common method used in route surveying based on measuring perpendicular distances off the centerline to locate features along the route.

Riparian: Belonging or relating to the bank of a river.

Rod: 1. A measure of length equal to 5.5 yd or 16.5 ft. 2. The corresponding square measure. Also known as a perch or a pole.

Rood: 1. A square measure equal to one-fourth of an acre, or 40 rod^2, in England and Scotland, and equal to 17.07 yd^2 (14.28 m^2) in the Union of South Africa. 2. A linear measure, varying locally from 5.5 to 8 yd.

Root-mean-square error (RMSE): A measure used to test the accuracy of measurements by comparing them to independently measured values from a higher-order survey. It is calculated in the same way as the standard deviation, except that the independently measured values are used instead of the mean value and the denominator is the number of variables rather than that number minus one. This measure is frequently used to evaluate topographic elevations established by photogrammetry or LiDAR.

S

Sag curve: A vertical curve connecting two tangents with a point of intersection lower than the points of curvature and tangency of the tangents.

Scale factor: A measure of the lineal distortion associated with the projection of ellipsoidal distances onto a plane surface.

Secant projection: A plane coordinate projection system where the intermediate surface slices through a small portion of the spheroid. Two lines of contact exist with the surface of the spheroid, creating two standard parallels or meridians.

Semimajor axis: One-half the length of the east-west axis of an ellipsoid.

Semiminor axis: One-half the length of the north-south axis of an ellipsoid.

Senior and junior rights: The doctrine whereby, when an overlap of land descriptions occurs, the holder of the earliest grant (the senior awardee) retains ownership of the overlap area, and the holder of the latter grant (the junior awardee) loses the area.

Shore: The place lying between the line of ordinary high tide and the line of lowest tide.

Shrinkage: The change in volume of earth from its natural state to its compacted state, expressed as a percent decrease from the natural state.

Slope stake: A stake used to indicate the intersection of the natural ground and the proposed cut or fill for excavation and embankment operations.

Span: A distance equal to 6 in.

Spatial interpolation: A process used to develop contour lines from a series of irregularly spaced ground elevation points.

Spirit level: A leveling instrument equipped with an elongated, slightly curved glass leveling tube usually filled with alcohol.

Standard deviation: A measure of variation in a data set that is commonly used as a standard for the acceptance or rejection of data. It represents the most probable error in a single observation and is defined as the square root of the sum of the squares of the differences between each variable in the data set and the mean of the set divided by the number of variables in the set minus one.

State Plane Coordinate System: A plane coordinate system widely used in the United States, Puerto Rico, and the Virgin Islands, comprising a series of multiple zones, which minimizes distortion because of its projection process.

Static GNSS survey: A geodetic survey that uses survey grade satellite receivers to collect satellite data on a fixed point and requires post-processing to determine position.

Subdivision laws: Regulations that define requirements for the subdivision of land, including types of streets and their dimensions, arrangement of lots and their sizes, land drainage, sewage disposal, protection of nature, and various other details.

Submerged land: In tidal areas, land that extends seaward from the shore and is continuously covered during the ebb and flow of the tide.

Swell: The change in volume of earth from its natural state to that after excavation, expressed as a percent increase from the natural state.

APPENDIX 1.B *(continued)*
Glossary

Systematic errors: Errors caused by physical laws. Such errors may be predicted and sometimes corrected for by observational procedures.

T

Tangent distance: The straight-line distance between the point of curvature and the point of intersection for a circular highway curve.

Tangent projection: A plane coordinate projection system where the intermediate surface is tangential to the spheroid. A single line of contact exists around the spheroid, which becomes the standard meridian or parallel for the projection.

Thalweg: The deepest part of a channel.

Three sigma error: The margin of error within which 99.7% of the observations (or residuals) fall within 2.968 standard deviations of the mean of the data set.

Three-wire leveling: A leveling process where the average of three readings (the top and bottom stadia hair readings and the center horizontal crosshair reading) is used for each reading of the rod.

Topographic map: A map providing a plan view of a portion of the earth's surface showing natural and constructed features such as rivers, lakes, roads, buildings, and canals. The shape, or relief, of the area is shown by contour lines, hachures, or shading. Also known as a contour map.

Topographic survey: A survey made to determine the relative positions of points, objects, and elevations so that their positions can be accurately represented on a map.

Total station: A digital theodolite integrated with an EDM to allow the measurement of angles and distances with one instrument.

Tracing contours method: A method of locating topographic contours for an area by following a specific elevation while making frequent measurements of the horizontal position of the line. This method may be used when mapping a specific elevation, such as that of the mean high water line.

Transverse Mercator projection: A projection that uses a cylinder with an east-west orientation. As used in the U.S. State Plane Coordinate System, the cylinder cuts the sphere along two meridians, parallel to the central meridian. Since the scale is true along the two standard meridians and is compressed between those lines and expanded outside of them, this projection is best for regions of greater north-south extent.

Trapezoidal rule: A method of calculating the area under a curved line defined by a series of straight lines.

Traverse: A series of straight lines connecting successive instrument stations in a survey. The relative position of each station in the traverse is determined by measuring line lengths and angles between consecutive lines.

Trigonometric leveling: A process for determining elevation differences using measurements of the vertical angle and distance to a distant point, then calculating the elevation of the point using basic trigonometry.

Trilateration: A method of determining the coordinates of points based on distance measurements from two points with established coordinates. Also known as the two-distances method.

Triangulation: A method of determining the coordinates of points based on angle measurements from two points with established coordinates on a baseline. Also known as the two-angles method.

Two sigma error: The margin of error within which 95% of the observations (or residuals) fall within 1.96 standard deviations of the mean of the data set.

U

Universal Transverse Mercator (UTM) system: A worldwide plane coordinate system using 60 zones, each covering a width of six degrees of longitude.

V

Variance: The difference between a specific value in a data set and the mean of that set.

Vector data model: A type of data file assuming a continuous coordinate space, which allows positions, lines, and area boundaries to be defined by x- and y-coordinate pairs, as opposed to grid cells as in a raster model.

Velocity of sound correction: A correction to fathometer soundings to adjust for the prevailing speed of sound in water, which varies with temperature and salinity.

Vertical curve: A curve, usually a parabola, connecting two tangents in vertical alignment.

W

Watercourse: A channel for water fed from permanent or natural sources running in a particular direction and having a channel formed by a well-defined bed and banks.

Witness stake: A stake used as a reference to a nearby hub or other type of construction stake. Also known as a guard stake.

World Geodetic System of 1984 (WGS 84): An earth-centered ellipsoid used for geodetic control by the United States Department of Defense.

Support Material

Index

Geodetic
meridian, 9-1
network standards, 12-3
Reference System, 23-1
System of 1984, World, 23-2
Geographic coordinates, 24-3
Geoid, 14-3, 23-1
separation, 15-7
Geometric, figures, drawing, 1-6
Geometry
plane, 1-5
satellite, 15-7
solid, 1-5
Geopotential Model 1996, 23-2
Giving line and grade, 28-1
Global
Navigation Satellite System, 15-6
Positioning System, 15-6
GPS
autonomous, 15-6
differential, 15-7
leveling, 14-7
real-time differential, 15-7
Grad, 9-1
Grade, 22-1, 28-1
giving line and, 28-1
point, 29-2
rod, 28-5
stake, 28-1
Gradient, 22-1, 28-1
Graduated staff, 18-2
Graph, 8-1
of the equation, 10-1
Grid
declination, 9-3, 23-5
meridian, 9-1
method, 16-3
system, 16-2
system, spherical, 23-2
system, spheroidal, 23-2
Guard stake, 28-1

H

Hairs, stadia, 14-3
Half-tide level, 18-5
Haul, free, 29-6
Hectare, 3-1
Height
of instrument, 14-4
orthometric, 15-7, 23-2
slant, 1-5, 1-6
High-definition surveying, 15-8
Horizontal
accuracy, 12-4
control, 16-1
spatial accuracy, 12-4
tie, 16-1
Hub stake, 28-1
Hydrographic
mapping, 17-5
surveying, 18-1
surveys, 12-3, 12-4

I

Imaging, three-dimensional, 15-8
Inch
conversion to decimals of a foot, 3-2
conversion to decimals of a foot (tbl), 3-2
Inclination, angle of, 7-3
Index
contour, 16-6
error, 14-10
Inequality
diurnal high water, 18-5
diurnal low water, 18-5
Inscribed angle, 21-1
Instrument, height of, 14-4
Intensity factor, 17-5
Intercept
form, 10-5
form, slope-, 10-5

x-, 10-3
y-, 10-3
International
Bureau of Weights and Measures, 3-1
System of Units, 3-1
Interpolation, spatial, 27-2
Intersection, 18-4
of two lines, 8-4
point of, 19-6, 21-2, 22-1
Interval, contour, 16-6
Invar
rod, 14-2
tape, 13-1
Inverse distance weighting, 27-2
Invert, 28-13, 28-14
Irregular network, triangulated, 27-2
Isosceles
triangle, 1-3
triangle (fig), 1-4

J

Junior
awardee, 24-4
right, 24-4

K

Key point, 16-3
Kriging, 27-2

L

Lambert conformal conic projection, 23-4
Land conveyance, 30-2
Laser scanning, 15-8
Latitude, 19-2, 23-2
Law
of cosines, 7-5
of sines, 7-5
platting, 25-1
Snell's, 17-5
subdivision, 25-1
Lead line, 18-1
Length
error, 14-10
of an arc, 4-2
of chord, 21-2
of curve, 21-2, 21-4
of long chord, 21-2
Level
bull's-eye, 14-1
digital, 14-2, 15-5
half-tide, 18-5
mean sea, 18-5
mean tide, 18-5
rod plumb error, 14-9
self-leveling, 14-1
spirit, 14-1
Leveling
balancing angles for, 19-4
barometric, 14-7
circuit, single-run closed, 14-7
differential, 14-4
GPS, 14-7
level, self-, 14-1
profile, 14-5
reciprocal, 14-9
three-wire, 14-5
trigonometric, 14-7
LiDAR, 17-4, 27-2
bathymetry, airborne, 18-7
Light detection and ranging technology, 27-2
Line
adjustment, lot, 30-2
agonic, 9-3
and grade, giving, 28-1
bearing of a, 9-1
directed, 8-1
direction of a, 9-1
intersection of two, 8-4
lead, 18-1
midpoint of a, 8-3
principal, 17-2
Linear equation, 10-2

Liter, 3-1
Locating the center of a circle (fig), 1-7
Long chord, 20-5
Longitude, 23-3
Loose-measure, 29-1
Lot line adjustment, 30-2
Lower frequency, 18-2
Lowest possible low water, 18-4

M

Magnetic meridian, 9-1
Manhole, 28-14
Map
contour, 16-1
parcel, 30-2
tentative, 30-1
tentative tract, 30-1
topographic, 16-1
Mapping
aerial, 17-1, 17-7
angle, 9-3, 23-5
hydrographic, 17-5
Mark, bench, 14-4
Marking pin, 13-2
Mass diagram, 29-7
Matching, cloud-to-cloud point, 15-9
Mean, 11-2
high water, 14-4, 18-5
higher high water, 18-5
low water, 14-4, 18-5
lower low water, 14-4, 18-4, 18-5
sea level, 18-5
tide level, 18-5
Measure
bank-, 29-1
compacted-, 29-1
loose-, 29-1
of angles, 1-1
Measurement, 2-1, 11-1, 27-2
electronic distance, 15-1
error, 11-2
Median, 11-2
Mercator system, Universal Transverse, 23-7
Merger parcel, 30-2
Meridian, 9-1, 23-3
distance, 20-1
distance, double, 20-1
distance method, double, 20-1
geodetic, 9-1
grid, 9-1
magnetic, 9-1
prime, 23-3
Meter, 3-1
Metes and bounds description, 24-2
Method
angle and distance, 16-2
average end area, 29-2
azimuth-stadia, 16-2
bearing-bearing, 19-7
bearing-distance, 19-8
Bowditch, 19-5
controlling points, 16-4
coordinate, 16-6, 20-3
cross-section, 16-4
deflection angle, 21-3
distance-distance, 19-9
double meridian distance, 20-1
grid, 16-3
prismoidal formula, 29-2
protractor, 16-6
right-angle offset, 16-1
tangent, 16-6
tracing contours, 16-4
transit, 19-5
two-angle, 16-2
two-distances, 16-2
Metric system, 3-1
Middle ordinate, 20-5, 21-2
Midpoint of a line, 8-3
Minute of an angle, 1-1
Mode, 11-3